FIRST ON THE MOON

FIRST ON THE MOON

A Voyage with

NEIL ARMSTRONG
MICHAEL COLLINS
EDWIN E. ALDRIN JR.

written with GENE FARMER *and* DORA JANE HAMBLIN

Epilogue by ARTHUR C. CLARKE

with photographs

KONECKY & KONECKY
72 AYERS POINT RD
OLD SAYBROOK, CT 06475

THIS EDITION PUBLISHED BY ARRANGEMENT WITH
LITTLE, BROWN
AND COMPANY INC.

PHOTOGRAPHS COURTESY OF NASA AND TIME, INC.

ISBN: 1-56852-398-X

PRINTED IN THE USA

To the eight astronauts who
did not live to share the
experience of the first
lunar landing

THEODORE C. FREEMAN

ELLIOT M. SEE JR.

CHARLES A. BASSETT II

VIRGIL I. GRISSOM

ROGER B. CHAFFEE

EDWARD H. WHITE II

EDWARD G. GIVENS JR.

CLIFTON CURTIS WILLIAMS

Contents

GLOSSARY xi

PROLOGUE: Quiet Here in Mission Control 1

BOOK I: Into the Vasty Deep

1. M Plus 2,974 13
2. T Minus and Counting 32
3. "I vunder vere Guenter Vendt?" 51
4. "He doesn't owe me a cup of coffee" 71

BOOK II: Where the Stars Don't Twinkle

5. "There isn't any magic selection" 101
6. NOUN, VERB, DSKY and REFSMMAT 121
7. The Ecliptic 140
8. Earthshine — and "Over the hill" 161

BOOK III: Eagle and Columbia

9. "See that ridge? That can be five hundred feet high!" 195
10. "An expression of man's self-determination" 220
11. "Don't forget one in the command module" 245
12. "Magnificent desolation" 280
13. "Open up the LRL doors, Charlie" 306
14. "It seems appropriate to share with you . . ." 336

EPILOGUE: Beyond Apollo *by Arthur C. Clarke* 371

ACKNOWLEDGMENTS 423

NOTES 425

Illustrations

The astronauts and their families (Ralph Morse for *Life*) 91

Neil Armstrong at home after the flight (Lynn Pelham for *Life* from Rapho-Guillumette) 92

Janet Armstrong watching liftoff (Vernon Merritt for *Life*) 93

Buzz Aldrin at home after the flight (Lynn Pelham for *Life* from Rapho-Guillumette) 94

Joan Aldrin watching liftoff (Lee Balterman for *Life*) 95

Mike Collins at home after the flight (Lynn Pelham for *Life* from Rapho-Guillumette) 96

Patricia Collins at time of liftoff (Bob Peterson for *Life*) 97

Neil Armstrong fishing with his sons (Ralph Morse for *Life*) 98

Astronauts leaving spacecraft after preflight test (Courtesy NASA) 183

Armstrong and Aldrin in preflight practice (Courtesy NASA) 184

Apollo 11 liftoff (Ralph Morse for *Life*, Courtesy NASA) 185

Liftoff seen from press site three miles away (Ralph Morse for *Life*, Courtesy NASA) 186–187

Earth from 112,000 miles in space (Courtesy NASA) 188

Lunar surface from sixty-three miles high (Courtesy NASA) 189

Back side of the moon seen from Columbia (Courtesy NASA) 190

Landing site of Eagle (Courtesy NASA) 191

Footprints on the moon (Courtesy NASA) 192

Aldrin descending from the LM to the moon (Courtesy NASA) 269

Aldrin setting up lunar scientific experiments (Courtesy NASA) 270–271

Aldrin on the moon (Courtesy NASA) 272

Armstrong in LM after moon walk (Courtesy NASA) 273

Eagle preparing to dock with Columbia (Courtesy NASA) 274–275

Recovery of the astronauts and Columbia after splashdown (Courtesy NASA) 276–277

Columbia being hoisted onto deck of *Hornet* (Courtesy NASA) 278

Astronauts in quarantine (Lynn Pelham for *Life* from Rapho-Guillumette, Courtesy NASA) 279

Glossary

This is not a complete "space dictionary"; no such thing exists, since acronyms and abbreviations are constantly being added. This list does identify the shorthand expressions most frequently used in the flight of Apollo 11.

ACQ. Acquisition.

AOS. Acquisition of signal from the spacecraft.

APOGEE. The highest point reached by a spacecraft in earth orbit.

APOLUNE. The highest point reached by a spacecraft in lunar orbit. Also called apocynthion.

APS. a) Ascent propulsion system of the lunar module. b) Auxiliary propulsion system of the S-IVB.

ATTITUDE. The position of a spacecraft in flight relative to the horizon or another fixed reference.

AZIMUTH. The bearing of an object measured in degrees from a fixed point, usually true north.

BASELINE. Standard or average performance, determined by preflight testing.

BEAM WIDTH. The angular separation between the right and left edges of the radar or radio beam.

BURN. A noun or a verb: the firing of engines or to fire the engines.

CAPCOM. Capsule communicator. The console position in the Mission Control Center (always manned by an astronaut) for relaying directions to, and getting information from, the spacecraft crew.

COAST. The flight of a spacecraft when the engines are not thrusting.

COMM. Communications.

CONSOLE. A term applied to the grouping of controls, indicators, etc., for monitoring and controlling a particular sequence of action.

CRYO. Cryogenic, supercooled, or −195 degrees Centigrade or less.

DELTA V. Velocity change.

DOI. Descent orbit insertion.

DOWNRANGE. Direction beyond the planned target point or landing site.

DPS. Descent propulsion system of the lunar module.

DROGUE. The guiding device of the lunar module into which the probe is inserted during docking with the command module.

DROGUE CHUTE. Small parachute which stabilizes the command module and then pulls out the main parachutes during reentry.

EGRESS. To exit the spacecraft.

ELLIPTICAL ORBIT. An orbit having the shape of an ellipse, i.e., an elongated circle.

EVA. Extravehicular activity, "space walk" or "moon walk."

F STOP. A specific lens aperture setting on a camera.

FAM AND OPS. Familiarization and operations manual.

FUEL CELL. An electro-chemical generator in which the chemical energy from the reaction of oxygen and a fuel is converted directly into electricity.

G. The force of the earth's gravity, approximately thirty-two feet per second in terms of acceleration.

GIMBAL. As a noun, a device with two perpendicular and intersecting axes of rotation on which an engine can be mounted and swivel in two directions. As a verb, to swivel a motor to change the angle of an engine or antenna.

HYPERGOLIC. An adjective to identify a fuel which ignites spontaneously upon contact with its oxidizer, eliminating the need for an ignition system.

INERTIAL GUIDANCE SUBSYSTEM. A subsystem which, utilizing gyroscopic devices and a computer, automatically determines spacecraft attitude and location without reference to external sources.

INGRESS. To enter the spacecraft.

LAUNCH WINDOW. The limited time period during which a launch can take place if the spacecraft is to arrive at the proper time and place on the moon.

LEVA. Lunar extravehicular activity, or "moon walk."

LOI. Lunar orbit insertion.

LOS. Loss of signal from the spacecraft.

LOX. Liquid oxygen.

MCC. Midcourse correction.

MCC-H. Mission Control Center, Houston.

MSC. Manned Spacecraft Center.

MSFN. Manned Space Flight Network, radio communications.

NAUTICAL MILE. 6,076 feet, about 1.15 statute miles.

OMNI. Omnidirectional radio antenna, used for sending and receiving.

PDI. Powered descent initiation.

PERIGEE. The lowest point reached by a spacecraft in earth orbit.

PERILUNE. The lowest point reached by a spacecraft in lunar orbit. Also called pericynthion.

PGNCS. The spacecraft's primary guidance, navigation and control system.

PITCH. The attitude movement of a spacecraft where the nose tips up or down, rotating on the "Y" axis.

POO. Program zero zero, an "idle program" used for clearing a spacecraft computer or returning it to zero for another computation.

PROBE. The device on the command module which engages the drogue on the lunar module during docking.

PTC. Passive thermal control, or "barbecue mode" of rolling the spacecraft to distribute solar heat.

PYRO. Pyrotechnic, or explosive. Pyrotechnic batteries are used to provide power to explosive devices.

ROLL. The rotation of the spacecraft around the "X" axis in the command module and "Z" axis in the lunar module.

S-BAND. A 2100–2300 megahertz band which relays voice, telemetry, television, data and tracking information during all phases of the flight and especially in deep space.

SPS. Service propulsion system, the large engine of the service module.

SQUAWK BOX. A small speaker, connected by telephone line to Mission Control, for home or office reception of live conversation between the spacecraft and earth.

STATE VECTOR. The calculation of the spacecraft's position and velocity from the earth or moon, from which calculations on board the spacecraft can be made.

TELEMETRY. The system of measuring anything from astronaut blood pressure to the angle of the spacecraft, and relaying the information to earth in coded signals.

TORQUE. A sideways force on the spacecraft.

TRAJECTORY. The flight path traced by a spacecraft.

YAW. A sideways swing of the spacecraft, rotation around the "Z" axis of the command module and the "X" axis of the lunar module.

PROLOGUE

Quiet Here in Mission Control

It's a strange, eerie sensation to fly a lunar landing trajectory — not difficult, but somewhat complex and unforgiving.

NEIL ARMSTRONG

IT was a steaming Sunday afternoon in Texas. From the air-conditioned Houston Manned Spacecraft Center the voice of NASA public affairs officer Douglas K. Ward described the scene, working up to good news the whole world hoped soon to hear: "This is Apollo Control at 101 hours 54 minutes. The time until the ignition for the powered descent is 38 minutes 55 seconds. I believe back in the viewing room we probably have one of the largest assemblages of space officials that we've ever seen in one place. Included among the viewers are Dr. Thomas Paine, NASA administrator; Jim Elms, director of the Electronic Research Center at Cambridge; Dr. Abe Silverstein, director of NASA's Lewis Research Center; Rocco Petrone, director of Launch Operations at Kennedy Space Center. From Marshall Space Center, we have Dr. Wernher von Braun, the director, and his deputy Dr. Eberhard Rees. Also a large number of astronauts, including Tom Stafford, Gene Cernan, Jim McDivitt and John Glenn. Here in the control room proper, down on the floor, are a number of other astronauts, including Pete Conrad, Fred Haise, Jim Lovell and Bill Anders, and Donald K. Slayton, director of Flight Crew Operations at the Manned Spacecraft Center. Sitting beside us in the back row of consoles here is Dr. Robert Gilruth, director of the Manned Spacecraft Center. Further down on the line is General Sam Phillips, director of the Apollo program. Also Chris Kraft is here, director of Flight Operations at the Manned Spacecraft Center, and George Low, the Apollo spacecraft program manager. We also see in the back viewing room Secretary of the Air Force [Robert C.] Seamans. . . . Ignition for the powered descent to the lunar surface is 36 minutes 30 seconds away."

PUBLIC AFFAIRS OFFICER (Ward): We're now 2 minutes 53 seconds from reacquiring the spacecraft, 21 minutes 23 seconds from the beginning of the powered descent. It's grown quite quiet here in Mission Control. A few minutes ago flight director Gene Kranz requested that everyone sit down, get prepared for events that are coming, and closed with "Good luck to all of you." We have a number of big plot boards which will be used to keep track of the burn progress. One will show the performance of onboard guidance systems, both the primary and the backup guidance systems, and compare the guidance systems with the Manned Space Flight Network tracking. These displays, by the time this is all over, will look a great deal like a combination Christmas tree and Fourth of July. We're now 1 minute 39 seconds from reacquiring the command module Columbia. Acquisition of the lunar module [Eagle] will come a little less than 2 minutes after that. At the time we acquire the LM it should be at an altitude of about 18 nautical miles descending toward the 50,000-foot pericynthion from which the powered descent to the lunar surface will be initiated. If for any reason the crew does not like the way things look as they are coming across the pericynthion, simply by not initiating the maneuver they will remain in a safe orbit of 60 miles by 50,000 feet, and if they desired they would be able to attempt the powered descent on the following revolution at a ground elapsed time of about 104 hours 26 minutes. We're now coming up on 30 seconds to acquisition. . . .

CAPCOM (astronaut Charles Duke): Columbia, Houston. We're standing by. Over. Columbia, Houston. Over.

COLUMBIA (command module pilot Michael Collins): Houston, Columbia. Reading you loud and clear. How me?

CAPCOM (Duke): Rog. Five by, Mike. How did it go? Over.

COLLINS: Listen, babe, everything's going just swimmingly. Beautiful.

CAPCOM (Duke): Great. We're standing by for Eagle.

COLLINS: Okay, he's coming around.

CAPCOM (Duke): We copy. Out. . . . And Columbia, Houston. We expect to lose your high gain [communications antenna] sometime during the powered descent. Over.

COLLINS: Roger. You don't much care, do you?

CAPCOM (Duke): No, sir.

EAGLE (lunar module pilot Edwin E. Aldrin Jr.): Houston, Eagle. How do you read?

CAPCOM (Duke): Five by, Eagle. We're standing by for your burn report. Over.

ALDRIN: Roger. The burn was on time.

PAO (Ward): Guidance says we're go. Gene Kranz just advised his flight controllers: "We're off to a good start. Play it cool." Twelve minutes now until ignition for powered descent. Everything still looking very good at this point. We presently show the LM at an altitude of 12.9 nautical miles and descending.

ARMSTRONG (lunar module commander Neil Armstrong): Horizon check right on time.

CAPCOM (Duke): Roger.

ALDRIN: Did you copy the star — I mean the sun check, Charlie?

CAPCOM (Duke): That's affirmative. We did, Buzz. Out.

PAO (Ward): Eagle is now at 10.7 nautical miles, 7 minutes 37 seconds from ignition.

ARMSTRONG: Our radar check indicates 50,000-foot perilune. Our visual altitude checks are steadying out at about 53,000.

CAPCOM (Duke): Roger, copy.

PAO (Ward): Coming up on 5 minutes to ignition. Gene Kranz getting a go/no go for descent.

CAPCOM (Duke): Eagle, Houston. If you read, you're a go for powered descent. Over. . . . Columbia, Houston. We've lost them on the high gain again. Would you please — we recommend they yaw right 10 degrees and reacquire.

COLLINS: Eagle, this is Columbia. You're go for PDI [powered descent initiation] and they recommend you yaw right 10 degrees and try the high gain again.

CAPCOM (Duke): MARK, 3:30 till ignition.

ALDRIN: Roger, copy. Balanced couple on, TTCA throttle, minimum. Throttle, auto, CDR. Stop button reset. Stop button. Check abort — abort stage reset. ATT control, 3 of them to mode control. PGNCS mode control is set. AGS is reading 400 plus 1. Standing by for arming.

CAPCOM (Duke): Eagle, Houston. If you'd like to try high gain, pitch 212, yaw 37. Over.

ALDRIN: Roger. I think I've got you on high gain now.

PAO (Ward): Coming up on 1 minute to ignition.

ALDRIN: Say again the angles. Over.

CAPCOM (Duke): Roger.

ALDRIN: We want to put them into use before we yaw around.

CAPCOM (Duke): Roger. Pitch 212, yaw plus 37.

ALDRIN: Copy. Over.

PAO (Ward): Current altitude about 46,000 feet, continuing to descend.

[*Below fifty thousand feet now. . . . Ignition time. . . . Time for Armstrong to fire the lunar module's descent engine which will land them on the moon. . . . Touchdown twelve minutes away. . . .*]

ALDRIN: Light's on. . . . One, zero. Ignition, 10 percent.

CAPCOM (Duke): Columbia, Houston. We lost 'em. Tell 'em to go aft OMNI. Over.

COLLINS: They'd like you on the OMNIs, Eagle.

ARMSTRONG: Okay, reading. . . . Stay with us, Mike.

COLLINS: Say again, Neil.

ALDRIN: I'll leave it in SLEW.

ARMSTRONG (to Collins): Relay to us.

ALDRIN: See if they got me now. I've got good signal strength in SLEW.

COLLINS: Yeah, you should have him now, Houston.

CAPCOM (Duke): Eagle, we've got you now. It's looking good. Over.

ALDRIN: Okay, rate of descent looks good.

CAPCOM (Duke): Eagle, Houston. Everything is looking good here. Over.

PAO (Ward): Two minutes 20 seconds; everything looking good. We show altitude about 47,000 feet [corrected].

ALDRIN: Houston, I'm getting a little fluctuation in the AC voltage now.

CAPCOM (Duke): Roger.

ALDRIN: Could be our meter, maybe — huh?

CAPCOM (Duke): Stand by. It's looking good to us. You're still looking good at 3, coming up 3 minutes.

ALDRIN: Everything looks real good. Altitude . . . right on.

ARMSTRONG: Our position checks downrange show us to be a little long.

CAPCOM (Duke): Roger. Copy.

ALDRIN: AGS . . . AGS is showing about 2 feet per second greater rate than it ought to be.

ARMSTRONG: I show us to be about . . . Stand by.

ALDRIN: Altitude rate is 3.2 [*static*].

ARMSTRONG: Roger, that's [*static*].

ALDRIN: I think it's gonna drop.

CAPCOM (Duke): Eagle, Houston. You are go.

ALDRIN: You are looking forwards.

ARMSTRONG: Yeah, we need to be backwards.

ALDRIN: Take it all at four minutes.

CAPCOM (Duke): You are go — you are go to continue powered descent. You are go to continue powered descent.

ARMSTRONG: Roger.

PAO (Ward): Altitude 40,000.

CAPCOM (Duke): And Eagle, Houston. We've got a data dropout. You're still looking good.

ALDRIN: PGNCS we got good lock on. Altitude lights out. Delta-H is minus 2,900.

CAPCOM (Duke): Roger. We copy.

ALDRIN: And we got the earth right out our front window.

PAO (Ward): Good radar data. Altitude now 33,500 feet.

ARMSTRONG: Give us the reading on the 12 02 program alarm.

CAPCOM (Duke): Roger. We got you. We're go on that alarm. . . . Six plus 25 throttle down.

PAO (Ward): We're still go. Altitude 27,000 feet.

ARMSTRONG: Throttle down!

ALDRIN: Throttle down on time! Throttles down better than in the simulator.

PAO (Ward): Altitude now 21,000 feet. Still looking very good. Velocity down now to 1,200 feet per second. . . . Seven minutes 30 seconds into the burn. Altitude 16,300 feet. . . . Altitude 13,500, velocity 9,100 feet per second. . . . Correction on that velocity, now reading 760 feet per second. . . . FIDO says we're go, altitude 9,200 feet. . . . [*Two thousand feet from High Gate — start the final landing approach*]. . . . Descent rate 129 feet per second.

CAPCOM (Duke): Eagle, you're looking great, coming up 9 minutes.

PAO (Ward): We're now in the approach phase of it, looking good. Altitude 5,200 feet. . . . [*Two thousand feet past High Gate, one statute mile above the moon*]. . . . Altitude 4,200. . . .

CAPCOM (Duke): Eagle. Houston. You're go for landing. Over.

ALDRIN: Roger, understand. Go for landing. Three thousand feet. Program alarm!

CAPCOM (Duke): Copy.

ALDRIN: 12 01.

ARMSTRONG: 12 01.

CAPCOM (Duke): Roger. 12 01 alarm. We're go. Hang tight. We're go.

ARMSTRONG: Two thousand feet.

ALDRIN: Two thousand feet. Into the AGS, 47 degrees.

CAPCOM (Duke): Roger.

ALDRIN: Forty-seven degrees.

CAPCOM (Duke): Eagle looking great. You're go.

PAO (Ward): Altitude 1,600. 1,400 feet. Still looking very good.

ALDRIN: Thirty-five degrees. 35 degrees. 750, coming down at 23. 700 feet, 21 down, 33 degrees. 600 feet, down at 19. [*Nearly to Low Gate, five hundred feet, two minutes of fuel left*] . . . 540 feet, down at 30 — down at 15. We're at 400 feet, down at 9, 48 forward. 350, down at 4. 330, 3½ down. We're pegged on horizontal velocity. 300 feet, down 3½. 47 forward. [*Static*] . . . Down 1 a minute. 1½ down. Got the shadow out there. 50, down at 2½. 19 forward. Altitude-velocity lights. 3½ down, 220 feet. 13 forward. 11 forward, coming down nicely. 200 feet, 4½ down. 5½ down. 160, 6½ down, 5½ down, 9 forward. Looks good. 120 feet. 100 feet, 3½ down. 9 forward. 5 percent. Quantity light. 73 feet, things looking good. Down a half. 6 forward.

El Lago, Texas

Jan Armstrong had been hunched on her knees before a television set in the bedroom, pencil in hand, studying the powered descent graph and checking landmarks on a detailed map of the moon. "We're coming up on Dry Gulch. Then Apollo Ridge. Twin Peaks. Off in the distance will be Smoky Ridge." The dog Super wandered in, his tail wagging. He was banished under the bed. At the 15,000-foot mark Jan sat up on her heels when she heard the words "Go for landing," and said, "Come on, come on, Trolley!" (Armstrong was flying the lunar module standing up, like an old-fashioned trolley driver.) At 220 feet she put her arm around her son Ricky's shoulder and kept it there.

CAPCOM (Duke): Sixty seconds.

Jan Armstrong leaned down, one hand over her mouth, her eyes a little brighter than usual.

At 3 P.M. Pat Collins was reading *Peanuts* and drinking a Coke. At 13,000 feet (four minutes to lunar landing) Pat shook her head and clenched a fist. When Charlie Duke said everything was looking all right (she heard him, as did the other crewmen's wives, over the "squawk box," a small loudspeaker connected with central communications in Mission Control) she encouraged him, "Charlie, you're a good boy, keep it up." Then, 3,000 feet: "Oh God, I can't stand it." At 2,000 feet she was biting her lip. Then, 1,600 feet: "Can I bear it?" Down to 700 feet; now she was biting a finger.

When the lunar module had come around the moon and reestablished wireless communication, Joan Aldrin had crossed over to the fireplace and bowed her head on her hands, resting on the mantelpiece. There was something very solitary about her, and everyone else in the room respected her privacy, crisscrossing around her. She said, "Talk about killing time . . ."

Nobody could find the ham sent by the ladies of the Webster Presbyterian Church (it eventually turned up in the refrigerator); Joan was concerned about Missy the dog getting her digitalis pill on time, and about the noise: "Girls, will you *please* try to cut it down to a low roar?" Astronaut Rusty Schweickart was monitoring the squawk box and making notes in the margins of the flight plan. His wife Clare asked him, "Will you say 'That's bad' if anything's bad?" Rusty said, "I will."

The Aldrins' son Michael was suddenly not around. "Where's Mike? Where's Mike?" asked Joan. Someone went upstairs and sent him down: "What's the matter? I'm watching upstairs." Joan said, "I know you were, dear. I know you were." She had a rather blank look on her face.

She gripped a door and huddled against the frame. She put one hand on the top of the lampshade. Nobody spoke. The lamp beneath her hand began to shake; it was the only thing moving in the room. Her eyes were now large with tears; she sniffed and dabbed at her nose with a tissue. *Sixty seconds to go . . .*

ALDRIN: Lights on. Down 2½. Forward. Forward. Good. 40 feet, down 2½. Picking up some dust. 30 feet, 2½ down. Faint shadow. 4 forward. 4 forward, drifting to the right a little. 6 . . . [*static*] . . . down a half.

CAPCOM (Duke): Thirty seconds.

ALDRIN: [*Static*] . . . forward. Drifting right . . . [*static*] . . . Contact light. Okay, engine stop, ACA out of detent.

ARMSTRONG: Got it.

ALDRIN: Mode controls, both auto, descent engine command override, off. Engine arm, off. 413 is in.

CAPCOM (Duke): We copy you down, Eagle.

ARMSTRONG: Houston, Tranquility Base here. The Eagle has landed.

CAPCOM (Duke): Roger, Tranquility. We copy you on the ground. You've got a bunch of guys about to turn blue. We're breathing again. Thanks a lot.

ALDRIN: Thank you.

Jan Armstrong hugged her son Ricky and smiled when she heard that they had had only thirty seconds of fuel left when they landed. Jan's sister, Carolyn Trude, leaned against a wall and said, "Thank you, God."

Pat Collins had her face screwed tight, head on her hands, when she heard the words "contact light." Then she smiled. It was the first time anyone had seen her smile in more than an hour. She said, "The Eagle has landed. We're on the moon and they say, 'Thanks a lot.' Why aren't they cheering? I guess that's why they don't send a woman to the moon — she would jump up and down and yell and weep. . . . Oh, Mike didn't hear it!" Then she heard Mike Collins's voice: "Yeah, I heard the whole thing." She said, "Oh, you did hear it! Good, honey! Now all we gotta do is get them out of there." Her seven-year-old daughter Ann cut in: "Tomorrow we're staying home from camp, aren't we, Mommy?"

In Buzz Aldrin's home the words "Okay, engine stop" rang out with surprising clarity. Everyone started clapping — everyone but Joan Aldrin, who was still shaking, her head buried against the wall. Then she fell into the arms of Uncle Bob — Robert Moon, the brother of Buzz Aldrin's mother. "Fantastic, fantastic," Rusty Schweickart was muttering through the hubbub. Joan Aldrin walked straight out of Uncle Bob's arms into Buzz's study and into the master bedroom without looking at anyone. Jan Aldrin, age eleven, was visibly shaken. Michael Archer, Joan's father, took little Jan into his arms and gently pushed open the study door for her to go to her mother. Later that night Joan Aldrin recalled, "My mind couldn't take it all in. I blacked out. I couldn't see anything. All I could see was a match cover on the floor. I wanted to bend down and pick it up and I couldn't do it. I just kept looking at that match cover."

Touchdown: it was 3:18 P.M. in Houston, where the event had the most shattering personal impact; it was 4:18 P.M. in New York, where they stopped a baseball game to announce the news and sixteen thousand people stood in Yankee Stadium to sing, joyously, "The Star-Spangled Banner"; it was 6:18 A.M., July 21, at the Honeysuckle Creek tracking station in Australia, where a staff of about one hundred was hanging onto the touchdown by radio circuits. But the date that would live in the history books was July 20. That was the calendar date in the United States, the nation which had underwritten the incredible voyage with much of its treasure and a little bit of its blood. July 20: the aviator Alberto Santos-Dumont was born on that day in 1873; Guglielmo Marconi, inventor of wireless communications, died on that day in 1937. But as long as western civilization endured men would remember July 20 for another reason: that was the day three American astronauts named Armstrong, Collins and Aldrin put man on the moon.

BOOK ONE

Into the Vasty Deep

OWEN GLENDOWER: *"I can call spirits from the vasty deep."*

HOTSPUR: *"Why so can I, or so can any man;*
But will they come when you do call for them?"

SHAKESPEARE, HENRY IV, PART I

Just how long can you hang this operation on the end of a limb? . . . Okay. We'll go.

GENERAL DWIGHT D. EISENHOWER
on the eve of D day, June 1944

It's like the first heart transplant. Until you try it you don't know what it's like.

CHRISTOPHER COLUMBUS KRAFT,
June 1969

1

M Plus 2,974

M STANDS for mobilization. It is a military expression. In the days when European wars were fought by conventional armies, mobilization (M day), either covert or overt, usually preceded a declaration of war. Covert mobilization, when possible, was a handy device to enable one general staff to get the jump on another. But every general staff had its own M plan and a timetable for achievement reckoned in terms of M plus days. In the age of space flight we would now label as a "timeline" the German general staff plan to envelop and annihilate the French Army in 1914 by M plus 39.[1] This one almost worked, and if it had the history of the next fifty years would have been written a bit differently. But there was no way to put a precise timeline on the mobilizations undertaken by the Soviet Union and the United States in the 1950's and the 1960's for the purpose of conquering space. Space: a big, all-encompassing word. Where did one begin? With the moon, of course. That is what one could see in the sky above, a tantalizing crescent, quadrant or full circle (depending on the day of the month) colored neuter white, warm yellow or fulsome orange (depending on the state of the filtering terrestrial atmosphere). Go to the moon? Such an operation could not be planned with exactitude out of experience; the experience did not exist, and there was no "fleet in being" made up of operational spacecraft. The "reserve divisions" that a chief of staff looks for when he makes his calculations existed only in the minds and persons and theories — some of them a little far-out — of the technological and scientific élite which had come into its own during and following the Second World War. But the idea was there, and it would not go away. Indeed the idea had been there for thousands of years: GO TO THE MOON! Set foot on it. Climb its lower foothills, then its higher foothills, then its mountains. Gaze, upon occasion, from it toward the earth; and finally return with an enrichment of man's accomplishment. Man is such a strange creature, lazy and energetic by accident or design; capable of heroic deeds and full of potential personal cowardice: capable, also, of accomplishment *within the means he knows about.* But something within man binds him to accept the idea of challenge: accomplishment with means he does not know about, or does not know enough about. Columbus accepted the challenge. He sailed from Palos, now a mud flat on the southern coast of Spain, with a roughly correct idea of the shape of the world and an absurdly incorrect notion about the size of it. His dead reckoning navigation was a marvelous piece of precision, but he never got where he wanted to go, and he never knew where he had been. Charles Augustus Lindbergh, in 1927, knew where he wanted to go, and he too did it by dead reckoning: from New York to Paris. It took him nearly thirty-four hours, but he got there, and he was surprised that thousands of Frenchmen were around to greet him. New York to Paris — it had been only forty-odd years ago, and now, every twenty-four hours, at least thirteen planes flew the Lindbergh route (the Great Circle), carrying to Paris twelve hundred people daily during the tourist season. The trip took them about six hours, and mostly they were bored. It was only forty-odd years ago, and now . . .

The moon! Did anybody care? The answer was yes. Aside from the fact that it suddenly seemed no more remote than Paris once did, every man of letters from the time of the ancient Greeks had written or spoken about the moon: Keats, Shelley, Coleridge, Shakespeare, Schopenhauer and Whitman in lyrical language; Sir Richard Burton, circa 1880, in words frosted with scorn and frustration: ". . . a globe burnt out, a corpse upon the road of night"; the anonymous author of "The Ballad of Sir Patrick Spens" in words reflecting a primitive sense of fear:

> *. . . I saw the new moone,*
> *Wi' the auld moone in hir arme,*
> *And I feir, I feir, my deir master,*
> *That we will cum to harme.*

The "green cheese" joke about the moon goes back at least to Erasmus in the fifteenth century, and probably farther back than that; Samuel Butler picked it up in the seventeenth century with a line in *Hudibras* ("And prove she is not made of Green Cheese"). But "go there"? To the moon? The superstitious ragamuffins who shipped with Columbus (many of them with misgivings) would have refused the Spanish equivalent of the king's shilling. (Or would they have done so? They were men too, men who knew about the idea and the challenge. In any event they did go. Why? Possibly because they were selected to go.) If one were about to go to the moon, the choice of ways and means was narrow, and the tools to do the job did not exist. They simply *did not exist.* So what did exist? There was a whole body of scientific and pseudoscientific literature on the subject, much of it woolly and worse than useless; some of it — as in the case of Jules Verne — surprisingly prescient. But all of it titillated man's natural and hereditary response to the challenge: what Admiral Richard E. Byrd called "desire to know that kind of experience to the full." [2]

Until the late 1950's nobody knew what the earth looked like; nobody had ever seen it with photographic perspective. Was it really round as Columbus (and the Greeks) thought? How did it appear from 240,000 miles out? By 1969 we knew that it was certainly not flat, despite the last-ditch protestations of the Flat-Earth Society in Britain, still in existence as a quaint anachronism; but it was not really round either. It was not a true cue ball — it was shaped more like a slightly squeezed orange, and because of its huge water masses, anyone living on Mars would call earth the Blue Planet. Could a voyage to the moon be undertaken with reasonable physical safety?

Clearly the answer to that question had to be in the negative.

Was it worth the prodigious effort and the stupendous cost? President John F. Kennedy agonized over the decision. "The cost," he said. "That's what gets me." [3] He was thinking: Forty billion bucks! And we don't even know if the damned thing will work! Is it worth that? Is anything worth that kind of gamble?

This was in the spring of 1961. President Kennedy was a scholar, or at least a well-read man; he knew about Erasmus, and he must have known about the reference to the moon being made of green cheese. Was it worth the gamble to check out Erasmus?

A few men had no doubt that it was worth the gamble; indeed, they thought it was not that much of a gamble. One of them was an Air Force major named Edwin Eugene Aldrin Jr., the son of a distinguished early birdman who had studied physics under a shadowy figure in American history named Robert Goddard, who even at the time of the First World War had some curious ideas about rocket engines. The elder Aldrin knew people named Orville Wright and Charles Lindbergh. The younger Aldrin, in May 1961, was thirty years old and working for his doctorate, on leave from the Air Force, at the Massachusetts Institute of Technology. ("I had some original ideas about piloting problems and rendezvous. . . .") Buzz Aldrin often wondered about President Kennedy's use of the word "if": ("Wasn't it a question of 'when'? Didn't he know that we could build these big boosters?") Perhaps he was not sure; this was a month after the Bay of Pigs, and at this time John Fitzgerald Kennedy was notably suspicious of professional expertise. But the question of "if" or "when" had already been answered: in 1903 by Konstantin Tsiolkovsky, sometimes called "the father of Soviet rocketry." He said, "The earth is the cradle of humanity, but mankind will not stay in the cradle forever." And there is a meaningful postscript on Tsiolkovsky's gravestone at Kaluga in the USSR: ". . . in the pursuit of life and space [man] will emerge timidly from the bounds of the atmosphere and then advance until he has conquered the whole of circumsolar space."

Tsiolkovsky died in 1935, twenty-six years before President Kennedy had to make a decision on whether the damned thing would work.

Even accepting the safety risk, could the tools be built? Could the job be done? By the spring of 1961 the answer had to be a qualified yes. More than three years had elapsed since the Russians astounded mankind by firing the first Sputnik into orbit, and shocked the Americans into putting their own space effort — a fairly casual thing until Sputnik went up in October 1957 — into overdrive. A little too much overdrive, as it turned out; when a Vanguard missile blew up on the pad at Cape Canaveral (now Cape Kennedy) in December 1957, Europeans exhibited some of the latent fright which always seizes them when it appears that the Soviets really *are* ahead of the United States in some critical area. "Do you Americans *really* know what you are doing?" a pro-American British politician,[4] indeed a future foreign secretary, asked an American visitor at that time. He was obviously somewhat shaken, all the more so because at the moment no American could give him an honest affirmative answer.

The Soviet achievement may have been a freak, indeed a kind of timeline gun-jump brought about successfully because they had been building powerful rockets for military reasons almost certainly having to do with a delivery system for their new nuclear weapons. This obviously started during the last years of Stalin, when — by most subsequent accounts, including that of his daughter Svetlana — the man had become a certifiable mental case. The hydrogen bombs which the Soviets began testing in the summer of 1953 (Stalin died in March of that year) were not only "dirty" in terms of radioactive fallout, but they were physically big. They required enormously powerful rockets if they were to be a threat against that distant country — the United States — whose overseas commitments, particularly in Europe, represented the main impediment to the Soviet Union's proximate territorial aspirations. Therefore the Russians had in being huge booster rockets which the Americans had only on drawing boards. At the time we did not need them for military purposes. By the time the Soviets had successfully tested their first H-bomb, the Americans had succeeded in building, with equivalent "yield," lighter nuclear weapons which could be delivered by smaller rockets which were already on the pads, but not so readily adaptable to space probes as the Soviet monsters. It would seem now that the lack of Soviet sophistication in nuclear weapons, and the requirements to deliver them, postulated the necessity to build, for military purposes, the huge boosters which the Americans did not then need. True, the Americans had the Atlas and the Titan, later to be so useful in the Mercury and Gemini programs, but by and large the development of American rocketry was proceeding at a slower pace. The appearance was a bit false, but the Soviet boosters did make the Americans seem to be lacking in rocket technology.

We do not know how much this enormous Russian effort in the late 1940's and early 1950's further impoverished an economy devastated by the Second World War; the Soviets keep their books differently, but it is always safe to assume that they have Great Russian motivations which have at least as much to do with Catherine the Great as with Marxist ideology. For the moment the Americans felt silly, and when they took a hard look at their own space program they felt no better. It was underpowered, undernourished and badly fragmented; the Army, the Navy and the Air Force each had a piece of the action, only vaguely coordinated by something called NACA — the National Advisory Committee for Aeronautics, founded in 1915. The situation *was* embarrassing.

In January 1958 the United States began to recover lost ground — and prestige — when an Army Redstone rocket put the first American satellite, Explorer 1, into orbit. The main thing this proved was that the Americans had a rocket which would work — and perhaps knew what they were doing, after all. It was also a triumph of sorts for the brilliant scientist Wernher von Braun, an émigré from Germany to Huntsville, Alabama; the man largely responsible for the V-2 missiles of the Second World War; the man of whom the entertainer Mort Sahl once said, "He aimed for the stars and often hit London"; the man who took the critical decision to transfer the German V-2 expertise (fifteen hundred people, fourteen tons of paper, consisting mostly of arcane mathematical equations) to the American theater of operations when the Third Reich was being obliterated by the American and Soviet armies in the spring of 1945.

We still had a long way to go, but things were beginning to move. A new agency, the National Aeronautics and Space Administration — NASA — took over the old NACA, plus selected facilities of the Army, the Navy and the Air Force. On April 9, 1959, NASA announced the selection of the first American astronauts, who would forever afterward be known as the Original Seven. Their average age was 34.5 years. Their average height was 5 feet 9.5 inches. Their average weight was 164 pounds. They were all married and they all had children. They were all military men and highly skilled pilots. Their names were Alan B. Shepard Jr., Lieutenant Commander USN; Virgil I. Grissom, Captain USAF; John H. Glenn Jr., Lieutenant Colonel USMC; M. Scott Carpenter, Lieutenant USN; Walter M. Schirra Jr., Lieutenant Commander USN; L. Gordon Cooper Jr., Captain USAF; Donald K. Slayton, Captain USAF. Four of these men had a "Jr." after their names to indicate they were named for their fathers. One would die tragically; in later years Deke Slayton would say meaningfully, "Of course, Gus Grissom is always with us." One would not fly at all in the program they decided to call Mercury — Deke Slayton himself, washed out by a minor heart condition.

But the decision concerning how far we would go beyond Mercury — and how fast — had not really been taken; and it was President Kennedy who had to take it. The evidence that it was possible to go to the moon was accumulating fast. On April 12, 1961, Russia's Yuri Gagarin, later killed in an ironically conventional plane crash, became the first man to orbit the earth in a capsule the Soviets called Vostok 1. The trip lasted 1.8 hours; it was the first time man had traveled at the speed of 17,500 miles an hour. Still Kennedy hesitated. Then on May 5, Commander Shepard put the United States in space — or almost in space! — with a fifteen-minute "suborbital" flight from Pad Five at the Cape, wedged into a 2,300-pound Mercury capsule shaped like the top of an ice cream cone (Shepard called it Freedom 7). When he parachuted, capsule and all, into the Atlantic Ocean, 302 miles away, words like "splashdown" and "downrange" were in the vernacular — along with "A-OK," an expression coined by a public affairs announcer, and one which the astronauts never did like, and never did use.

The Redstone rocket which powered Shepard's first flight was strictly a Model T; everyone in the United States felt a little better, but that Redstone was not going to take Shepard or anyone else to the moon. Neither was that little one-man Mercury capsule. A go-ahead had been given to build the massive rocket which would be called Saturn, but not to build a three-man spacecraft. Until almost the last minute, Kennedy drew back; he wished the decision would go away. He kept asking the scientist Jerome Wiesner, "Can't you fellows invent some other race here on earth that will do some good?" [5] So far, no cigar. But Kennedy smoked cigars, and twenty days after the Shepard flight he bit one — hard. In a second State of the Union message to Congress, he said: "I believe that this nation should commit itself to achieving the goal, before this decade is out, of landing a man on the moon and returning him safely to earth."

That was M day, and a good one to remember: a date that will live in honor, not in infamy. It was May 25, 1961. The industrial and technological mobilization Kennedy was proclaiming for peaceful purposes miniaturized every military mobilization undertaken by any nation — in the qualitative sense at least. The audacity of the man! He was truly summoning Shakespearean "spirits from the vasty deep," with no sure knowledge — except, perhaps, an intuitive knowledge — as to whether they would come. And he even put a timeline on the accomplishment: *before this decade is out.*

On May 25, 1961, Neil Armstrong (blond hair, blue eyes, 5 feet 11 inches, 165 pounds) was a civilian test pilot for NASA at Edwards Air Force Base in

California. He came from a small town in Ohio called Wapakoneta. There, as a boy, he had had a recurrent dream . . .

« I could, by holding my breath, hover over the ground. Nothing much happened; I neither flew nor fell in those dreams. I just hovered. They can't have been bad dreams. But the indecisiveness was a little frustrating. There was never any end to the dream . . . »

He had flown Panther jets off carrier *Essex* during the Korean war: seventy-eight combat missions. He was in one of the earliest all-jet squadrons, and his squadron took more casualties than Neil Armstrong liked to remember . . .

« I got three air medals. Apparently I caused a lot of damage to bridges and trains and things, but really they handed out medals there like gold stars at Sunday School. »

Once he coaxed a badly crippled jet back to the deck of *Essex*; another time he did not get his aircraft back. Flying a low-level mission, he clipped a wing on a cable, then nursed the plane far enough to parachute into American-held territory. But he was pure pilot; indeed he was known in the Navy as a "hot" pilot. So he resigned from the Navy in 1952 to fly the hottest stuff he could find. He had been a lieutenant (j.g.), but now he was Mr. Neil Armstrong. He went to work as a pilot and engineer for the old NACA's Lewis Flight Propulsion Laboratory in Cleveland, Ohio, a hundred and forty-five miles from his hometown, then transferred to NACA's high speed flight station at Edwards Air Force Base in California as an aeronautical research pilot. He knew about airplanes; he had started building models out of scraps when he was nine years old. Now he was flying the hottest stuff in the world: the B-47 jet bomber, the F-100, the F-101, the F-102, the F-104, the F-5D, the X-1 rocket airplane, and the X-15 — that one to more than 200,000 feet, at 4,000 miles an hour. He pushed his flight log to more than 4,000 hours of flying time. He lived with his wife Jan in a primitive cabin five thousand feet up in the San Gabriel mountains. When they moved in it had neither electricity nor plumbing, but it had a great view; on clear days Jan could see her husband flashing through the sky in the most exotic aircraft of the era.

The Mercury people made some tentative overtures to the test pilots at Edwards, but Neil Armstrong was not much interested — then — in becoming an astronaut . . .

« The X-15 experimenters were not basically in agreement with the Mercury approach. We weren't sure that was the right way to go; some of us felt a winged vehicle represented a better approach. Worse than that, perhaps, we tended to think of the space task group people as babes in the woods. Then, in February 1962, John Glenn orbited the earth three times in a little less than five hours, and we began to look at things a bit differently.

It was a hard decision for me to make, to leave what I was doing, which I liked very much, to go to Houston. You don't have to be in any particular program or wear a particular color of shirt to find research questions that need answering. It wasn't a question of do you want to be an astronaut or do you want to sweep streets. But by 1962 Mercury was on its way, the future programs were well designed and the lunar mission was going to become a reality. I decided that if I wanted to get out of the atmospheric fringes and into deep space work, that was the way to go. »

Meanwhile, at Edwards, Neil Armstrong had struck up a nodding acquaintance with a slim Air Force captain named Michael Collins, an experimental flight test officer.

Mike Collins (brown hair, brown eyes, 5 feet 11 inches, 165 pounds) was a fellow who looked and talked all of his Irish. He disliked gobbledygook and he did like the inherent simplicity of the English language. Much later, when the special (or "astronymic") language of space seemed to him to be getting out of hand, he had a way of interrupting a conference by asking, with a simplistic look on his face, the innocent question: "Do you mind explaining that in agricultural terms?" Mike was also an Army brat. He never had a hometown. He was born in Rome, the son of a general and the nephew of another one — "Lightning Joe" Collins, one of the American Army's finest corps commanders in Europe during the Second World War, later Army Chief of Staff. Naturally, Mike drifted into the military; but he never thought of himself as entirely committed to it . . .

« I went to West Point to get a free and excellent education. Then I had to do four years, in the Army or in the Air Force. I chose the Air Force because it sounded more exciting and more innovative. I did think the Army might be better for a career as such, but I wasn't really thinking about careers then. I have a tremendous admiration for the military, but I guess I'm more of a civilian at heart. In fact, after I had done my four years, I thought seriously about chucking the whole thing. But I was in a fighter squadron in France, and I liked what I was doing. I truly enjoy flying. I do it well and I enjoy doing something that I do well. It infuriates me to hear somebody say, 'Well, it all counts for twenty, you know . . .' The translation of that is: they don't care if they are doing their job well or not; they don't care about anything; it's just another month gone on the calendar, another month closer to their twentieth year in the service when they can drop the whole thing and retire. Now that attitude really gets my back up.

Besides, I *like* fighter pilots. I really do. They're good guys. As a group, I like them better than I like any other group. They're very independent people. They're not just talkers; they're doers. They have their lives on the line in a lot of ways, and they're good people. They say what they mean. There is no sham or pretense about them. The good ones you know not by what they say but what they do, and when you fly with a guy you can tell what he is. You can be irresponsible and you may get away with it being a fighter pilot; but you most certainly cannot as a test pilot. Fighter pilots can be impetuous; test pilots can't. They have to be more mature, a little bit smarter. They have to give more thought to what they're doing, or they're going to — well, maybe not kill themselves, but, even worse, they'll come to wrong conclusions about airplanes and others will kill themselves later when the aircraft reach squadron service. They have to be more deliberate, better trained — and they're not as much fun as fighter pilots. People think of a test pilot as a fellow with a white scarf trailing out of the cockpit as he puts the plane into a power dive to see if the wings fall off. Fighter pilots may have a little of that in them, but not test pilots. The test pilot has to be more of an engineer, and a lot of his work is drudgery, studying charts and graphs, determining what is safe for the airplane to do and what isn't. And the test pilot has to be *older* . . . »

In France Mike Collins felt older. He also felt a commitment to flying per se, of which test piloting is a logical extension. He applied for the Air Force test pilot school at Edwards Air Force Base. He didn't make it until 1961, but he did make it. Although he never flew the black and rather fearsome looking projectile called the X-15, which now shares the entrance hall of the Smithsonian Institution in Washington with the original plane flown by the Wright brothers and Lindbergh's *The Spirit of St. Louis*, Collins did test-fly practically everything else.

In that arcane world which test pilots can share with no one, neither Neil Armstrong nor Mike Collins had ever heard of Edwin Eugene Aldrin Jr. In fact hardly anybody knew him by that name; he called himself Buzz Aldrin. The subject of his doctoral thesis at MIT was rendezvous and docking in space. This technique had to be perfected if man was to go to the moon, and the moon was where Buzz Aldrin wanted to go: not just because his mother's maiden name was Marion Moon, but because he was a man intrigued by the ultimate challenge. And this was it for Buzz Aldrin (blond hair — what there is of it; blue eyes, 5 feet 10 inches, 165 pounds) . . .

« When you try to answer the question of motivation — 'Why do you do it?' — you have to ask another question: 'What type of people are you talking about?' In our business we are talking about airplane pilots, fighter pilots. If you are a fighter pilot, you want to get hold of the hottest thing you can find. Being able to solo really puts you on that pad. Then you ask yourself, 'What else can I fly? What else is available?'

So you come down to the ultimate, the space program. It's there. Some people don't think they have a chance to make it and don't bother to try. Others say, 'I'll accept that challenge.' So they try. Some make it and some don't. But there's a treadmill that's going, and if you make it, you're on it. You have some control over it by your own performance. A lot of fate determines where you fit into the puzzle. But why do you do anything? Maybe because you were selected to do it. »

As a "leaning Presbyterian," Buzz put a touch of Calvinist predestination into that remark; but, aside from any significance an astrologer might read into his mother's maiden name, Buzz Aldrin did seem to have been "selected." There was, for instance, the matter of his father, who held a commission in the Coast Artillery Corps in the spring of 1918 and was due to go to France to command a battery when he was suddenly detached to the Aviation Section of the Army Signal Corps to teach a cram course in aeronautical engineering set up by MIT for the Army and the Navy. One of his students ("They were all quite a nice bunch") was LeRoy Grumman, founder of the Grumman Aircraft Engineering Corporation, which eventually would make the lunar module, Eagle, from which a man named Buzz Aldrin would step onto the moon.

In those days hardly anyone knew anything about aerodynamics, but the elder Aldrin and Orville Wright knew a little. They met when Aldrin was organizing the Army's Air Service Engineering School at McCook Field in Ohio following the end of the First World War: "I lived right across the street from the president of the National Cash Register Company. The old gal with whom I stayed was a widow, very close to the Wright brothers. So I went out and called one Sunday, and from then on I was in touch with Orville Wright whenever he came to the field." At McCook the elder Aldrin lectured majors and lieutenant colonels who had flown in combat in France and even commanded airfields but who knew nothing of aerodynamics ("the difference between direct current and alternating current, things of that kind"). He learned to fly himself and was posted as an aide to the airpower prophet Billy Mitchell in the Philippines. Back in the United States, he got the Army to send him to MIT to get a doctorate ("I said, 'I paid for my education. I'm not like you West Pointers who got it for free' "). But there was not much future in the Army of the 1920's for a man with a Sc.D., and in 1928 Buzz's father went to work for Standard Oil of New Jersey. The next year his new boss, Vice Admiral Emory S. Land of the Guggenheim Foundation, who knew of Colonel Aldrin's earlier association with Robert Goddard, said to Aldrin, "Slim Lindbergh is running around here with his tail between his legs." Aldrin said, "Let's send him up to see Goddard." Lindbergh then went up to Clark University in Massachusetts, and that is how he made contact with the new world of rocket fuels. (On March 16, 1926, Goddard had fired the first successful liquid-powered rocket in the United States. It traveled one hundred eighty-four feet in a Massachusetts farmyard.) Aldrin's main job was to promote commercial aviation; only the oil companies had money for that sort of thing in those days, and in 1930 he flew a single-engine Lockheed Vega over the Alps at 14,000 feet — to the horror of the Swiss military, who tried to talk him out of it. That was in the same year Buzz was born, in Montclair, New Jersey. Three years later Buzz had his first plane ride — again in a Vega, to Miami, Florida, with his father as pilot . . .

« It was a ten and one half-hour trip, with a fuel stop in Charleston, South Carolina. I remember the trip very well, because I went down by plane and I got sick. We went back by boat and I did not get sick. I remember also that in Florida, down in the Biscayne area, I wandered away and got lost. Of course I didn't know I was lost. Then a car came cruising by to fetch me. It was the folks. I also remember falling face down in the water at some beach. I was afraid of water after that, and I stayed afraid of it for years. »

He went to West Point because he wanted to fly, and if you wanted to fly the academy route (West Point or Annapolis) was, at that time, the road you traveled. It led Buzz Aldrin to Bryan, Texas, where he got his pilot's wings, then to Korea, where he flew sixty-six missions (two MIGs destroyed) before he came home with the Distinguished Flying Cross. Later, in Germany, he renewed his acquaintance with another Air Force pilot named Edward White, whom he had known at the Point and who was interested in getting into "the space program" — something then called Mercury. Buzz Aldrin began to think about that too . . .

« I accepted a long time ago the fact that there are things I can't do better than a lot of other people. When I was at West Point I accepted the challenge to try for a Rhodes scholarship, but I found out that I wasn't the kind of guy they were looking for. I'm an introvert, I guess.

But what is challenge? What determines 'excellence'? In grade school it's the marks that you get, what you do on the athletic field. At West Point the name of the game is, 'Do what people tell you to do, keep your nose clean and work out your academic progress.' I fitted into that pretty well.

I knew I needed more formal education. Not because I wanted to know more for the sake of knowing; I needed knowledge that I could put to useful work. And I did have these ideas . . .

So I asked the Air Force to send me to MIT, and that is where they sent me. I also tried for the 1962 class of astronauts. That's the year Ed White made it. But test pilot experience was required, and I didn't have it. I tried for a waiver. I knew darned well I wouldn't get it, but I wanted the application in the record. Of course I was disappointed, and I even considered leaving MIT to get the test pilot experience. But I finally decided, 'Here I am. I might as well get the best out of it.' So I stayed in Cambridge. Then, the next year, they dropped the test pilot requirement, and I qualified for the program. »

On January 9, 1969, three men were named to what NASA called the G mission — and everyone else started calling Apollo 11. Their names were Neil Alden Armstrong, flight commander; Michael Collins, command module pilot; Edwin Eugene Aldrin Jr., lunar module pilot. They were going to the moon.

But when? One now could forget about the word "if"; finally we *were* going. In fact, we had already been fairly close to the place. The whole Apollo program, however, came a tragic cropper in January 1967 when three astronauts, one of them Mike Collins's close friend and West Point classmate Ed White, the same Ed White Buzz Aldrin had known in Germany, lost their lives. A flash fire asphyxiated them inside their spacecraft during a simulated launch at Cape Kennedy. For nearly two years nobody flew; indeed the whole spacecraft had to be redesigned, taken apart and put together again. Men might still lose their lives by catastrophe in space, but that particular kind of disaster must not be allowed to happen a second time.

The delay actually allowed more than enough time to perfect and test-fly the giant Saturn V booster which was to be Apollo's workhorse. The first big V thundered aloft from Cape Kennedy, unmanned, on November 9, 1967, and another was fired on April 4, 1968. By the summer of 1968 the redesigned spacecraft was ready, and in October that year Walter Schirra, Walter Cunningham, and Donn Eisele flew the first manned Apollo mission — the C mission, or Apollo 7. Their task was to check out the new craft in earth orbit. To hurl them only that far, the big Saturn V was not needed. They rode up on a smaller, uprated Saturn I and their flight was a dizzying success: 163 earth orbits in 260.2 hours; the Apollo spacecraft was a "dream to ride."

The next step, according to NASA's master plan, should have been to launch a D mission, putting the Apollo spacecraft and the spindle-legged lunar module, the LM, into earth orbit so crewmen could practice docking the two together, practice extraction of the LM from its protective cover, practice flying the funny-looking LM in the relative safety of earth orbit.

But by the summer of 1968 it was apparent that the LM, which was ultimately designed to touch down physically on the moon, would not be ready to fly that year. Something else had to be done meanwhile. What about an intervening mission? Should we do the same thing that Apollo 7 had done? Or the same thing with variations? That would have been the safe decision, and it was considered. But bolder minds began turning to the more "aggressive option" — to forget about the lunar module for the time being and go for broke by flying around the moon. Around the moon? *Around the moon.*

Unmanned craft had already done it — Russia's Luna flights, the United States' Surveyors and Orbiters, but men had never done it.

One advocate for the "aggressive option" was Lieutenant General Samuel Phillips, NASA's Apollo program director. He liked the idea, but it meant designing a whole mission from scratch. It was like trying to cram all the preparations for the 1944 Normandy invasion into two or three months. True, there was another crew in training for a high-orbital flight with the lunar module; possibly it could be moved up a slot. This was the "E mission" crew, headed by Colonel Frank Borman. It had trained for months to fly the third manned Apollo on a deep space probe to test equipment and navigational capabilities several hundred miles out from earth. Could that crew perhaps go from "several hundred miles" to about 240,000 and circumnavigate the moon — without the LM? Dr. Thomas Paine, deputy administrator of NASA, had some doubts. The training schedule would be terribly tight; after all, nobody had flown such a mission before, no man had ever flown on the big Saturn V (both earlier tests were unmanned), and the horror of the 1967 pad fire was still fresh in everyone's memory. Trained crews were not expendable. At one point Dr. Paine telephoned the E mission commander, Borman, and asked him bluntly if he thought a circumlunar flight at this point was a good idea. Borman said he did think so, and he thought the risk was minimal. But even Borman was not sure that he, along with his fellow crewmen James Lovell and William Anders, had a circumlunar "go" until Dr. Paine made the announcement a few days later. Then, after the decision had been taken to go circumlunar, the E mission was redesignated C prime.

A few months after the flight Borman admitted, with inelegant but telling eloquence, that "We had a horseshoe up our backside." Whatever they had, the flight of Apollo 8 was stunning: a perfect 147-hour trip to the moon and back, including ten lunar orbits and a Christmas Eve reading, live from seventy miles off the moon, which no earthbound listener will ever forget: *In the beginning God created the heaven and the earth . . .*

It was at the time of the Borman-Lovell-Anders flight that the crew for G mission — Apollo 11 — was picked. For ten years the manned space program had gone forward much as a skyscraper is built, each mission adding a few floors at a time. Now three names went on the top floor: Armstrong,

Collins, Aldrin — and it was Aldrin who once said, "A lot of fate determines where you fit into the puzzle."

Still the lunar module had to be checked out. In March 1969 James McDivitt, David Scott and Russell Schweickart did that for the first time — the D mission, Apollo 9. They took a lunar module into earth orbit. McDivitt and Schweickart got into the weird contraption — in a sense, the first true spacecraft — and flew off by themselves, then came back to be picked up by Scott, who had been flying the command module alone. This test of the lunar module was critical; so was the "spacewalk" of Rusty Schweickart, who was testing the portable life support system (the PLSS backpack that would be used when another lunar module crew actually stepped on the moon). Everything worked; also, all the docking and spacewalk experience accumulated during the Gemini program (ten manned flights between March 1965 and November 1966) was now paying rich dividends.

The success of Apollo 9 suddenly offered a tempting possibility. Should F mission, for which an able crew had been picked, land on the moon? This was Apollo 10, scheduled to go in May. The lunar module assigned to this mission could not land on the moon; it was too heavy to land and satisfactorily return a crew into lunar orbit. But there was still time to unstack the Saturn V assembly and replace its lunar module with a lighter version which was considered to be moonworthy. The public half-expected that to happen; by now the astronauts were making the most difficult business in the world look almost too easy. Computer-minded people, even in NASA itself, were certainly tempted. But Colonel Thomas P. Stafford, the commander of F mission, was against it, even though that would have meant, for him, the fame of being the first man to step on the moon. "There are too many unknowns up there," he kept saying. "Our job is to eliminate as many of them as we can, and the only way we can do that is to take this thing down to nine miles or less and see how it behaves that close to the moon. And we'll map the landing sites with our cameras. We can't get rid of the risk element for the men who will land on the moon, but we can minimize it and that's our job — to find out everything we can so that only a small amount of unknown is left."

General Phillips and Dr. Paine, now administrator of NASA, agreed with Stafford's estimate. In Houston, so did Christopher Kraft, director of Manned Space Flight Operations, and George Low, Apollo program manager. So Stafford, John Young and Eugene Cernan took the command module and the lunar module through the exact trajectory planned for Apollo 11 in July. Then, while Young orbited the moon in the command module, Stafford and Cernan flew the lunar module down to less than 50,000 feet off the surface and put the whole show on thrilling color television, seen live 240,000 miles away. Having done everything called for in the Apollo 11 flight plan but a final powered descent to the moon's surface, they rendezvoused and docked with Young in the command module. "Snoopy and Charlie Brown are hugging one another!" shouted Tom Stafford with a boyish delight he did not bother to disguise. Then they sent the descent stage of the lunar module, its job done, spinning into the vast nothingness of a solar orbit — and came home. Now it was up to Armstrong, Collins and Aldrin.

Not all of the unknowns had been solved. It appeared that the mass concentrations ("mascons") on the lunar surface — that is, areas of extreme mass density which occur in or near the big flattish *maria* or "seas" of the moon — were still doing odd things as seen by the ground tracking radars. The moon's gravity had long since been calculated to be one-sixth of earth's, as applied to anything resting on the lunar surface, but that figure represented an extrapolation of data about a place no human being had ever visited, and there was now some doubt that it was entirely consistent. The new data indicated gravity variations, presumably because of the mascons, which were probably meteoric in origin but which had properties we could only guess at. Three new men had to go the rest of the way and find out what they could. Their flight plan, down to 50,000 feet, was identical to that of Apollo 10, although it once occurred to Buzz Aldrin that there was a slight difference . . .

« In the back of their minds they [Stafford, Young and Cernan] knew they didn't have to land on the moon, so they didn't count on it. Up here in the front of our minds we know we are going to land, so we may wind up doing a few things differently. »

By the end of May, Armstrong, Collins and Aldrin were tired men. They were spending ten, twelve, fourteen hours a day in their simulators — the gigantic earthbound machines which John Young called "the great train wreck." All week they would work in those machines at Cape Kennedy in Florida, then fly in T-38 jets to their homes near Houston to mow the lawn and cope with household matters like television breakdowns and leaky plumbing and childhood injuries. They saw little of their wives, and what little Jan Armstrong saw of Neil Armstrong during that period bothered her a good deal: "Neil used to come home with his face drawn white, and I was worried about him. I was worried about all of them. The worst period was in early June. Their morale was down. They were worried about whether there was time enough for them to learn the things they had to learn, to do the things they had to do, if this mission was to work." Neil Armstrong, as flight commander, had the extra chore of flying the LLTV (lunar landing training vehicle), a fractionally different version of another machine in which he once had almost lost his life. But the hardware was ready; nothing had gone wrong on Apollo 9 and Apollo 10 that could not be fixed. Could the men be ready for liftoff on July 16? That decision had to be taken, because during the second week of June it was time to start putting hypergolic propellants into the fuel tanks of the lunar module and the command module — they were already on the pad for G mission. The members of the crew could not take the decision; they were not necessarily the best judges of their own physical and mental state. The man who had to take it was General Samuel Phillips. This lean, gray-haired man had to say, "Okay. We'll go" — or push the mission back to August.

Washington, D.C.
June 12, 1969

Sam Phillips's office had a green slate board in it. Sometimes he picked up a piece of chalk and drew things on the slate board. He liked to draw "curves" which indicated where you probably were, where you might get to be, and where you might never get to be — all in human terms, of course. There was a curve which he drew, rather freely, which went up and across to the right — rather like the left side of the moon's outline when it is at full. He was aiming for what might be called "full performance" — that is, when the men reached the top of the curve and leveled out, you might as well go ahead. They might become fractionally better, but not all that much better.

General Phillips knew about making decisions. He had made a pretty big one himself — back in 1962: "I was the director of the Minuteman program. I had to make the decision on going operational with this program. That meant unattended rockets, with nuclear warheads, remotely controlled by wires. For me, this was a very personal decision. It had to be approved by President Kennedy — and it was. But it still was a very personal decision."

So was the decision he had taken to fly the C prime mission — Apollo 8, ten times around the moon and back home: "I had to decide whether we were ready. That isn't the kind of decision you plug into a computer. You are dealing with human beings, different human beings, and other human beings have to weigh and decide."

He decided affirmatively; Dr. Paine endorsed the decision, and Borman, Lovell and Anders went.

General Phillips remembered another decision — the one taken by General Eisenhower in June 1944: do we invade Normandy or not? An RAF wing commander named Stagg kept coming in with gloomy weather reports; under those circumstances Sir Trafford Leigh-Mallory did not think he could provide the necessary tactical air support; General Sir Bernard Law Montgomery was willing to say the hell with the weather and let's go; and a man named Eisenhower had to decide. He postponed the invasion for twenty-four hours, and ships in the English Channel, loaded with troops, had to be called back; then Stagg brought good news instead of bad about the weather. Leigh-Mallory was still dubious, but Eisenhower finally said, "Okay. We'll go." [6]

In June 1944 Sam Phillips was a captain in the Army Air Corps. ("I flew a P-38 that day.") Now he had to make a kind of D day decision himself. The chain of command linked General Phillips to George Mueller, associate administrator of NASA, to Dr. Paine, administrator of NASA, a Democrat promoted to the top job by a Republican President named Nixon. And, at the very top, there was the President himself.

On June 12, 1969, Tom Paine was not in Washington at all. He was in Paris, but he was standing by a telephone. Tom Paine is a man who does not want to run things all the way down: "I have these squawk boxes in my office and in every room in my house. When they are in space I hear every word spoken, from air to ground. At no time do I try to run a mission from Washington. I'm not going to tell anyone to turn left or turn right to avoid a boulder. That's their job." And he had left General Phillips certain specific instructions before he went to Paris: "I told Sam that if he had any reservation, *any* reservation, about the men, about the equipment, about the launch-pad facilities — then defer the whole thing and go in August." This was the kind of language Sam Phillips understood. He knew that the decision was up to him, and that George Mueller and Tom Paine would stand behind him. Paine knew that the lives of three men, three rather particularly trained men, were hanging in the balance. He was concerned about that, but he was not going to interfere with what Sam Phillips was doing.

So it came to pass that at 12:30 P.M. on June 12, 1969, a four-way conference telephone call took place. It was all done with squawk boxes. Sam Phillips and his staff were in Phillips's office — room 5043 in a slab-sided building three blocks around the corner from the Smithsonian in Washington. George Low and Gene Kranz, chief of Flight Control, were on the line in Houston, along with Dr. Charles Berry, director of Medical Research and Operations. Then there was Lee James, Saturn V program manager in Huntsville, Alabama, and — at Cape Kennedy, Florida — Deke Slayton, director of Flight Crew Operations, and Rocco Petrone, director of Launch Operations. A dozen or so men, all surrounded by staff expertise, had a voice in the decision: go or no go?

Sam Phillips had the opening statement to make, and he made it: the purpose of this conference call — a "meeting" of sorts — was to decide whether the preparations already made and in train supported the July 16 launch date. Sam Phillips was very tough on this point: "I am fully prepared to delay if something is not ready, or if we are pushing these men too hard. If we are doing that, then we will reschedule for August."

This launch date — July 16 — had been on the Apollo calendar for about a year. It could still be "slipped" — that is, changed. But General Phillips had to stand ten feet away and think about that. How much is enough? Another year of training would make them only marginally better. Besides there are severe limitations on this mission. We have limited them to one extravehicular activity on the moon. We won't do the full-blown exploration package we once talked about. We're going to go there more than once. How long can you keep this thing *hanging on a limb?* There is the psychological factor. Astronauts are not supermen; they're people, like athletes. And like athletes, they can go stale . . .

But Sam Phillips did not want to go into the morning of July 16 with a bunch of tired astronauts, and that was really what he was getting at on June 12. There did not seem to be many questions to ask about the hardware, about the flight controllers, about the paper work; all that was in order, and Gene Kranz in Houston and Lee James in Huntsville, Alabama, checked out everything quickly enough. The only question left concerned the human beings — the kind of question Sam Phillips could not put into a computer. He did not want a lot of advice, but he did want the best advice there was, because he had to take the decision; and in this area he had to rely on Deke Slayton and Dr. Berry.

Chuck Berry was skeptical. He necessarily took a medical view of things, and he was worried about the heavy pace of the training. He would have liked Armstrong, Collins and Aldrin to have a little more time.

The astronauts were not on the squawk box. They had not made a recommendation for or against slipping the flight, and they had not been asked to make one. Neil Armstrong had reported, through Deke Slayton, that the training was very tight — but they were up to date.

Now Deke Slayton had to give his judgment. He reviewed the experience of the astronauts with simulators. They had given a lot of trouble on C mission — Apollo 7, Schirra, Cunningham and Eisele. Sam Phillips remembered that "those simulators gave us fits." But the simulators had since given the astronauts progressively fewer fits; during the training of the Apollo 11 crew, which was the whole point of the discussion, they had been functioning at something like ninety percent productivity. True, the men had become very tired. But they were over what looked like the worst part, and possibly going into a period when they could plane down a bit in their training.

Deke Slayton said he had no reservations. He recommended: "Go."

Chuck Berry did not resist the decision.

After an hour and a half it was finally up to Sam Phillips. He said: "Go." That was it.

The squawk boxes fell silent in Sam Phillips's office in Washington, at Cape Kennedy in Florida, at Huntsville in Alabama, and at the Manned Spacecraft Center, twenty-five miles out in the prairie from Houston, Texas.

When she next saw her husband a few days later, Jan Armstrong noticed a difference immediately: "The turning point came with the decision to go.

After that everything seemed to get better. They knew they were going, and the knowledge seemed to take the weight off their shoulders. That last weekend they were home [the July 4 weekend] they were in great shape. When I talked to Neil on the Monday [of launch week, two days before liftoff] he was ready to go. When I talked to him on Tuesday he was *really* ready. They were ready to go *then*."

Jan Armstrong had only one worry — that the men would get near the moon and then, for some unpredictable reason, be unable to accomplish their mission. If they had to abort, three disappointed men would have a long and depressing ride home. But, as July 16 approached, she had decided not to worry even about that: "It doesn't do any good. They're up there on their own. There isn't anything we can do to help. They have to hack it. If they can't hack it — well, that will be awfully sad, but there isn't anything we can do. So there's no point in my worrying."

July 16: this was the date the explorer Roald Amundsen was born in 1872; the date Nicholas II, last Czar of all the Russias, was executed by the Bolsheviks in 1918; the date the first atomic bomb was exploded in New Mexico in 1945. It was, dated from Kennedy's second State of the Union message in 1961, M plus 2,974.

Meanwhile Tom Paine had returned from Paris, and he decided there was one more thing he had to do. He wanted to see the three men once again. On July 10 he flew down to Cape Kennedy to have a private dinner with them. He wanted to tell them something: If you get into trouble up there, don't hesitate to abort. Come on home. Don't get killed. If you do have to abort, I promise you that this crew will be slipped ahead in the mission sequence. You will get another chance. *Just don't get killed*. . . .

To go places and do things that have never been done before — that's what living is all about.

MICHAEL COLLINS

2

T Minus and Counting

Cape Kennedy, Florida

IN any Apollo mission the fifth day before flight is a big one. It is called F minus five; only in the latter stages of the final countdown does the nomenclature change to T minus hours and minutes (T stands for "time before liftoff" or "time to liftoff"). F minus five, in this case, was Friday, July 11: the day of the last comprehensive medical examinations for the three Apollo 11 astronauts. There would be another examination on the morning of launch day, July 16, but there would not be enough time for that to amount to much; as Mike Collins once said, "The doctor looks in one ear and if he doesn't see out the other, you go." Today, July 11, was the day to determine if that "ingenious assembly of portable plumbing" which Christopher Morley[1] once called the body of man was ready, in the human forms of Neil Armstrong, Mike Collins and Buzz Aldrin, to undertake the grand voyage which had required ten years and twenty-four billions of dollars' worth of preparation. There could be no thought of crew substitution at this point. There was, to be sure, a backup crew consisting of Jim Lovell, Bill Anders and Fred Haise — all highly competent men. But, since Armstrong, Collins and Aldrin were named in January to fly Apollo 11, too much specialized training had been theirs and mostly theirs alone. For at least two months it had been clear to all that if Armstrong, Collins and Aldrin did not go on July 16, nobody went for at least a month. It was much too late to play platoon football; indeed, one member of the backup crew, Bill Anders, was in the process of separating himself from space flight to take an administrative job with the National Aeronautics and Space Council in Washington.

The medical responsibility was Dr. Berry's. He had been skeptical about the July 16 launch date when the decision was taken on June 12; now he had to make sure that, from his point of view, the decision had been the right one. Certainly the astronauts were in better overall shape physically than they had been a month earlier; everyone had noticed that, particularly their wives. But a minor case of the sniffles, a telltale redness in the throat, an earache, a stomach upset, an unpredictable intestinal disorder — any of these things could bring the whole enterprise to a sudden, shuddering stop. There had been a situation like that once before, and not too long ago. Back in March, about a week before Jim McDivitt, Dave Scott and Rusty Schweickart were due to fly Apollo 9 and give the lunar module its first test in earth orbit, all three men developed sore throats. It was at this time that the expression "tired astronauts" came into wide usage; the expression was Dr. Berry's, and it summed up a worry he had to live with all the time. Berry strongly recommended that Apollo 9 be postponed for three days, and it was.

A three-day slip in this case was out of the question. It was a practical thing to do in the case of Apollo 9 because that flight was not going anywhere near the moon. Given acceptable weather conditions on earth, actual at time of liftoff and predictable at time of splashdown, a day or two did not matter.

Even minutes mattered in the case of Apollo 11. The time of launch — 9:32 A.M. eastern daylight time — had been chosen with great care. The date and the hour had been fixed to take full advantage of the so-called "lunar window," which is a shorthand expression to identify a complex combination of circumstances. The lunar window is actually a limited period of time in which a mission like Apollo 11 — or Apollo 10 — can do its job. The time is limited by rocket propulsion capabilities, by the time of dawn or sunset at the launching site and in the downrange recovery and abort recovery areas, and above all by lunar landing visibility conditions. These are governed by the positions of the earth and the moon vis-à-vis one another on a specific day and at a specific hour. The moon's distance from the earth is not constant; it varies from 252,000 to 221,000 miles and averages out at about 240,000 miles. The moon requires twenty-seven days, seven hours and forty-three minutes to make one revolution around the earth, and its position relative to the sun results in a lunar day and a lunar night of about fifteen days each. It is not that easy to hit the moon at all, let alone a specific landing site; as recently as 1965 an unmanned Soviet moon shot (Luna 6) missed the whole place by one hundred thousand miles and went wandering off into space. One does not get into a rental car with a road map and aim haphazardly for a destination which itself is moving at an average rate of 2,287 miles per hour. At the time of launch one does not aim for where the moon is, but where it will be a few days later. If Armstrong and Aldrin were to touch down at the landing site that looked most favorable — Landing Site 2 on the moon maps, in the Sea of Tranquility — at a time of optimum visibility, Apollo 11 had to go at 9:32 A.M. or very close to that time. In the larger sense, this time had been fixed a few billion years earlier — whenever it was that the earth and its moon satellite settled into their orbital relationship. (If, walking at a fixed gait, a person swings over his head, at a constant rate, a ball attached to a piece of string, he will never see the back side of the ball; therefore we never see the back side of the moon.) There was some margin for error, in the case of a slightly delayed launch, which could be corrected by changing the Saturn V's azimuth — that is, its horizontal bearing from a fixed point or direction, in this case true north. But other things in the flight plan then would have to be changed, especially the timing of translunar injection — TLI — the big burn that had to take place at just the right moment to eject the spacecraft from earth gravity and send it coasting to the moon. In any event the margin was painfully small, no more than four hours, and a last-minute mechanical malfunction could still scrub the mission for a whole month.

So could a sore throat. Since their last trip home to Houston over the Fourth of July weekend the Apollo 11 astronauts had lived a semi-isolated existence. Except to go into their simulators and practice for the hundredth time, or perhaps the five-hundredth time, the mechanics of landing on the moon and getting off, they rarely left their windowless, and unearthly quiet, crew quarters: small and immaculate private bedrooms, small but adequate private offices, names on the doors, names to be stamped indelibly into the history books; a conference room where they could go over the flight plan and the landing charts once again. The conference room had perforated walls at each end to filter the air and keep it germ-free, but Mike Collins could never quite convince himself that it worked; the air always seemed, to him, to be flowing in the wrong direction. Lew Hartzell, the cook, saw them every day; he had to prepare, three times a day, the plain but substantial fare which satisfied everyone — everyone, perhaps, except Mike Collins, who had become fond of dishes like coq au vin and boeuf bourguignonne when he was a fighter pilot in France and, in some areas, definitely could outcook Lew Hartzell. That was not important now; it was important that Lew Hartzell did not catch cold himself. He did catch cold once, just before a Gemini launch, but the only available replacement was more ill than Lew Hartzell. So Hartzell wore a mask when the crew was around. This time he had a medical checkout a few days before the flight; his fondness for beer showed in a large midriff, but he had not caught cold. Neither had NASA's Tom Paine; otherwise he would have been the first to cancel his private dinner with Armstrong, Collins and Aldrin on the night of July 10, and give them his final message by telephone.

It was the day after that dinner that Dr. Berry had to determine whether somebody, anybody, in spite of all the "clean room" precautions, had communicated a germ or a virus to three men who had to be at their physical peak. The actual physicals were done by three other doctors: Al Harter, Jack Teegan and Bill Carpentier; Carpentier would pick them up on carrier

Hornet. They had the assistance of a person all the astronauts loved — nurse Dee O'Hara.

It was 4:30 A.M. when Dee O'Hara walked into crew quarters on the third floor of the Manned Spacecraft Operations Building at Cape Kennedy. She had left the Holiday Inn two hours earlier to drive to the preflight examination area in the same building; it took that much time to get ready. She had slept in the same room, number 216, which the Holiday Inn always blocked off for her two weeks or so before launch day, knowing that she would be coming over from Houston about that time. She had had that room since the middle of the Gemini program, and she had become a little superstitious about it. Once, when she arrived at the Cape for Apollo 7 (Schirra, Cunningham and Eisele, the first Americans to fly in space since the fatal pad fire twenty months earlier), she ran into a night clerk who did not know the background — it was the first time Dee O'Hara had been there since the fire. She insisted on putting Dee into another room, number 245. It overlooked the ocean and actually had a better view than 216, which only overlooked the swimming pool with a big palm tree in between, but that night Dee O'Hara had a bad case of the shakes: "They didn't like me anymore. I got paranoid. I was in a strange place, and all of a sudden I was just destroyed." Next morning, over coffee, she ran into a friendly face at the desk and was immediately moved into 216. Even Dee O'Hara could not tell you what was so special about 216: "It's upstairs, which I usually don't like. The view isn't all that much. But it's *my* room. And they had just redone it and repaneled it. I was the only one who had a newly paneled room. They had even put in a new bathroom for me. I said, 'Who did you have in there? The bathroom was in good shape when I left.' They laughed and said they were redoing sections at a time, and since they knew I was coming in they went ahead and did my room first." That same morning, when Dee drove to work, she barely noticed that someone was putting a new set of letters on the big white sign outside the motel entrance. The man had got as far as WELCOME, and Dee paid no attention. That evening, when she was fighting her way back to the motel (the left turn off North Atlantic Avenue to the Holiday Inn is a tricky one), she saw the full lettering out of the corner of her eye and was so startled that she got halfway into a wrong lane and nearly caused an accident. It said: WELCOME BACK HOME DEE O'HARA. Inside the parking lot, finally, she circled the sign a half-dozen times just to look at it.

On July 11, inside the preflight examination area, another kind of welcome awaited her. Neil Armstrong came in, and they chatted about the flight as Dee checked him in. "I suppose," Armstrong said, "they are going to make a big deal of all this."

Outside the examining room Mike Collins was drinking coffee, relaxed, his legs crossed as he waited his turn. Dee walked by in her crisp white uniform, her short dark hair, wearing her vivid pink-grape lipstick. She playfully kicked Mike on the bottom of the foot to make him look up.

"Hey, Mike," Dee asked. "Where are you going to watch the launch from?"

"Gee," Mike said. "I don't know. I hope not from the beach."

"Well, I'll use my influence and try to find you a spot."

"Great, could you get the other two guys one too?"

"I know this is your first launch. It's best to get up high."

"I'll remember that, thanks," Mike said.

It was the first good look Dee O'Hara had had at the men since she arrived at the Cape on June 27: "They were just so busy. I guess I saw them two or three times, just in passing. Usually I don't speak to them, don't bug them at a time like this, unless they seem to want to talk to me." But what she saw that morning looked great to Dee. The atmosphere was easy and casual, although perhaps not so much fun as back in the Mercury days when the original astronauts had their quarters in another building and used to hide the urine samples from her: "It would be in the shower stall, the air conditioner, the heater — anywhere." Once Wally Schirra, during Mercury, brought his sample in a gigantic container the size of a water cooler. Another time, during Gemini, there was a solemn procession of astronauts down the hall to the examination area. Dick Gordon led the march, carrying a plastic container with a lace doily draped over it and a little lace butterfly hanging off the top. After a day in the simulators the astronauts would bring back slim brown briefcases, stuffing in with the flight papers the little bottles Dee wanted. One day John Young opened the briefcase and a strange expression came over his face. Dee looked and burst out laughing: "I'll never forget him sitting there on the chair, and he opened the briefcase and he looked in there and the thing had tipped over. He hadn't put the lid on right. His papers were ruined. He had a big cigar in his mouth. . . . Of course some of the doctors didn't think it was so funny." In earlier days, before the Apollo flights, Dee also used to take blood samples, although she really did not want to. She had been proficient in laboratory work at one time but was out of practice. However, "The crews got on this thing about nobody doing it but me. What those guys didn't realize was that I was the worst person in the world to do it. God was good to me. I never missed a vein. Now I don't draw the blood. John Potter does it, and he's a hundred times better than I am. I had to talk an arm and a leg off to Dave Scott to convince him. He was having nobody but me."

Now there was a research Ph.D., Carolyn Leech, to run the urine study program, and Dee's big job was the paper work, which seemed to get more massive for each flight. Come F minus five it was twelve pages, per man, just for the physical. There was the ENT — ear, nose and throat. The ophthalmology part included a visual examination and looking into the eye with a funduscope; the internal medicine examination dealt with all major parts of the human system — lungs, abdomen, especially the heart. Space flight makes the heart behave oddly when astronauts are asleep. Gordon Cooper's heart, and Rusty Schweickart's, dropped down to a beat in the thirties when they were sleeping; by normal terrestrial standards they were barely alive. (They were finely trained men, but the spacecraft atmosphere was different.)

Back in the Mercury days Dee O'Hara had only one little form for her records and comments, with the history on the back of the sheet. For Apollo 11 she started the examination with six forms per man, which would expand to eighteen pages per man when she typed up her report — all that plus the labs: "You can't punch a tape as fast as you can type on a typewriter. Working on this 1050 data processor is something I enjoy as a change of pace, but I'll just get started and here come the doctors: 'Would you mind filling this prescription for me. . . . Get such and such a drug. . . .' I don't dare turn my machine off because that wipes out all the codes and I have to start over again.

"Then there are those other forms, the 1042's, the 88's, the 89's and so on. Some go to the Air Force in Washington, some go to Aircraft Operations and some go to the Navy. To do a physical as such isn't a big thing, but all the paper work just about undoes me. We do make quick copies of the F minus five physicals so the surgeon can take them out on recovery and will know what's what, just as a review. But some of the other stuff — late, late, late. God, if I could just have a slave."

Early that afternoon the men were pronounced as fit as they looked. Medically speaking, they were ready to fly. Dee O'Hara could not relax; she still had all that paper work to do. But over the weekend the men could relax — a little. They would put in some more time in the simulators, go over the flight plan again, but all this was in the nature of final review. The main work was done, and there would be a little time in those last few days to read or jog on the beach. As a practical matter they knew all they would ever know about the mission until they flew it. Buzz Aldrin had once said, "We could spend another year isolating these little problems one by one and we would never really get them all. We could spend too much time doing that, so much that we could forget what this mission is all about. The idea is to go to the moon, land there, take off and get back." Now the medical examination showed them at the top of that performance curve which General Phillips had been drawing in his office — right where Sam Phillips wanted them to be. After months of confinement to what Buzz Aldrin had called "the treadmill," months which saw them become, against their wishes, a bit dehumanized, they were now ready for the challenge — another word Buzz Aldrin used frequently. The medical fitness hurdle was a big one for all three men, but perhaps a little bigger for Mike Collins. He was the handball champion of the astronauts, and early in 1968 his game had gone off: "My legs felt peculiar, as if they didn't belong one hundred percent to me. I had heard prizefighters talk about their legs going and I thought, well, instant old age." It turned out that he had a bone spur and a loose disc in his neck that were pressing on his spinal cord. A complex operation was fully successful and restored him to flight status but cost him a position on Apollo 8 — the first circumlunar flight. Now he was back in business, cleared for a key job on the biggest mission yet — command module pilot, in some ways the loneliest job of all. Mike Collins would not set foot on the moon, but he would orbit it all by himself for twenty-seven hours, and it would take all of his skill and know-how to get Neil Armstrong and Buzz Aldrin home.

Probably Armstrong, Collins and Aldrin were the only three people in Brevard County, Florida, who had a reasonably relaxed weekend. In Cocoa Beach, and the adjacent community, Cape Canaveral, the tension was becoming indistinguishable from physical pain. Cocoa Beach and Cape Canaveral are visual incongruities. For ten miles northward along Atlantic Avenue, between Patrick Air Force Base and the Kennedy Hilton, just short of the point where a main highway peels off westward and a restricted-access highway veers right to the entrance of the eighty-eight-thousand-acre expanse of land occupied by the Kennedy Spacecraft Center, motels and cocktail lounges bearing names like Gemini and Satellite nest alongside gleaming new buildings marked with the mightiest names in American technology: Boeing, North American Rockwell, Chrysler, McDonnell Douglas, General Electric, IBM, Grumman Aircraft — dozens more. There is something unreal, something unearthly, about the place; it is as if all its creature comfort facilities are maintained on standby, like Dee O'Hara's room number 216, which in fact they are. Cocoa Beach is a rather dull place between "shots," and the motels are almost always glad to welcome the odd tourist who shows up without an advance reservation. But "shot time" is a wild time in Cocoa Beach. A special kind of hysteria seems to seize the place; it is as if a whole mass of people had gone on LSD for a predictable number of days. People drink abnormal amounts of hard liquor, eat irregularly or not at all and hardly ever seem to go to bed. The bars are populated in twos and threes by an extraordinary number of unescorted women, the vast majority of them quite respectable, drawn to the scene, perhaps, by a force or an urge which could stand some Freudian analysis. Private parties proliferate to lavish proportions; most of them are sponsored by contractors or prime news media in the interest of "image" public relations, but some are given by private citizens of means. At shot time Cocoa Beach exerts its own gravitational pull.

Yet the scale of the parties distorted the full picture. Many thousands of people drove or flew more than a thousand miles for Apollo 11, to share a moment of history, and did not go to parties at all. One of them was a Texas automobile dealer named Jay Marks, who was widely known for putting his bald head on billboards around the Houston Manned Spacecraft Center to advertise himself as a "topless dealer." He was a casual acquaintance of the Neil Armstrongs'. Marks had done the cocktail circuit when Apollo 8 flew in December 1968; this time he and a friend from General Motors loaded their sons into a camper, drove to Cocoa Beach, camped out, fished and surfed as they waited for the moon shot to go. "We were sober the entire time and I never enjoyed a week more," Marks said later. "There were some pretty swinging parties at the Cape but believe me, there were more of *us* than there were of *them.* Apollo 11 gave a lot of nice people a chance to get acquainted." Such people, on the night before the shot, ate sandwiches and drank soda water or a few cans of beer as they waited, with quiet impatience, for the happening which had so stirred their emotions. Of course those who lived it up in the cocktail lounges that night were also emotionally moved, and it may be that they were venting their emotions in the only way they knew how.

By the evening of Tuesday, July 15, there was not a vacant hotel room within fifty miles of Cape Kennedy. For thirty miles west to Titusville, on both sides of a highway more or less paralleling the launch facilities eight to ten miles distant across the Banana River, automobiles, house trailers, bicycles and motorcycles were parked nose to rear: hundreds of them, thousands of them. The same thing was happening along the road north to the point where a pass gate shut off public access. Perhaps the most extraordinary thing of all was the fact that Jules Verne had described the scene, fairly accurately, in fictional prose set down one hundred years earlier:

> As may be readily imagined, very few people enjoyed their slumbers the previous night. . . . The weight of protracted suspense many a breast felt to be rather an unpleasant burden, which became only more irksome as the last moment approached. . . . From an early hour in the morning, a countless multitude covered the vast plains extending all around Stony Hill as far as the eye could reach. . . . In fact the arrivals at Tampa latterly, by land and water, by horse, steam and human power, had been so enormous that the *Daily Truth Teller* estimated the number actually present within a two-mile radius of Stony Hill, to be very little short of five million souls.[2]

Verne even had the location almost right, and he did have the right American state; in *From the Earth to the Moon* the spacecraft carrying the explorers Barbican, Ardan and McNicholl was fired in Florida from a giant cannon only a hundred miles west of Cape Kennedy. Verne named his spacecraft Columbiad, and the Apollo 11 astronauts had that in mind when they named one of their two spacecraft Columbia. He also sang admiringly of the wondrous American mechanical genius long before the expression "technological gap" entered the language. And Verne had the saloon atmosphere about right: "Who said a Rocky Mountain sneezer? All right, sir. Root beer?

No, sirree. White beer? Coffee? No such stuff here! Get them in the temperance house around the corner. Eye-opener? Yes, sir. Rum punch for two! One glass Plantation Bitters! California champagne! Who said California champagne?" Drinking habits changed, of course; at Cocoa Beach in July 1969 the big seller was the $1.25 "liftoff martini," and for the venturesome souls there was a "moonlander" consisting of crème de menthe, crème de cacao, vodka, soda and a squeeze of lime. ("We stick an American flag atop the whole mess. If you can't drink it, you salute it.")

All week the Cocoa Beach regulars hoped that Jules Verne had miscalculated one statistic — the one about five million visitors. Aside from 3,497 accredited newsmen, including 812 from fifty-four foreign countries, ordinary people were coming from as far away as the Union of South Africa to see the launch if they could, or to watch it on television, which as a matter of state policy did not exist in South Africa. Verne did overestimate when he used the five million figure; even so, more than one million people did come to the Cape, and that was enough to create the most appalling traffic jam Florida had ever seen. It was freely predicted that, despite advance stocking, there would be shortages of food, gasoline and liquor, but this did not happen. What Cocoa Beach did run out of was alarm clocks. By the middle of Tuesday, the day before launch, there was not one to be had in the stores.

Even the Cocoa Beach regulars, the people who had been involved directly or indirectly in the space program since Vanguard and Redstone and Atlas, since Mercury and through Gemini and now Apollo, people who presumably should have been case-hardened by now, could not escape the fact that there was a different sensation about this flight. It was the one they had been waiting for, some of them for more than ten years. Charles R. Johnson, manager of the convention and visitors bureau for the Cape Kennedy Chamber of Commerce, felt as if he were a child again, looking forward to Christmas: "Hell, this is it!" The excitement manifested itself in odd ways. Dr. Burton Podnos, a thirty-eight-year-old psychiatrist who directed Brevard County's mental health center, found that he was getting an unusual number of telephone calls from parents; the standard report was "Johnny is uncontrollable this week." The county library had a record number of reservations — twenty-six, presumably all made by adults — for the novel *Portnoy's Complaint.*[3]

Then there was the weather. It was abominable. Cocoa Beach is tolerable in the springtime, when sea breezes lower the temperatures sharply from early evening until midmorning; in mid-July, when the thermometer regularly hits one hundred degrees, the seaside humidity and the heat combine to approximate the climate of an African swamp. The sun, when it is out, reflects with blinding intensity off boiling hot concrete and the white loamless soil which will support palmetto scrub but not much else. Even with dark glasses people tend to develop a squint. The sun is not always out; it rained on Sunday, July 13, and for a few merciful hours the temperatures were bearable. All day Monday and Tuesday black clouds moved overhead; it was the thunderstorm season, but weather forecasters stubbornly insisted, as they had been insisting for days, that weather conditions for a manned rocket launch — which required a minimum ceiling of 2,000 feet — would be satisfactory Wednesday morning.[4]

At 5 P.M. Tuesday the Apollo news center at Cape Kennedy issued a sobersided "status report":

> The Apollo 11 countdown will resume tonight at 11 P.M. when launch crews will begin chilling down the systems which will be used to flow the cryogenic [ultra-low temperature] fuels into the Saturn V launch vehicle. The cryogenics, liquid hydrogen and liquid oxygen, are stored and transferred at extremely low temperatures and the systems must be chilled down to accept the propellants. . . .
>
> The prime crew for Apollo 11 spent most of the day relaxing in the crew quarters. Apollo 11 commander Neil Armstrong and lunar module pilot Edwin Aldrin did spend some time in the lunar module simulator.
>
> Tonight the crew will have an early dinner consisting of combination salad, broiled sirloin steak, mashed potatoes, buttered asparagus, tomato purée, cottage cheese, fruit bowl, bread, butter and a beverage. After dinner the crew plans to retire early. [The flight crew skipped the salad, the asparagus and the fruit bowl.]
>
> The weather outlook for a 9:32 A.M. launch tomorrow continues to be satisfactory. There will be a broken cloud cover at 15,000 feet. Visibility will be ten miles, winds light and variable, and the temperature near eighty-five degrees.

Neil Armstrong would have preferred something else to eat; he had confided to his wife Jan, "I'm sick of steak." But steak and vegetables it would be on Tuesday night, and still another steak — with eggs — early Wednesday morning. That was the way it had always been before a launch, and Lew Hartzell the cook was not about to challenge the gremlins of fate by changing the menu — not for this flight.

The astronauts had invited President Nixon to dine with them that Tuesday night in the crew quarters, and he had accepted. Then everything went wrong. Dr. Berry seemed to think it was an unnecessary risk; suppose the President had an incipient cold and did not know it? The story got out and the President deferred; he intended, in any event, to be present on carrier *Hornet* at the time of splashdown. Everyone wound up embarrassed, angry or both. The astronauts were upset, not only because of an apparent slight to the President, but because they considered the objection based on a false premise. In a strict medical sense they were not "isolated" and could not be. "The really ridiculous part of it," Mike Collins said later, "was the fact that every day we were in contact with twenty or thirty people and had to be if we were to complete our training — the secretaries, the suit technicians, the simulator technicians and so on." Hardly anyone bothered to conceal his irritation with Dr. Berry. NASA administrator Thomas Paine moved in and renewed the dinner invitation, but by this time there was such a flap that the President felt it prudent to decline with thanks. Astronaut Frank Borman, who had been detailed to be the President's technical aide-de-camp for Apollo 11, was visibly upset; he had just returned from a highly successful goodwill visit to the Soviet Union, and he thought the whole affair "totally ridiculous . . . damned stupid." But what had been done could not readily be undone; as Borman said, "If one of them even sneezed now on the lunar surface it would be because President Nixon was there [for dinner]." So the astronauts had their Tuesday night dinner with Deke Slayton, with the three members of the backup crew (Jim Lovell, Bill Anders and Fred Haise), and with a half-dozen other astronauts who had worked closely on the eleven mission.

By 10 P.M., an hour before the final countdown resumed, Armstrong, Collins and Aldrin were sound asleep.

Hardly anyone else, for thirty miles around, was asleep. Every motel within driving distance had a countdown party going. In fact, there was so much partying that not a few people wondered about the size of the national hangover should something go sour the following day — or the following Sunday, when Armstrong and Aldrin were due to land on the moon.

That Tuesday a former President of the United States, Lyndon Johnson, flew in from Texas to be the Gannett newspaper chain's luncheon guest of honor, along with former NASA administrator James Webb. Johnson's presence was almost mandatory; as John F. Kennedy's Vice President he had chaired a council to produce documented answers to the questions of cost, manpower and administrative responsibility involved in putting a man on the moon "and returning him safely to earth." Mr. Johnson said, "I have ridden on every flight and I doubt a human could be as concerned or troubled until splashdown as I am or have been."

Wally Schirra, the only astronaut to fly in all three programs (Mercury, Gemini, Apollo), dropped in on the Johnson-Webb luncheon. "I'm sorry you couldn't eat with the crew," he told the former President. "That seems to be bad form these days."

Everybody who was anybody, or had been anybody in the space program, was at the Cape for this one. Dr. Wernher von Braun came from Huntsville, Alabama, literally dropping in on a private party — a helicopter landed him and Mrs. von Braun on a golf course. A surprise guest at the same party, in Titusville, was Dr. Hermann Oberth, one of the great, half-forgotten names of rocketry. It was Oberth, a Transylvania-born German citizen, who along with America's Robert Goddard and Russia's Konstantin Tsiolkovsky, all working independently, had come to conclusions about space travel which had long since sketched out the basic working formulas of the Apollo 11 mission. A bent, white-haired man of seventy-five years, the only one of the three pioneers still alive, he got a standing ovation.

Off in a nearby hotel room was another name from another age — T.

Claude Ryan, chairman of the board of the Ryan Aeronautical Company. He had been with a little California company called Ryan Airlines when, back in 1927, it built *The Spirit of St. Louis*. The corporate descendant of that company had built the landing radar for Apollo 11's lunar module, and tomorrow Mr. Ryan would watch the launch with an old pioneering friend named Lindbergh.

Past midnight, now: M plus 2,974 — July 16. It was NASA associate administrator George Mueller's birthday. It was as if the astrologers had been looking over shoulders. Now, ten miles from the din and the revelry, a Saturn V, only the fourth such vehicle to be flown with a three-man crew, stood out on Pad 39A, where it had been moved on May 20, before Tom Stafford, John Young and Eugene Cernan had completed their Apollo 10 flight. Illuminated by floodlights, it stabbed 363 feet upwards in the gloomy night like a gigantic fluorescent pencil, looking just a little bit different from the ones which had flown before because everyone knew that this "big boy" had the biggest job of all to do.

There was a warm sprinkle of rain, and the moon itself, the object of this enterprise, was nowhere to be seen in the heavily overcast sky. It was on the wax, not on the wane, which seemed a good omen. But there were angry stabs of lightning in the northeast and northwest. Were we back in the time of many gods? Did Somebody Up There not like this whole idea?

In Cocoa Beach the bars closed at 3 A.M.; the Cape bars had special permission to stay open on this night of the long party until 5 A.M. — forty-five minutes after Armstrong, Collins and Aldrin were to be awakened by a knock on the door which meant that it was time to get on with the job.

The knock on the door — three doors — came from Deke Slayton. Dee O'Hara had been in the building for nearly two hours, having risen at 2 A.M. to drive past all the countdown parties still going full blast. From her car she could see the floodlighted Saturn V, gleaming this night of all nights with a kind of supernatural intensity: "I had seen it there so many times, but this time it was so — just so exquisite. It really *was* different from all the others, because of what it was supposed to do. . . ."

To save time, Dee O'Hara had set up a special little examination area near the astronauts' bedrooms. It was about 4:50 A.M. when Buzz Aldrin came in: "He looked like the same old Buzz but he had a sort of pink, tender look. He wasn't exactly smiling but he looked relaxed. A soft look. He was wearing a white terry cloth bathrobe and slippers. He said 'Hi' and walked into the adjoining room to register his nude weight. I have to fill out the forms on that, plus recorded sleep and heartbeat — all that. Then Buzz came back out and smiled and went into his bedroom to put on shirt and slacks. I remember he said to me, 'I hope you don't mind taking these things back . . .' and I said no, of course not." Aldrin was handing Dee his personal belongings — billfold, clothing, etc. — which he had been using at the Cape but which had to get back to Houston somehow. Normally these things are turned over to the backup command pilot; this time both Aldrin and Armstrong turned their things over to Dee. Mike Collins had already mailed his home.

The man Dee called "darling Mike" came in next. He was smiling and he said, very formally, "Good morning, Nurse O'Hara, what are you doing up so early?" Dee said, "Well, I didn't have anything else to do." They both laughed. Collins is a slight man, but from inside the weighing room he yelled "One hundred eighty-eight pounds!" — it would amuse Dee, of course. Then he came out and said, "Nurse O'Hara, I want you to behave yourself."

Neil Armstrong was the last one in. He was just out of the shower; his hair was standing straight up. He picked up Mike Collins's line: "What are you doing here at this hour?" But he was smiling and Dee O'Hara thought he looked "so cute."

The routine over, Dee O'Hara was left once again with a pile of forms to fill out. She heard Mike Collins cry out: "Man, let's get going. What are we waiting for? I'm *hungry!*" That love Dee had for them all welled up in her again, as it always did on launch day. Once again she had that old schizophrenic feeling — half exuberant, half sad. She stood by the door and waited for somebody's last words: Give my love to the wife and the kids. . . . Dee, I want you to behave yourself. . . . This time, "Neil came by and gave me a big smile and a big wink. I don't think we said any more."

Nobody said goodbye. Nobody ever does. Not on launch day.

Lew Hartzell was employed by ARA Services, Inc., the company that contracted for all food services at the Kennedy Spacecraft Center. He had learned to cook in the Marines. Later he operated a bakery, then joined the Merchant Marine — he held an ordinary seaman's Z card — and began to cook on ships, mostly tugboats. Al Shepard hired him at the time of Gemini 4 (June 1965): "Eleven men before me had failed either the physical or the security check, and I passed both. I cooked Captain Shepard a meal on my own time and he offered me the job. I told him, if they want to eat plain and simple, okay — but look, maybe I'm not good enough. He said the astronauts were meat-and-potato men. Okay. I told him I could cook everything from biscuits to cornbread. Oh, I picked up a little fancy stuff when I was working on yachts. You don't slice a tomato there; you make it look pretty — like a rose. And how to make hors d'oeuvres. I don't do that here. Give these boys an hors d'oeuvre and they'll throw it at you." The crew kitchen was closed down after the 1967 pad fire, and Hartzell spent most of the interim cooking up and down the East Coast on a yacht owned by the head of the Vanity Fair Paper Company. But he stayed in touch, and when they began to get ready for the manned Apollo flights in 1968 he leaped at the opportunity to come back — not because of the money, which was not great, but because, like hundreds of others, he had become part of the program and did not want to let go. He got an apartment at the Cape, but often as not, particularly when the astronauts were working late, he slept at the Space Center in the Chevrolet camper which he liked to take down to the Florida Keys between shots to look for gold doubloons. He did the food shopping personally, switching mysteriously from store to store because he bought in big quantities and did not want anyone to know who was going to eat the food. "There are a lot of kooks in this country, and somebody could spray a little DDT on the vegetables or needle the meat. I got twenty-four pork chops one time and at the checkout counter someone said, 'Can we help? Is this a restaurant order?' I said no, I've got a tug tied up at the dock. One time I said I was buying for the Girl Scouts."

It had not been a difficult Tuesday night for Hartzell — only a dozen for dinner. (Once, for Apollo 10, when Vice President Spiro Agnew was there, he had crowded in thirty people for dinner.) And there would be only five for breakfast — the three Apollo 11 astronauts, plus Deke Slayton and Bill Anders. After cleaning up the dinner dishes he went out to his camper about 9:30 to sleep, but it was too hot. He went back to the crew quarters and slept in astronaut Gordon Cooper's room. He got up at 2:30 A.M. and wondered if he should not stop sleeping in that camper. He was too big for the bed and he had already broken the window once with his feet. As usual, his back hurt; once he had fallen down a yacht hatchway in the dark and broken a hip, and later, on a supply boat bound for South Africa, he fell against a big metal rivet and fractured the lower part of his spine. Breakfast was scheduled for 5 A.M., and Hartzell used the intervening time to get the meat and the eggs ready, set the table and make up a batch of lemonade for the VIP viewing stand. He never started cooking steak and eggs until the astronauts arrived in the dining room, which they did shortly after 5 A.M. There would be no biscuits this morning — just "straight-dough" bread. The astronauts were now on a low-residue diet, which meant that for several days Hartzell had been cutting down on butter and cutting out milk entirely. They were never on a special diet at any other time. They were too well trained for that to be necessary. The conversation over the food, as always during a pre-launch breakfast, was studiedly casual. "The atmosphere *is* a bit different," Mike Collins once said. "We all know why it is different, but it's hard to explain. There's no drama to it. It's just that this is the day we go, and that knowledge does make the *feeling* different. . . ."

By then Neil Armstrong's wife Jan was on a boat in the Banana River. Alone among the three Apollo 11 wives, she had elected to come to the Cape to watch liftoff. She stayed in sealed-off quarters on South Atlantic Avenue; reporters tried all day Tuesday to find her, but none succeeded. Late

Tuesday night she had herself driven to the Space Center; her pass would take her to within three miles of where the Saturn V, her husband's Saturn V, stood perched on its pad. She was there at midnight; it was just something she wanted to do. Then she went back to her quarters and slept three hours. At 4:15 A.M. it was time to awaken her sons — Ricky, twelve years old, and Mark, six years old — and go get on the boat. Mark understood it all. "My daddy's going to the moon," he said. "It will take him three days to get there. I want to go to the moon someday with my daddy." Jan Armstrong had wanted to be present for liftoff when Neil last flew (Gemini 8), but Neil did not want her to come. This time she refused to stay home. On the boat she had good company, good friends: David Scott, who had flown with Neil on Gemini 8, and his wife Lurton; Pat Spann, an El Lago (near Houston) neighbor whose husband was in NASA's Mission Support Office, and another space wife, Jeanette Chase (her husband was in the Recovery Division) who with Jan coached a synchronized swim team at the El Lago Keys Club. As liftoff approached, the Armstrong boys were playing a game called Peggity.

Lola Morrow was with Dee O'Hara in the astronaut viewing area — north of the VIP stand, the best view in the ball park. Until recently she had been, in effect, the astronauts' backup secretary — at the Cape. She got hotel rooms for them and their families when there were no hotel rooms to be had. She saw that a car was waiting at Patrick Air Force Base when one of the astronauts flew in from Houston — or from Downey, California, or from Bethpage, Long Island, where Grumman was building the lunar module. On the NASA organizational chart, her job was impossible to describe. Once, when there was some kind of strike by the maids, she personally made up the beds and changed the linen in the crew quarters. When Gus Grissom called, the day before his Gemini 3 launch in March 1965, to say his bed was too narrow (he did not want to fall out on the floor the night before he was going up into space), it was Lola's problem. He also wanted a softer mattress. Then John Young called and so did Wally Schirra and Tom Stafford; none of them wanted to sleep in those narrow, hard-mattressed beds. By this time it was midafternoon. Lola called up a furniture company and said she wanted four beds, with softer mattresses, in four hours. She got Deke Slayton to sign the purchase orders, and the beds arrived at 8 P.M. Lola got the sheets and made them up herself. Then someone said, "Aren't you going to tuck us in?" Lola said, "No, gentlemen. Your beds are ready for you and good night." Later the astronauts, or some of them, decided the mattresses were too soft. Lola went out and got new mattresses — hard on one side and soft on the other. She did all this for a take-home pay of about eighty-eight dollars a week.

She went to work for NASA in 1962 as a travel clerk. In 1964, when the

Gemini training program was being organized in a serious way, the astronauts had no secretary at the Cape. They did not even have an office of their own with telephones. One of them — Lola Morrow does not remember which one — said, "Be our secretary for a while until we find one." She does remember that it was Gus Grissom who said, "Just *be* our secretary." Lola was scared of the prospect, and she turned down the job. Then, one Monday morning, Deke Slayton left word that she was supposed to have an astronaut office open by Tuesday morning. She went downstairs from the fourth floor to the third, put a sign on the door and waited for things to happen. On Tuesday morning a mail boy came around wanting to know what the mail symbol would be: "I said it would be H O — the sign said H O Astronaut Office, for 'Ho, ho, here we go.' "

That was all behind Lola Morrow on the morning of July 16; she was preparing to leave the space program and find a public relations job. She had intended to do that anyway, after Apollo 11; but under no circumstances was she going to miss this liftoff. Not the one that was going to go all the way! And she had her memories — some good, some bad. One of the good ones was about the night of October 24, 1965. It was her birthday. At one o'clock in the afternoon of that day Al Shepard called and said there was a problem with the mission rules books — the big black loose-leaf notebooks containing checklists for every crew for every stage of a flight; and a Gemini flight was coming up. Shepard said they needed to be updated, but Lola knew they already had been updated; she had stayed at the office until seven o'clock the night before doing it. Shepard said, "I'll come and get you."

Shepard walked her to the briefing room in crew quarters. An updated rule book did not seem to be in sight. Shepard said, "I know there's one. It must be on the table in back of us." Lola turned around to look at the biggest birthday cake she had ever seen. ("My husband wasn't the kind to remember birthdays. I couldn't remember the last time when anybody gave me a birthday cake.") Then somebody gave her cuff links; cards came every day, and so did two more birthday cakes and flowers. Finally Deke Slayton said, "Lola, how long is your birthday going to last?" Later she thought it had lasted about ten days.

Then there was the bad memory. It was about something that had happened on January 27, 1967 — the worst day of all in the Apollo program. Lola did not like to use the word intuition to characterize her feelings that day, but it was one of those days when nothing seemed to be going right. Apollo 1 — the first manned flight, the one due to go on February 21 — was on the pad, and that day Gus Grissom, Edward White and Roger Chaffee were to rehearse a launch countdown, strapped in their couches in a spacecraft on top of a Saturn I-B. (Gus Grissom had once hung a lemon on that spacecraft — what was wrong with it?) Grissom normally came into the office acting rather chipper, but Lola thought he was not so chipper on the morning of January 27. Edward White seemed very preoccupied, and Roger Chaffee "sort of shot by." Finally Lola went to Ed White and asked, "Is everything all right? Are you all right?" He had been writing letters and he said, "Lola, I'll take care of everything when I get back — when we finish this test." He was sitting at a desk in his bathrobe and slippers — "little scuffs I had picked out for them, white terry cloth with a little blue."

As she listened to the squawk box that afternoon, hearing everything said between the pad and the crew, Lola Morrow became more apprehensive: "I went in to Chuck Friedlander [chief of the astronaut support office] and said, 'It doesn't seem to be going right.' And it didn't. For one thing communications were bad. And one thing seemed to be going wrong and then another. There were too many problems. I kept thinking to myself — 'Project Apollo. We should call it Project Appalling.' Then Mr. Friedlander said to me, 'You've had a hard day. Go home.' "

She called astronaut Stuart Roosa, who was in the blockhouse "Stony Position." Monitoring the rehearsal countdown was part of Roosa's training. Lola Morrow said to him, "Major Roosa, I don't know how to say this but something's wrong down there. I'm staying here until I hear something from you." Roosa said they were having a meeting then to decide whether to abort the countdown; then he called back and said, "Lola, it's go. The problems are ironed out. You can go home."

Lola Morrow stopped at the Holiday Inn for fifteen minutes to see some out-of-town friends. Then she called her daughter Linda, who was then eleven, and Linda said Chuck Friedlander had called. She called Friedlander. He asked her, "How soon are you going to be home?"

"Right away," Lola said.

"Then call me when you get there."

Lola went home and called the office. Chuck Friedlander said there had been an accident. . . . "We don't know the details. Can you come out?"

"What kind of accident?"

"We've had a fire. Somebody's got hurt out there. We don't know who."

"Who is it? The guys? How bad is it?"

"We don't know yet," Friedlander said.

"Can we find out?"

"It's pretty bad. . . . At least one of them."

Linda Morrow was standing by the television set and suddenly shouted, "Mommy, the astronauts!" Lola went to the television and the talk was about the fire. She went back to the phone and asked, "What's going on here?"

"Lola, it's all of them," Friedlander said.

Lola put the telephone back on the hook and drove back to the office. Every telephone was ringing there — the whole world was calling: "I looked at Chuck and he looked at me — that was all. Deke Slayton came in, and he was shaking like a leaf. He had a cigar and I tried to light it for him. He couldn't hold it and I couldn't light it. I started helping Deke call wives — Betty Grissom first. Wally Schirra answered, so he told her. Ed White's line was busy and Roger Chaffee's line was busy — Martha was calling the Holiday Inn at that moment to see if Roger was in his room. . . ."

It had been three seconds after 6:31 P.M. eastern standard time, when someone — probably either Chaffee or Grissom — yelled something like "Hey!" through the communications static. Then came another cry: "We've got a fire in the cockpit." Then more noise; something like "Let's get out of here. Open her up." Then, no noise at all.[5]

It was 6:45 P.M. before they got the hatch open for the doctors. Grissom, White and Chaffee were dead. They had probably been dead for about ten minutes. They had certainly lost consciousness no more than thirty seconds after the first cry. They did not burn to death, but they were asphyxiated when the fire raced through the pure oxygen atmosphere of their spacecraft and sent carbon monoxide, in the form of heavy black smoke and soot, bursting through their breathing apparatus and into their lungs.

So many things turned out to be wrong with the spacecraft — a hatch that took too long to open in an emergency, flammable plastic materials vulnerable to the slightest spark in a pure oxygen atmosphere — that some people later wondered whether the fire, horrible as it was, did not in the long run serve a useful end. Tom Stafford thought a lot about that. "It was terrible," he once said. "But we had to do so many things to that spacecraft — literally hundreds of things. If we had gone on using it as it was, which we probably would have done had it not been for the fire, we might have lost two or three crews up there." And Dave Scott, talking about the tragedy a year later, said: "Everyone says how awful. It would have been worse if it had happened in space. Then everyone would have thought it was a micrometeorite puncture and we would have been constrained to fly an armored car instead of a fire engine."

Nobody ever forgot Gus Grissom's remark: "If we die, we want people to accept it. We are in a risky business. . . ."

But people like Dee O'Hara never could quite "accept it"; she heard the news over the radio in the laundry room of her Texas apartment, and her first reaction was that it was all quite impossible: these men seemed to her to have the stuff of immortality in them; how could they die? Nor did Lola Morrow ever get over it; she had seen Apollo come back all the way in a little more than two years, but now, watching the big rocket, waiting for the first streaks of dawn to lighten the ominously purple Florida sky, she could only hope and pray that nothing had been left out of the planning for this mission. She knew that the men who would fly this day were magnificently trained and supremely confident; but it was still "a risky business" and something could go wrong. . . .

It was a few minutes past six-thirty in the morning when Armstrong, Collins and Aldrin walked out of the MSO Building to get into a white van which would transport them about five miles to Pad 39A. Each wore a bulky white space suit which had cost, tailormade, one hundred thousand dollars — ten times what it had cost, in 1927, to build *The Spirit of St. Louis* for Charles A. Lindbergh. They carried portable oxygen ventilators, and they waddled a little as they walked. It was difficult to tell who was who; but then one of them jerked a "thumbs up" signal which had a certain style to it. That was unmistakably Buzz Aldrin. Give or take a few minutes, it was T minus three hours.

At the viewing stands, three and one-half miles from the pad itself, early arrivals could already hear the metronomic chant of the loudspeaker: *T minus . . . and counting . . .*

It is a monster, that rocket. It is not a dead animal; it has a life of its own.

GUENTER WENDT

3

"I vunder vere Guenter Vendt?"

As Apollo 7 lifted off in October 1968, a voice from the spacecraft came crackling through the static: "I vunder vere Guenter Vendt?" The pun was credited at the time to Walter Schirra, but the voice actually belonged to Donn Eisele, a man with a talent for imitating accents. Guenter Wendt's German accent was so thick that even he laughed about it, but that day he had gone nowhere and all three men in the spacecraft were glad of it. Wendt worked for North American Rockwell, and as "pad leader," his job title was a little vague; the astronauts knew him affectionately as "the fuehrer of the pad." John Glenn had given him the nickname, and he was not offended by it; in one particular area Guenter Wendt was a martinet, and he did not pretend to be anything else: "There's no reason to say I am narrow-minded. Just do it my way and you will have no problem at all." It was his responsibility to make sure that the spacecraft itself, as distinct from the stack of booster rockets beneath it, was ready for flight. Once it arrived at Cape Kennedy and went into major testing and manned runs, he supervised everyone who touched the spacecraft, be it an engineer, a quality control man, a flight technician. "It's easy to get along with Guenter," Pete Conrad once said. "All you have to do is agree with him." Installation of every component, every piece of equipment, came under his jurisdiction. Nothing, absolutely nothing, was added to the apparatus until Wendt said: "Okay, *now.* Do it." He had launched every Mercury flight, virtually every Gemini. He had even launched Ham and Enos, the two chimpanzees. At that time he worked for McDonnell Douglas. When North American Rockwell got the contract to manufacture the Apollo spacecraft, Wendt thought of switching over. But he feared he could not get the authority he needed, and at the time of the January 1967 fire he was in Titusville, Florida, with McDonnell, as supervisor of the test range there. He had nothing to do with the spacecraft in which Grissom, White and Chaffee died, but he always felt uncomfortable when people said (as they did), "Oh, we wish you had been there." He thought it would be presumptuous for him to suggest that maybe he would have noticed something, done something, in time to prevent the tragedy. "On the other hand," he said, "maybe it was meant for me not to be there because I would have taken it very hard." Guenter Wendt was not a man to guarantee anything. Before John Glenn's Mercury capsule, Friendship 7, took off for three earth orbits in February 1962, he told Mrs. Glenn: "Annie, we cannot guarantee you safe return of John. This would be lying. Nobody can guarantee you this — there is too much machinery involved. The one thing I can guarantee you is that when the spacecraft leaves it is in the best possible condition for a launch. If anything should happen to the spacecraft, I would like to be able to come and tell you about the accident and look you straight in the eye and say, 'We did the best we could.' My conscience then is clear and there is where my guideline is."

After the fire, some of the astronauts knew who they wanted back in charge of the pad when manned flights were resumed. One of them was Wally Schirra, who liked the fact that Guenter Wendt literally terrorized everybody. They all remembered the time, during Gemini, when a project engineer decided he was going to make a pad fix on the spacecraft, with or without Wendt's permission. He got himself up on the mobile service structure, and there he had the misfortune to run into "the fuehrer."

"You had better remove yourself from the structure," Wendt said. "You can't do that," the engineer said.

Wendt called downstairs to security and said, "Do you want to pick up a man?"

Then, Wendt recalled, "The guy comes up on the elevator and he says [to the engineer], 'You like me to put handcuffs on you, or are you going to go by yourself?' The engineer dropped his jaw, but he left. Maybe this system is wrong, but I have had pretty good success with it. If I don't do a good job, I get out. I can't compromise."

It was Deke Slayton who got Wendt transferred to North American Rockwell, and Wally Schirra who got him changed from the midnight shift so that he would be in charge when Schirra, Cunningham and Eisele got into the Apollo 7 spacecraft on launch day. Schirra went to the office of Buz Hello, North American Rockwell's vice-president and general manager for launch operations, and said, "Buz, you are going to have to change the rules. Guenter Wendt is going to be the pad leader when we get into the spacecraft. After we lift off you can put anybody there you want, but he is going to see us off." Wendt's shift was changed, and he saw Schirra's crew off.

Wendt had not belonged to von Braun's group of German rocket scientists. He had studied mechanical engineering in Berlin and had flown night fighters in the German Air Force during the Second World War as a flight engineer. He had learned the trade of building aircraft; in his time apprenticeship in Germany was a four-year business. He also learned how to handle any kind of machinery and hand tools. But immediately after the war, before the German recovery miracle had taken place, there was not much opportunity in his homeland for a man of Guenter Wendt's training. His parents were divorced; his father had emigrated to the United States and was living in St. Louis. Wendt joined his father there. McDonnell wanted to give Guenter a job immediately but could not; the company was working on Navy contracts, Wendt was a German citizen and America was still officially at war with Germany. So Wendt got a job as a truck mechanic, and although he had never worked on trucks before, he was bossing a shop within a year. After five years he had his American citizenship, and he changed over to McDonnell at once. He liked living in St. Charles, outside St. Louis: "It was a small community, small-town USA. The biggest event was Friday night when you went downtown and just walked around. You didn't shop. You just met everybody. I liked it better than a big town." In 1958 and 1959 he started going to Florida, where he worked first on ballistic missiles. He met people like Kurt Debus, one of the von Braun group who would become director of the Kennedy Space Center, along with Christopher Columbus Kraft Jr., Robert Gilruth, Kenneth Kleinknecht and a half-dozen others who would, along with Guenter Wendt, bear heavy responsibility for the flight of Apollo 11. Wendt took very seriously his new American passport: he waved the flag, or rather he was looking for a flag to wave. He thought that the South Gate entrance to the Kennedy Space Center, the one the astronauts most frequently used, ought to have an American flag. He asked for one, and got a letter explaining that according to Air Force regulations only one flag could fly officially at each base. At Patrick Air Force Base that flag was outside the headquarters building, No. 425. Wendt got on the telephone and protested so loudly that finally a colonel flew to Washington to iron things out. Guenter's flag flew — until a new base commander took it down about six months later. Wendt went into action again and threatened to write congressmen. After two days the flag flew again, making Patrick probably the only U.S. Air Force base in the world with two official flags.

In the summer of 1969 Wendt still remembered those early days at the Cape with nostalgia: "It is still as exciting today, but then there was the personal contact — you could bank more on somebody else's word. Somebody would say, 'I'll do that for you' — and it was done. Today you can have it in writing and the guy says, 'Oh, it was on the schedule, but we skipped it.' " He was still exasperated, after twenty years, about his difficulty with colloquial English, especially because his wife, who was reared in Hamburg, spoke the language perfectly, with a slight English accent. But in one way Wendt found his accent an advantage; when he went on the network which linked the spacecraft with Mission Control, the first thing he always heard was "Hey, Guenter . . ." And Wendt would reflect, "I never have to say who I am!"

Everyone at the Cape — and at the Houston Spacecraft Center — knew who he was. By July 16 Wendt had been working on spacecraft No. 107, the Apollo 11 spacecraft, for five and one-half months — ever since January 22, when it was flown from North American Rockwell's Downey, California, aerospace plant to the Cape Kennedy "skid strip" in the Super Guppy, a slow-moving but big-muscled aircraft which was built to do this particular type of work. It looked a bit like a pregnant whale with wings — indeed the first version of the aircraft, a Boeing Stratocruiser which had been acquired by an independent developer named John Conroy, then modified and recertified, had been called the Pregnant Guppy. The Pregnant Guppy flew all the early Titan rockets to Cape Kennedy, along with the Gemini capsules. Its successor, the Super Guppy, was flown by Aerospace Lines to carry the third stage of the Saturn V rocket — the S-IVB.

Wendt's first task was to oversee the receiving inspection checks to make sure that there had been no damage in transit. Then he had to supervise the power tests in the altitude chamber, both unmanned and manned; next put the lunar module, the spacecraft adapter and the command and service module all together; then — on April 14 — move the whole mass into the Vehicle Assembly Building, cable it up and run a major systems check that would last nearly two weeks. There were ten different modes of aborts, and all of them had to be simulated — that is, tested.

As late as mid-March, when Apollo 9 flew, some of the Saturn V assembly which would power the Apollo 11 spacecraft in July was lying around the floor of the VAB in pieces — e.g., the leaf-shaped metal "skirts," each twice the circumference of an ordinary coffee table, of the five F-1 engines which would be clustered up to provide 7.5 million pounds of thrust at the moment of liftoff. By now, every statistic concerning Apollo had become difficult for the finite mind to assimilate. There was the size of the VAB itself. A symmetrical structure 716 feet long and 517 feet wide, it had a high bay of 525 feet. Both the United Nations Secretariat Building and the Statue of Liberty could have been moved into any two of its four main doors to rest on a concrete floor which had been ground down to perfect level with wristwatch precision. It required a ten thousand-ton air-conditioning system with a capacity to cool some three thousand homes; without air conditioning, the VAB's 129,482,000 cubic feet of space would develop its own atmosphere. Clouds would form inside, and it would rain.[1]

The VAB was not intended to be the biggest building in the world; it was designed to be big enough to assemble, vertically, the three stages of the Saturn V and the Apollo spacecraft — the command and service module, the lunar module, and the adapter which linked the two. But until Boeing needed a bigger building near Seattle to assemble the 747 "jumbo" jet, the

VAB was, in terms of cubic space, the biggest building in the world.[2] By mid-May it was time to "mate" the electrical systems of the rocket boosters with those of the spacecraft, then move the whole stack to Pad 39A, three and one-half miles distant, on the six million-pound transporter (or "crawler"), which is 131 feet long and 114 feet wide, traveling on four double-tracked crawlers, each shoe on the crawler track weighing about one ton. The actual stacking was done by crane operators 462 feet above the VAB floor, working from digital computers and listening over earphones to talkers beside each segment to detect movements and tolerances of as little as 1/128th of an inch; if vertical balance were lost, the whole thing would fall over and become a billion-dollar pile of junk. The trip to Pad 39A required six hours on a three-layer roadway nearly as wide as an eight-lane turnpike, topped and sealed with asphalt, then covered by river rock to reduce friction during steering — the steering being done by controllers who communicated with each other by radio from diagonally opposite cabs.

Finally, on the pad: time for Wendt and his crew, now grown to about sixty men, to run still another major systems check and a flight readiness test, then — after June 12, when Sam Phillips had given the word "go" — load the self-igniting hypergolic propellants into the CSM (command and service module) and the lunar module. This meant three hundred pounds of monomethylhydrazine fuel and nitrogen tetroxide oxidizer to feed each of the CSM's four "quads" of engines (each quad had four clustered engines fired separately or in combination, i.e., one engine in each of two quads, or in all four quads, or by itself, to achieve roll, pitch and yaw in flight); plus 41,000 pounds of propellants for the SPS (service propulsion system) engine which fired to make lunar orbit insertion, transearth injection and midcourse corrections, and another 23,245 pounds of the same for the lunar module's descent and ascent engines. This loading required a week; once in the tanks the hypergolic fuels would stay there until launch.

Timbercove, Texas

Guenter Wendt was not the only man who considered spacecraft No. 107 "his"; another was Kenneth S. Kleinknecht, who was stationed at the Houston Manned Spacecraft Center as manager of all command and service modules in the Apollo program office. He had been in the space business since there had been such a thing: engineering manager for the old NACA, project manager for Mercury, assistant manager for Gemini. In the last half of June he still had a lot to do, and the meetings started at seven or seven-thirty in the morning. Then there was one day a week to spend at Downey, California, where North American Rockwell built the command and service modules. There were all those daily "action assignments" on questions that needed answers. There had been that problem of gas getting into the astronauts' drinking water. During Apollo 10 Tom Stafford swallowed a dose of chlorine and was nauseated; Kleinknecht had people working on a filter device to remove the chlorine, which was there to kill bacteria, before the water came out of the water ports. (Activated charcoal did the job.) Then there was the matter of the fiber-glass insulation in the tunnel hatch of the spacecraft. The film which stuck it down came loose during Apollo 9, but the insulation itself did not blow out. They stuck down the film all over the edges of the insulation for Apollo 10, but that was not good enough either. The film came loose again. This time the insulation did blow up, and Tom Stafford and Eugene Cernan went into sneezing fits; at the moment it reminded Stafford of "plucking chicken feathers in Weatherford, Oklahoma." The question arose: was the insulation really necessary? Answer: negative; after all, Apollo 10 had got along without it during the long journey home from the moon. So it was taken out of the Apollo 11 spacecraft. And how about that high-gain communications antenna? If there were to be good communications from lunar distances, that antenna had to work. It had been flown on three earlier missions, but Kleinknecht still did not consider it fully "qualified"; there had been some kind of problem about facilities at a subcontractor's plant. All of the antenna's components were qualified, and the whole system had been through one mission "duty cycle" to the moon and back. Still Kleinknecht was not satisfied. He would put it through a second duty cycle to make sure.

He had had to turn down an Apollo 10 recommendation that the 11 crew carry rubber cords called "bungees" on their flight to hold down papers. Too risky; rubber was a flammable material. Kleinknecht and his people designed two kinds of clips with snaps to be put in various places around the inside of the spacecraft cabin: "You could stick a little card under a clip, stick a flight plan under it. The spring clip would even hold a camera magazine securely. Tom Stafford said they lost a camera once and it took them fifteen minutes to find it. It floated away in the weightlessness and they didn't know where to look for it."

There were also the "anomalies" that were still being checked out following the flight of Apollo 10. The matter of the glass insulation that came loose was only one of them; there were about fifty cases in all. Ten or so turned out not to be anomalies; they involved things that were working correctly but did not so appear at the time because all the information was not in hand. But there were other things — "like the vent plug, camera failures, the cooling pump for the fuel cell — I think we had an electrical short in one of the pumps. The fuel cell would still work but not under the full load. That's why we have three fuel cells; they're not the most reliable pieces of equipment." Kleinknecht did not keep an errors percentage count: "The reliability and quality people keep book, and they know how many failures there have been and how many are open and how many have been closed.

I'm not interested in bean counts. I'm interested in one bean only and that's a big fat zero." He still could not quite "close" the anomaly of why the lunar module had behaved the way it did on Apollo 10 — when the LM went automatic thirteen miles above the moon and "took off" unexpectedly — causing Eugene Cernan to exclaim over the open radio circuit an oath for which he later received a deluge of mail, which ran about twenty-five to one in favor of his having expressed himself freely. But two electrical systems had to fail simultaneously for the lunar module to do such a thing on its own, and that seemed most unlikely. Was there not the possibility of human error? That could never be ruled out. *There are all those switches to throw,* Buzz Aldrin had said, at one point. *We'll remember not to throw that one, but there are others we could throw at the wrong time.* . . . The lunar module itself did not come back to be checked out for a malfunction; having no heat shield, it could not be brought back into the earth's atmosphere without burning up. But the deduction of error or accident seemed sound enough, and Apollo 11 would go ahead.

It had been a long road for Kleinknecht since he was graduated from Purdue in 1942: engine research on airplanes used in the Second World War, work on the early experimental airplanes at Edwards in California after the war, then to the NASA Space Task Group at Langley in the 1950's, then to Houston with about twenty-five others from the Task Group which Robert Gilruth brought over with him in 1962. Somehow it had become more and more complicated since Mercury: "It's funny. You can't do the job without people, and then people make the mistakes." He could not even keep his mind entirely on Apollo 11; someone had put a water glycol mixture, which belonged in the coolant system, into the waste water system of spacecraft No. 109, for Apollo 13, to be flown in 1970. Kleinknecht was a tall, silver-haired man of forty-nine years, very slow of speech, and if everything worked all right splashdown day, July 24, would be his fiftieth birthday. . . .

North Amityville, Long Island, New York

Still another man had a right to consider Apollo 11 — or at least a piece of it — "his." His name was Herman Clark. He was a black man, born in Roanoke, Virginia, in 1932. He was a quality control inspector for Grumman in Bethpage, Long Island, and lunar module No. 5 — for Apollo 11 — was "his." He had had it from the very beginning: "We made all the bulkheads and the skins, the docking tunnel. We built it. We saw it when it was nothing but a big hulking piece of metal. On LM 5 I had a particular interest because I had looked at every damned hole on that thing, and every joint. The quality control inspector is a sort of nitpicker. We're the ball breakers, in plain English. We're the most unwanted people. The quality control man, the inspector, is a guy that not many people like because it's his job to criticize another man. And when you criticize another man, it's rough. There are ways of doing it where you can be diplomatic about it, but in this program, there is one way — you know. You do it by the book. There's no other way. You've got to have almost one hundred percent traceability, all items. Every discrepancy, every scratch, every dent that occurs during the manufacturing of this product has to be recorded and documented. The smallest scratch. We have to find it. To me it's money because if the product gets out — well, then, I get paid. But I also have to realize, as a QC, which is drummed into our heads, that the product is the important part. Yes, we have a schedule, but I don't have a schedule to follow. All I'm responsible for is that the product be right. If that product's not right, you can bet I'm going to hear about it.

"I am the only Negro in my area — in the inspection part — and I direct a number of white people, and I get damned good cooperation from these guys. My outlook on the racial problem is — I don't have to win your heart, but I will win your respect. I think the guys respect me. And the way a man respects a guy is because I fight for him. The way I fight for him is that whenever there is a disagreement I stick up for him. Our section of the plant is separated from the rest of the plant. We have over there a pretty good relationship going. The people who work on that project, on the LM, take pride in it. The guys there are not there to fight a racial problem. If you've got a problem, take it out of here. . . ."

Herman Clark was not too concerned about liftoff — rather, he thought in advance that he would not be too concerned. Then NASA asked him and his wife Rosa to be guests at the launch. They stayed at a motel, got up early each day and — like Jay Marks, the Texas automobile dealer — skipped the Cocoa Beach nightlife. On launch morning they wound up in a viewing site about a thousand feet from a man they recognized as former President Johnson. Rosa Clark struck up a conversation with a man next to her who was dressed in sports clothes and was fiddling with some camera equipment. Herman Clark did a double take and said to Rosa, "Do you know who you're talking to? You're talking to Barry Goldwater!"

Herman Clark had not checked out the spacecraft; that was up to other people. Nor had he checked out the Saturn V boosters. He was thinking ahead — to the time when Armstrong and Aldrin would separate from the mother spacecraft and fly Eagle, Herman Clark's LM No. 5, down to the surface of the moon. He was thinking about that ascent engine, the only one they had, which had to fire to get the two men off the moon again. Weeks in advance he had acquired goose pimples about that moment: *I mean, you've got two guys sitting there and just the idea. . . .*

Nassau Bay, Texas

Ted Guillory was a flight planner. In fact he was a "lead flight planner." It was his responsibility to sign a document of some 240 pages — it would be signed also by six other people, including Deke Slayton, George Low and Chris Kraft — which Neil Armstrong, Mike Collins and Buzz Aldrin would use as a celestial road map to take them to the moon and back. It would be timelined from 9:32 A.M. EDT — but more importantly, from liftoff, in ground elapsed time, the amount of time in flight, 195 hours and 40 minutes to splashdown in the Pacific. In the execution of the flight plan GET was the only kind of time that really meant anything, the only kind of time that related to the hour of launch, the moon landing itself and to splashdown. There would be two copies of the flight plan on board, loose-leaf books weighing about two pounds apiece, plus about twenty pounds of data and reference books. The whole mass would cost about five thousand pounds of fuel to send aloft — i.e., one pound of weight, whether it be paper or the raw flesh of man, required two hundred pounds of fuel to get into orbit and then be shot free of the earth's gravity.

Guillory's basic preparation had been done long ago; after all, this mission had been scheduled a year earlier. During the last six weeks before launch the task was to make the flight plan operational. It had been simplified somewhat; originally the mission had called for two moon walks, and that had been cut down to one. But again and again Armstrong, Collins and Aldrin would use the flight plan, in its penultimate stage, as a road map to simulate the entire G mission in the Cape Kennedy "train wreck." There were the navigational sightings, plus recalibrating the attitude of the earth, or whatever the astronauts used for a horizon. Before the circumlunar mission (C prime, Borman, Anders and Lovell) in December 1968, for which the flight plan — like everything else — had to be put together in a tremendous rush, Guillory and his men had found it useful to draw crude little pictures into the plan: "See here, what is this pitch 180, roll 180 and pitch 322? That's all nice, but here you drew a picture and that was the way the spacecraft should be with respect to the moon. Sometimes the simple things that you don't have a big computer program for turn out to be the most useful of all." There had been some difficulty about star sightings on that flight. Buzz Aldrin was on the backup crew and he suggested centering the star chart around the lunar equator; that would make it easier to figure things out. Guillory recalled, "We said yeah, that's a good idea, but we didn't know how to do it. So we had to query the Air Force. They said they could do it. Sure enough we got a star chart back but it was right before launch and we had to fix it up. Then Buzz came back and said it would be better if the star chart centered about the plane of the ecliptic. We redid it again, but this time it wasn't so hard because we had our program that tossed up the stars automatically and we just got the computer to reposition the stars." It was about this time, looking ahead to Apollo 11, that Guillory began to comprehend some of Buzz Aldrin's depth: "Boy, he's really something. He carried a slide rule for his Gemini flight on the rendezvous and I sometimes think he could correct a computer. I can remember hearing him say things like, 'If I get an answer that says I'm twenty feet out of plane I'll believe ten of that, but I won't believe all twenty of it.' He came in with something on C prime about 'If you plot the middle gimbal angle on the star chart it'll tell you all these things and you won't even need the ground.' Nobody could figure out what he was talking about so we did it. And sure enough, he had a way to figure out all these things with this little old sine curve — how to point to be perpendicular to the sun, for instance. To him it was just plain old simple."

This was a simpler flight plan, although — as in the case of Apollo 10 — something extra had to be built into it for the command module pilot: "What about the time Mike Collins is going to be flying the spacecraft by himself? That's quite difficult — to monitor all the systems, to fly the craft, to track everything. So on Mike's copy of the flight plan, for the period in which he would be alone, we had to build the checklist right into the plan itself so Mike didn't have to run around and look for his checklist while he was following the flight plan." And the flight plan had become more rigid than its predecessors, even by early May 1969, when Guillory said, "In C prime we could have gone twenty orbits in lunar orbit, or ten, and we had some flexibility. Here we're going to go and do a certain job, and we have very little flexibility in changing things. We can't even change a sleep period without going before a board!" There was indeed some argument about whether Armstrong and Aldrin should have a sleep period on the moon before doing their one EVA, and the continued discussion actually prevented publication of the "reference plan" — the normal first stage of any flight plan. The first published plan was called "final," but it had to be revised to put in an optional sleep period.

Like so many others, Ted Guillory sometimes yearned for the simpler days of Gemini: "It's harder now because our job is to know a little bit about everything, TV and all that, and with all these different systems I can't keep up with them. In Gemini just about everybody knew the systems and if you had a problem they could tell you switch by switch what you did. In Apollo the people know a little bit each — until the last minute when we all try to learn the whole flight plan so we can support the mission."

By July 15, the day before launch, Guillory and a twelve-man crew were ready to support the mission. The flight plan was aboard the spacecraft, and it was nearly time for Armstrong, Collins and Aldrin to get aboard.

That afternoon, in Nassau Bay, Patricia Collins got her hair done, picked up the three children at day camp and heated up some frankfurters and beans for supper. She was a Finnegan from Boston, a pretty, deceptively fragile-looking woman with a vivid smile. Her movements were quick; she seemed to flash rather than walk around the room. In the evening she had a final telephone conversation with her husband; he asked if he should call her from the Cape the following morning before launch, but Pat said no. In the evening she read a few chapters of Arthur Hailey's *In High Places*. She wondered idly what was happening to her application to enter the University of Houston for graduate study.[3] She had been an English major at Emmanuel College in Boston, and she was wondering if she was good enough to do some writing on her own. The last night Mike had been home, over the Fourth of July weekend, she had showed him her notebooks for the first time. Mike liked what was in them, which relieved Pat; she thought that if the contents had been rubbish he would have known it and said so.

Joan Aldrin also went to the hairdresser on July 15. She was conscious about getting ready for houseguests — her father and stepmother, Mr. and Mrs. Michael Archer of Ho-Ho-Kus, New Jersey. (Her mother, who had fostered the match with Buzz by inviting him to dinner back in 1952, was killed in the crash of a private plane while Buzz was in Korea.) Jeannie Bassett had already arrived from San Francisco; her astronaut husband, Major Charles A. Bassett II, had been killed in a plane crash in 1966. Buzz's uncle, Robert Moon, and Mrs. Moon were coming from Arcadia, California. No hotel space was available, and Joan had to find a place for them to stay with friends. She was a strikingly attractive blonde who had a master's degree in theater arts from Columbia University. Before she met Buzz she did some television acting, and she still acts and occasionally directs for the local community theater. When she learned, in January 1969, that Buzz had been selected for the moon landing crew, she hardly knew whether to cheer or weep: "I wished Buzz were a carpenter, a truck driver, a scientist, anything but what he is. Now I understand how some of the other wives felt — wanting to run and hide. I want him to do what he wants but I don't want him to . . ." A deeply sensitive woman, capable of great emotion, at one point she had made elaborate plans to throw herself into extra housework during the flight of Apollo 11 — wash the windows, paint the walls, anything to keep sane. Now it began to look as if those things would not get done. She started thinking about a pet to replace the monkey — the "smelly little beast," as Buzz called him — which caught cold and died after a bath in the swimming pool. Michael, age thirteen, was crazy about all animals. He seemed to want a snake, like the baby boa constrictor at the Dick Gordons'. Joan Aldrin would think about that.

Cape Kennedy, Florida

Guenter Wendt stayed at the pad until nearly six o'clock Tuesday evening, making sure his precious spacecraft was in good shape. Then he drove home and slept for an hour. Before midnight he was on his way back to the VAB. He had to round up all his troops and make sure the trucks were parked in the right places; then he could have a leisurely breakfast. As usual he found the rocket a little scary: "The whole thing moves and groans. It has its own noises you haven't heard before." He and everyone else had had a bad moment a week earlier when the pad technicians discovered the possibility of a leak in "the bird." Walter Delle, a thirty-seven-year-old quality control inspector for Boeing, was summoned. It took all day and half the night to determine that the leak was inside the liquid oxygen tank. On the morning of July 10 Delle entered the tank and climbed the walls, listening for a hiss which would locate the leak more specifically. He narrowed it down to the helium tank manifold, which Delle regarded as the most delicate piece of equipment in the tank. *A slip of the wrench,* he thought. . . . *It would take at least four days to replace the manifold; they would miss their window and the shot would have to be postponed.* He left the tank to talk with people at the Cape, at the Houston MSC and at Boeing's Michoud plant in New Orleans, where he was normally based; should they take the risk of the leak or take the risk of repairing it? Delle recommended the latter, and everyone finally agreed. He went back into the tank. Guided by charts, he tightened "B" nut on the manifold to maximum torque. It took him thirty minutes. He left the tank, had it pressurized with helium and returned for a final inspection. It was fixed. His wrench had not slipped.

Guenter Wendt was going over that in his mind, and a lot of other things; the launch control book for this mission had seventeen hundred pages in it. *So many things that could go wrong* . . . Guenter Wendt even had a few things on his mind which were not in the launch control book — things like an absurd little present for the flight commander.

Joe W. Schmitt was up and at work in the MSO Building by 3:30 A.M. on July 16 — forty-five minutes before Armstrong, Collins and Aldrin were awakened. Joe Schmitt was the lead suit technician. Indeed he had suited up every American astronaut who had gone into space — Mercury, Gemini, Apollo, the lot. As missions became more ambitious, the suits became more elaborate. The ones for Apollo 11 had five hundred parts in them. Essentially the suit was one patch over another: an inner layer of five-ounce Nomex, two outer layers of filament-coated Beta cloth, a fourth layer at the knees, elbows and shoulders. (A man *could* fall down on the moon, and a torn suit in that situation was not a pleasant possibility to think about.) Joe Schmitt had a FAM and OPS (Familiarization and Operations) manual, but it was usually about three months behind. And for this mission — the big one — there were two kinds of suits, both made by ILC Industries Inc. (International Latex) of Dover, Delaware. Mike Collins had the lighter "intravehicular" version; it weighed about thirty-five pounds. Armstrong and Aldrin, who were to walk on the moon, had the heavier "extravehicular" version, which weighed about fifty pounds. Complete with the PLSS, the Armstrong and Aldrin suits weighed one hundred eighty-three pounds, but lunar gravity would reduce that weight to a little over thirty pounds. The suits were to provide an artificial atmosphere of one hundred percent oxygen, adequate mobility and micrometeorite and visual protective systems. The probability of the command module getting hit by a micrometeorite, an object the size of a cigarette ash — the same statistic presumably applied to a man on the moon — had been calculated to be 0.000815. Any micrometeorite that did hit would be traveling at the rate of sixty-four thousand miles an hour, and as for that . . .

Well, that was a detail beyond Joe Schmitt's power to cope with. But he could make sure that all the gadgets on the suits were in place — the communications carrier on the helmet, the electrical connector on the chest, the sunglasses pocket on the sleeve, the pressure gauge on the forearm, the arm bearings between the elbow and the shoulder joint to provide more mobility in reaching overhead: plus the lunar boots, plus the LEVA (lunar extravehicular activity) covering helmet with two shades of visors, one of them coated with electrically charged gold. Armstrong and Aldrin could see outside, but in the event they did encounter a strange being on the moon, that being could not see them inside.

The first thing Joe Schmitt and his four associated technicians had to do on launch morning was unbag the suits and lay them out. They had last been used on July 3, and after the countdown demonstration test they had been taken apart and put together again — a four-day job. The component parts had to be examined for wear, deformation, deterioration, scratches, sharp edges that might catch and rip, cleanliness. The directions were maddeningly specific; a certain kind of screw had to be turned four times, not three times or five, with a certain kind of torque screwdriver. Four different kinds of cleaning solutions were specified for various parts of the suits. The technicians had actually been working on these suits since the crew was chosen at the time of Apollo 8, and for the lunar landing some modifications had to be made. Buzz Aldrin was concerned about his arm length and his leg length, because when the suit was pressurized the gloves had a tendency to move away from the fingertips. Armstrong requested an accessory pocket on his thigh, with a flap opening toward the knee. He also needed a "contingency sample pocket" so he could collect a small amount of lunar soil quickly in case it were necessary to abort the mission and leave the moon immediately.

At 4:30 A.M. the transfer vans arrived — one prime, one backup. They had to be checked out: the communications consoles, the technician headsets, the purge ventilators (three for the crew, six spares stored in each van), the gas masks (eight in each van, in case there should be a hypergolic fuel leak on the pad). The crew's wristwatches were wound and set to Houston time. There was no idle chatter when the men came in from breakfast about five-thirty. But as Schmitt was stowing checklists and data cards into the suit pockets he called the astronauts to come over: "I always make a point of letting them know where their things are. So they came over and each one had a sandwich made of rye bread. And they each had a little tube of some sort of fruit drink. They stood there and loaded up their own pockets."

Now it was 5:55 A.M. and time to suit up: legs in first, pull the suit up to the waist, then over the shoulders and put your head through the neck ring; then the communications carriers (the "Snoopy" hats), gloves and helmets. Then came a final series of checks for possible pressure leaks (an improperly closed zipper could cause a leak), plus another check on communications and still another on the amount of oxygen inside the suit, and finally a switch from the console to the portable ventilators. At one point Neil Armstrong dropped a film cassette on the floor. Joe Schmitt was relieved when he saw Armstrong bend over, quite easily, in his space suit to pick it up.

While all this was going on Deke Slayton went outside to tell newsmen, standing in the glare of television camera lights, that the crew would soon be out, that so far it was going just as it had gone on other missions, that it was going to be a good mission.

At 6:26 A.M. Joe Schmitt's number one backup, Ron Woods of International Latex, led the way out of the suiting room. Schmitt brought up the rear. There were signs on the wall. One said "The Key"; another said "To the Moon"; still another said "Is Located." Guenter Wendt had asked Joe Schmitt to put the signs there. As they passed by crew quarters Schmitt noticed someone standing there with a brown bag in his hand: "Mike Collins grabbed it. We got into the van and Neil said, 'Joe, reach in my pocket and get out this little card for me.' I found the little card — I didn't pay any attention to what it said — and stuck it under his watchband."

It was 6:37 A.M. when the flight crew and the "close-out" crew of six people were all assembled at the pad. The close-out crew consisted of Joe Schmitt and Ron Woods on the suits; John Grissinger, a mechanic from North American Rockwell, on the command module; "Lucky" Chambers, a NASA quality control man; Guenter Wendt, the fuehrer of the pad; and Fred Haise, the backup lunar module pilot. Haise would be the last man to leave the spacecraft. In fact he had been there for several hours before the Apollo 11 flight crew arrived. He had done the same job for Apollo 8; he went through a switch list, setting up the spacecraft for launch through a series of switches and checks, all those things in the count, timelined so that he would finish about the time the other people arrived.

Guenter Wendt was waiting at the 320-foot level when they all came up on the elevator. This was the "thirty seconds of recreation period" he always looked forward to. Armstrong, Collins and Aldrin had taken off the yellow galoshes they had worn over their boots when they left the MSO building. Neil Armstrong was the first man off the elevator, and Guenter Wendt had something for him that he did not have: "So I had a four-foot-long key to the moon. It was made of foam and covered with nonflammable tape, and it came out silver. Of course they couldn't take it along." Neil gave Guenter Wendt a ticket for a "space taxi" marked "good between any two planets." It was the little card Joe Schmitt had fished out of Armstrong's pocket and put under his watchband.

Mike Collins reached into that brown bag he had picked up at the last minute in crew quarters and handed Guenter Wendt a plaque with a mounted fish on it. The inscription was "Guenter Wendt, Trophy Trout." The fish was only seven inches long. Wendt did not know what to do with it: "Apparently nobody ever told Mike that when you mount a fish you stuff it first." He decided to take it home and put it in the deep freeze.

Wendt did not expect a serious present from anybody. But he and Buzz Aldrin were both Presbyterians, and Buzz suddenly handed him a modern version of the New Testament, one of the hundred and thirty copies that Buzz had given to the Webster Presbyterian Church in memory of his mother, Marion Moon Aldrin. It was inscribed "On permanent loan to G. Wendt."

At the 320-foot level Joe Schmitt was no longer in voice communication with the crew. In the van, after the switchover to a second set of portable ventilators (one ventilator will not last a man from suiting room "egress" to spacecraft "ingress"), Schmitt had wished them "a real good flight." Buzz Aldrin had replied, "You take yourself a good vacation when you get us all off." It was 6:52 A.M. when ingress started. Neil Armstrong was the first to enter the spacecraft, grasping the overhead handrail with both hands to swing himself through the hatch of the command module on top of the Saturn V, which was groaning and purring with strange noises as white streaks of vapor were emitted by the fuel tanks — which were constantly being "topped off," although the loading of liquid hydrogen and liquid oxygen, plus the high-grade kerosene called RP-1, had been completed about the time the men had left crew quarters. Armstrong eased himself into the commander's couch on the left. He had not been home in ten days, and he briefly wondered about the pile on the desk in his working den. *The bills are easy — no creative effort required. They come with their own envelopes, IBM cards. You just write out the checks and send them off. But that pile of personal mail. It always accumulates faster than I can deal with it.* There was no time to think about that now. Joe Schmitt was right behind Armstrong to start hooking him up — first the communications, because then the commander could say good morning to the ground, and test conductor Skip Chauvin could say the same and be heard.

Mike Collins was next. Normally he would have been last, since on other Apollo missions the command module pilot rode in the center couch. But Mike had been the last to join the Apollo 11 crew and Buzz Aldrin had already trained for the center seat. It seemed a waste of time to retrain him, so Mike simply learned the right-hand seat and began the flight there. That made Buzz the last to enter the spacecraft because it was the logical way; he did not have to climb over either of the other two.

The last thing Joe Schmitt had to do was the suit circuit gas sampling, to make sure the men had not lost their oxygen purge: "I did that and was ready to egress — and old Mike Collins reached up and grabbed my hand."

It was 7:22 A.M.: ingress completed. The suit hoses were hooked up — three of them. Two lines were for breathing (at least ninety-five percent oxygen) and one was for communications and biomedical data. Now there were only four men in the spacecraft — the flight crew and Fred Haise. The other five members of the close-out crew were in the "white room" on swing arm No. 9. For the next thirty minutes Haise would be the busiest man in the spacecraft: check the olive-drab lap and shoulder strapping for each man — fire-resistant material introduced since the pad fire; set the communication switches in front of the center couch; make sure that all loose gear had been removed — the tool boxes, the portable ventilators. Fred and Mike together reconnected the seat-strap strut lock to the ceiling of the spacecraft. Then Haise crawled out, closed and sealed the hatch. It was 7:52 A.M. — exactly one hour after ingress had begun. They were right on time — the liftoff plan called for closing the hatch at T minus one hour and forty minutes.

It was not a particularly harried time for the crew; they had been much busier in their simulators. But there was still work to do, starting with spacecraft-to-ground readouts to make sure that no switches had been bumped during the ingress procedure. Meanwhile the cabin atmosphere was being purged to achieve a combination of sixty percent oxygen and forty percent nitrogen, much less volatile than the pure oxygen atmosphere in use at the time of the pad fire, although the crew continued to breathe pure oxygen through their suit loops.

Suddenly that valve acted up again — the same one that had to be fixed during the countdown demonstration. But the intervening week had given the launch crew time to work up a system using a set of replenish valves — a kind of bypass. A crew went out to the pad, did a quick but intricate torque job — it worked. No "hold" necessary.

More to do: check out all the emergency detection procedures; Armstrong handled the spacecraft end of that job, and it took about thirty minutes. Then there were the gimbals on that big service propulsion system engine directly under the crew in the service module — big arms which functioned as a sophisticated steering mechanism in space; did they work?

T minus 61 minutes and counting. . . . As Neil Armstrong moved his rotational hand controller we assured ourselves that the engine did respond by swiveling or gimballing. . . .

At some point when all this was going on Jim Lovell, Armstrong's backup flight commander, got on the microphone. The preceding night he had said to Neil, "This is your last chance to tell me if you feel good. Because if you do, I'm going to have myself a party." Now he cut in to ask Neil one more time if he could go.

"You missed your chance," Neil said.

There was no apprehension inside the spacecraft; the men had confidence in themselves, in their hardware and in their ground support. Walter Cunningham once recalled that for a crew the dominant feeling on launch day was "Please launch and get us away from all that hand wringing." Yet there was unmistakable apprehension among the five thousand or so spectators three and one-half miles away on the ground. Men and women walked about restlessly in the sunlight which now blazed down with tropical intensity; the rain had long since stopped, and the cloud cover was a thin one at fifteen thousand feet — plenty of visibility. The forecasters had been right after all. Yet the mission could still be aborted, even after ignition at 8.9 seconds; that had happened in the initial seconds of a Gemini launch, when there was a malfunction and the engines of a Titan rocket cut out automatically before it left the pad. Tom Stafford and Wally Schirra were strapped in their spacecraft and wondering what had happened. Probably nobody would ever match the relaxation achieved by Gordon Cooper during the countdown for his Mercury flight; he dozed off on his couch.

We have just passed the fifty-six minute mark. . . . We are doing quite well, in fact some fifteen minutes ahead on some aspects. Armstrong replied that was fine as long as we don't launch fifteen minutes early.

On her boat Jan Armstrong was saying, having been up most of the night, "I wish they would hurry up and get this off so I can go to sleep!" As the time approached she took her two boys and a transistor radio out on the forward deck. "We're going to go, we're going to go!" she kept repeating.

Nassau Bay, Texas

In the Collins home Patricia Collins, wearing watermelon silk Chinese pajamas, stumbled groggily into the kitchen about two hours before liftoff and made a cup of coffee. Squinting at the television, she quipped: "I guess we don't have to worry about their getting off. They have to — none of them knows how to read the alternate flight plan." Then the launch pad flashed on the screen. "Oh God," she said, "I hope they get to do it." Then she brightened and said, "I'll just have to stand over it and push." The big topic of conversation in the Collins home was not the moon shot but the missing tree; the night before a violent thunderstorm had uprooted an old oak on the front lawn. Almost immediately a yellow car pulled up at the curb and a wiry man in boots and Levis sloshed through the rain and presented himself at the door: "I'm Mason, the tree artist. I see you've had some trouble here." Pat Collins said she would be grateful if he could remove the debris and chop it up for firewood. "But that's my wishing tree," wailed ten-year-old Kate Collins.

Joan Aldrin had set the alarm for 6 A.M. Houston time, two and one-half hours before liftoff, but she turned it off and went back to sleep. She awakened at ten minutes before seven — almost the exact moment the spacecraft hatch was being sealed at Cape Kennedy. The morning paper and the milk were still on the front porch. By seven o'clock a Nassau Bay Community Group van had arrived to clean up the damage caused by the thunderstorm the night before. Joan dressed herself in elegant vivid green, her blonde hair very neat, barefoot to walk on the cool flagstoning of the family room. Then she exclaimed: "My God, the flag!" She had forgotten to put the American flag outside the house, something she had been afraid for forty-eight hours she might forget to do. "Outside" was now full of press. She found someone else to put up the flag.

An hour before liftoff Jeannie Bassett and Joan were watching television in a haphazard way, cutting up coffee cake and taking the preceding night's dishes out of the dishwasher. At ten minutes of eight Joan said, "I think my kids should get up" — and Jeannie Bassett went off to call them. Trying to fix a wobbly television picture, Joan complained: "Every time they say something about the Aldrins the TV goes off!" She disappeared into the bedroom to call Pat Collins about the lost oak tree, then returned to sit in a new chair — a big round one — that even Buzz had not seen yet. (She would not sit in that chair again for eight days.) Andrew, age eleven, thought he would like to be at the firing room at the Cape; he wanted the best seat in the house. Joan Aldrin said, very calmly, "I think we have it right here." Astronaut wife Caroline Henize arrived and knelt down on the floor by Joan. Caroline was very pregnant. Joan said, "Caroline, don't you dare have your baby here."

T minus sixty seconds and counting. Fifty-five seconds and counting. Neil Armstrong just reported back. It's been a real smooth countdown. We have passed the fifty-second mark. Our transfer is complete on internal power with the launch vehicle at this time. Forty seconds away . . . all the second stage tanks now pressurized. Thirty-five seconds and counting. We are still go with Apollo 11. Thirty seconds and counting. Astronauts reported, feel good. T minus twenty-five seconds. Twenty seconds and counting. T minus fifteen seconds, guidance is internal. Twelve, eleven, ten, nine, ignition sequence starts. Six, five, four, three, two, one, zero, all engines running. LIFTOFF! We have a liftoff, thirty-two minutes past the hour. Liftoff on Apollo 11. Tower cleared.

In the last few minutes Pat Collins's gestures had become slower, more deliberate. She chose her words more carefully; she lighted her cigarette more slowly. Ann Collins, age seven, was absorbed by a story in the Houston *Post* which was all about the oak tree felled the night before. Pat Collins was going over the flight plan: "What's the apogee?" she asked astronaut wife Barbara Young. "I can't find the apogee." Ann asked permission to go fetch a friend; she wanted to show the friend where the tree had been. It was the wrong time to ask: "Honey, Daddy's going up right this second."

"Can I go after the launch, then?"

"All right, after the launch."

Pat by now was biting her lip: "It sits there a long time. It's sitting there forever. Maybe we should push. *Push.* . . . There it goes — *it's beautiful, beautiful.* . . ."

At one point there was a message from the launch crew to the spacecraft that had the word "Godspeed" in it. Joan Aldrin half sneezed and half coughed, on the verge of tears. At T minus 02:49 she began shredding a cigarette in the ashtray that rested on a small stool beside her chair; then she began twisting her handkerchief, controlling her emotion by constant flexing of her hands. *Forty-six seconds* . . . "Something's burning!" Joan said suddenly. Everyone started looking for cigarette ends, but the heat was coming from some camera lights that had been set up in the family room. Liftoff: Joan sat quietly, very rigid, swallowing hard, and said not one word. The room stayed silent for about three minutes. Then, with the rocket reported thirty-five miles downrange, Andy Aldrin said, "There it is." It was seven minutes after liftoff when Joan finally said something, still without moving: "It must be clear today; they can follow it so well."

Cape Kennedy, Florida

Beth Williams was with nurse Dee O'Hara. She is the widow of Marine Corps Major Clifton Curtis ("C.C.") Williams, an astronaut assigned to the crew of Charles ("Pete") Conrad. He was to have been on Apollo 12, the second moon landing. He died in the crash of a T-38 in a wooded area twenty-one miles northeast of Tallahassee, Florida, in October 1967. He was hurrying home to Houston because Beth had just discovered she was pregnant with their second child.

Watching the launch of Apollo 11, Beth Williams cried a little: "I was standing beside Dee and we were clutching each other's hands. I probably broke hers, I held it so hard. Tears streamed down our faces. No sobs or anything, just all this water." Dee O'Hara had "this terrible knot in my stomach, and I could feel myself stiffening. Then came the crescendo, the noise, and I felt that I was getting taller and taller and taller. . . . Everybody got his money's worth on this one."

Lola Morrow was with Dee O'Hara too, and she cried. But women were not the only ones who cried. At the press viewing site, situated three and one-half miles from the launch pad because someone had calculated that to be the maximum distance an exploding Saturn V could toss a one-hundred-pound piece of hot metal, people behaved very strangely. They jumped up and down. They yelled, and not very coherently. They cheered. And they cried — in fact they cried and cried and cried. A dark-complexioned man with a camera strung about his neck turned from the scene at just the moment he should have been taking pictures. There was a look of transported ecstasy on his face. Nothing like this had been experienced since John Glenn had done the first American earth orbit in 1962, perhaps not since Al Shepard rode in the first manned Mercury capsule the year before that.

On her boat Jan Armstrong heard the noise on the radio but could not locate the rocket with her eyes: "Fine, but where is it? Where is it?" Then, through the rocket's own launch cloud, she saw the first stage drop away and she exclaimed: "There it is! There it is!" The party then went back to the television set, but Jan Armstrong was glad she had seen it the way she had seen it.

Are there no men to think as earnestly as one climbs a mountain, and to write with their uttermost pride?

H. G. WELLS

"He doesn't owe me a cup of coffee"

FOR those on the ground, liftoff was that awesome sight which, within the limits of man-made phenomena, only a nuclear explosion could surpass: those searing stabs of horizontal flame at the base as the five F-1 engines ignited, consuming — in 8.9 seconds — more than eighty-five thousand pounds of fuel before the rocket even left the pad, thirty-three times the weight of the fuel carried by *The Spirit of St. Louis* forty-two years earlier on Lindbergh's New York–to–Paris flight; then, fifteen seconds later, the shattering sound waves, shaking the viewers' grandstand and the earth itself, accompanied by popping noises which sounded like the angry crackle of old-fashioned musketry. Strapped to their couches, breathing through their suit hoses and experiencing the first pressure of a "G" effect which within two minutes and forty seconds would build up to four and one-half times the force of earth gravity, Neil Armstrong, Mike Collins and Buzz Aldrin saw none of these fireworks. What they did experience was something like the slow, insistent lift of a powerful elevator, and they heard a noise which sounded like the distant rumble of an express train. At first there was little sense of movement, although they knew they had lifted off; launch control had said so. Then the altitude and velocity figures began to build up on the computer displays, and they knew that — in Donn Eisele's expression — they "were really hauling the freight."

It was the first thirty or forty seconds after liftoff that worried people like Kenneth Kleinknecht and Guenter Wendt. *There's not much of a chance if something goes wrong then,* Kleinknecht had always in mind. *Once they have ignition on the second stage, and once they have insertion, it gets a little easier. . . .*

Neil Armstrong reporting their roll and pitch program which puts Apollo 11 on a proper heading. Plus thirty seconds. Armstrong's heartbeat had now gone up to 110, Collins's to 99, Aldrin's to 88 — all lower than the figures for their Gemini flights, all satisfactory. Looking back to his Titan-powered Gemini flight (Gemini 10, with John Young), Mike Collins thought this was a very different kind of liftoff . . .

« It was, I thought, quite a rough ride in the first fifteen seconds or so. I suppose Saturns are like people in a way — no two of them are exactly the same. Ours was very rough at first. I don't mean that the engines were rough, and I don't mean that it was noisy. But it was very busy — that's the best word. It was steering like crazy. It was like a woman driving her car down a very narrow alleyway. She can't decide whether she's too far to the left or too far to the right, but she knows she's one or the other. And she keeps jerking the wheel back and forth. Think about a nervous, very nervous, lady. Not a drunk lady. The drunk lady probably would be more relaxed and do a much better job. So there we were just very busy, steering. It was all very jerky and I was glad when they called 'Tower clear,' because it was nice to know that there was no structure around when this thing was going through its little hiccups and jerks. But the jerkiness quieted down after about fifteen seconds. »

The rocket appeared to have taken off at a slight angle, which puzzled many who were at the Cape and many others who were watching on television, but prevailing wind conditions accounted for a kind of optical illusion. A NASA engineer explained it: "Let's say the wind tries to push the thing over. The booster does not push against it and try to fight it. If it did, it might break up. Rather, the booster says, 'OK, you pushed me over so far. I will compute a new trajectory and compensate for the push.' And we guide the engines according to velocity changes [Delta Vs], not according to elapsed time."

Downrange one mile, altitude three to four miles now, velocity 2,195 feet per second, about fifteen hundred statute mph. . . . Eight miles downrange, twelve miles high, velocity four thousand feet per second, more than twenty-seven hundred statute mph. Time for Mission Control in Houston to take over. . . .

But not time for the thousands of emotionally charged onlookers to leave their viewing sites; because of the high and thin cloud cover visibility at the Cape was excellent, and on this day they could *see* the Saturn V start to do its work. *This is Houston. You are go for staging.*

ARMSTRONG: Inboard cutoff. . . . Staging and ignition.

HOUSTON (CAPCOM Bruce McCandless): Thrust is go all engines. You are looking good.

ARMSTRONG: We've got skirt sep. . . . Tower is gone [confirming the engine skirt separation and the separation of the launch escape tower, which would have been used only in the event of an emergency abort during launch]. . . . Be advised the visual is go today. . . .

COLLINS: Yes, they finally gave me a window to look out.

Neither Armstrong nor either of the other two men in the crew could see the spectacular part of all this. Millions of other people did watch the 138-foot-high first stage, the S-IC, drop off into the Atlantic like a fiery tin can, having consumed in less than three minutes forty-five times the weight of fuel burned by a Boeing 707 in seven and one-half hours on an average flight from New York to Paris.

On her boat on the Banana River Jan Armstrong watched the last vapor trail disappear, then sat and listened while David Scott explained that some earth orbits were better than others for TLI, which was scheduled to take place in a little less than three hours. This mission, Scott said, had a good orbit. The boat's skipper asked Jan if the years of living through launches and test piloting hadn't begun to affect her. Jan laughed and pointed to the gray in her hair. "I haven't aged a day," she said.

Downrange 530 miles, altitude 95 miles, velocity 17,358 feet per second — nearly twelve thousand statute mph. Still go at 7 minutes 41 seconds. At 8 minutes 17 seconds a fuel-low sensor would be uncovered to start the next staging sequence; at 9 minutes 11 seconds the outboard engines of the second stage, the S-II, would cut off; one second later the S-II itself would cut off and drop, nearly eighty thousand pounds of dry weight, like the S-IC, into the Atlantic. This would leave the astronauts with one power plant to fire them to the moon — the S-IVB's single J-2 engine. Unlike the F-1 engines which boosted the rocket cluster off the pad, the J-2 could be shut down and restarted; but first it had to be fired for about two and one-half minutes to put the spacecraft into parking orbit as a preliminary to translunar injection.

ARMSTRONG: It's a nice day for it. These thunderstorms downrange are about all.

[*Altitude one hundred miles, downrange 883 miles, outboard engine cutoff.*]

ARMSTRONG: And ignition.

HOUSTON (McCandless): This is Houston. Predicted cutoff at 11 plus 42 [seconds].

ARMSTRONG: Shutdown.

Right on time . . .

Nassau Bay, Texas

Patricia Collins was poring over the flight plan. She read it better than most astronauts' wives, although occasionally she went to the blue "cheater" which she kept under an ashtray — a condensed version of the plan, written in more or less plain English. "There's Daddy talking," she said. "I wish we'd get confirmation of orbit." Then it came; the radar of the Canary Islands tracking station showed Apollo 11 in a circular orbit 103 nautical miles above the earth. "Very nice," Pat Collins said, nodding approval. Ann Collins asked, "Are they on the moon yet, Mommy?" Pat laughed and said, "Oh no, honey. It takes two days to get to the moon." Young Michael Collins corrected her: "Three days." He had had to be reprimanded several times for being too noisy; now he went away and reappeared with adhesive tape over his mouth. "Now can I go to Mary's to tell her about the tree, Mommy?" asked Ann.

Just before the second staging, leading to the separation of the S-II from the S-IVB, Joan Aldrin leaned back in her chair with a deep sigh. Still picking at her handkerchief, she asked, "I wonder if they're going to have TV coverage throughout TLI?" Bill Der Bing, a NASA protocol officer, said he didn't think so. Missy the dog jumped up on the chair beside her. Joan said firmly, "She's not allowed in the chair!" and shoved her down. Then Joan put on her glasses and went into the kitchen to study an abbreviated flight schedule pinned onto the wall. Suddenly the room broke up and the children wandered away; the tension was over — for the time being.

Before the flight Mike Collins and Glenn Parker, a flight crew training instructor at Cape Kennedy, had made a private bet. The bet concerned navigation. Shortly after liftoff Collins had to take a sighting on two different stars, using the telescope on the command and service module, to get an angle fix. Then he had to feed the angle information into the computer. The computer would check his angle and then come back to tell Collins how accurate his sighting had been. Glenn Parker bet Collins that his degree of inaccuracy would be .02. Mike Collins bet that the figure would be .00 — perfect. The bet was a cup of coffee.

This is Apollo Control at 52 minutes and the station at Carnarvon, Western Australia, is about to acquire Apollo 11. We'll stand by live for this pass.

ALDRIN: Houston, Apollo 11. Would you like to copy the alignment results?

HOUSTON (McCandless): That's affirmative.

ALDRIN: Okay, noun 71. . . . Check star 34. Over.

HOUSTON (McCandless): Roger, we copy, and the angles look good.

[*Another voice cut in*]
COLLINS: Tell Glenn Parker down at the Cape that he lucked out.
HOUSTON (McCandless): Understand tell Glenn Parker he lucked out.
COLLINS: Yes, he lucked out. He doesn't owe me a cup of coffee.

Collins's degree of inaccuracy was .01 — "one shade less than perfect," as Parker said at the Cape. . . . "And it was a fun bet."

Theodore H. White once wrote of the American investment in the Marshall Plan, which put Europe back on its feet after the Second World War: "A sum of money so huge ceases to be money. It becomes a force in its own right that spends itself, that creates its own magnetic fields, that grows out of its own vitality the organs, tentacles, fibrils and limbs needed to match its growing responsibilities." [1] Facts about the Saturn V invoked an analogy; this awesome vehicle also had "organs, tentacles, fibrils and limbs" — and it certainly had responsibilities.

Its very statistics defied the human brain. Including the boost protective cover and the launch escape system, which had a 150,000-pound thrust engine of its own, hopefully never to be called into use, the Saturn V stack — S-IC, S-II, S-IVB, the three-foot-high instrument unit, the lunar module, the command and service module, the escape bumbershoot at the top — stood three hundred sixty-three feet high, tall as a thirty-six-floor building, more than three times the height of the Titan stack for Gemini, three and one-half times the height of the stack for the Redstone rocket which flew four missions in the old Mercury program, fifty-eight feet higher than the Statue of Liberty from the bottom of its base. The number of parts in the whole was beyond counting; the command module alone, as distinct from the rest of the rocket stack, had more than two million parts. The upper two stages of the rocket were actually two giant thermos jugs, supercooled to minus 423 degrees Fahrenheit for the liquid hydrogen and to minus 297 degrees for the liquid oxygen. The walls had to be strong enough to take the tremendous thrust of the engines without buckling, tolerant enough not to turn brittle at extremely low temperatures or at the five thousand degrees of heat generated by the engines when they were fired. When the rocket was fueled on Pad 39A, great chunks of frost would form on the outside and occasionally drop to the ground. This Saturn V stack was the heaviest yet launched; fully loaded on the pad, it weighed almost six and one-half million pounds. But that figure included more than six million pounds of fuel — liquid oxygen and kerosene for the first stage, liquid oxygen and liquid hydrogen for the second and third stages, the self-igniting (and scarcely pronounceable) hypergolics which went into the separate control systems of the command and service module and the lunar module. When the two modules and the S-IVB went into earth orbit, the weight of the package would be down to a little less than three hundred thousand pounds; after translunar injection, which consumed still more fuel, the weight would be down to a little less than one hundred forty thousand pounds. The first stage, S-IC, alone burned in two minutes and forty seconds, at the rate of fifteen tons per second, a sufficient weight of kerosene — converted into its gasoline equivalent — to fuel, for three hundred years, the motorcar of an average citizen who drove ten thousand miles a year and used seven hundred fifty gallons of gasoline a year; a distance total of three million miles, well in excess of the round-trip mileage to the moon and back. The kerosene in the S-IC and the liquid hydrogen in the S-II generated thirty-five *billion* British Thermal Units, an energy output which would permit Consolidated Edison to run New York City for about an hour and fifteen minutes.[2]

The liquid oxygen and kerosene combination for the first stage was a relatively simple one. The F-1 engines which lifted the rocket off the pad were monstrously muscled but not particularly sophisticated; they could not be shut down and restarted, but that was not necessary because they had only one job to do and it would all be over in less than three minutes. The J-2 engines, burning liquid oxygen and liquid hydrogen, two very bad-tempered elements which roar to life at the touch of a spark, had to be quite sophisticated indeed. The S-IVB's single J-2 had to shut down on signal, then restart to send the men of Apollo 11 on the long coast to the moon. Always there was the weight problem; some sections of the S-II's fuel-tank walls had been machined down to 80/1000ths of an inch. This was a feasible thing to do because the extreme cold of liquid hydrogen actually added strength to the metal. The fuel's chill also would serve to keep the engines from melting, and the heat from the metal fuel tubes incorporated into the engine structures was just enough to warm the liquid hydrogen and turn it into gas as it entered the combustion chamber.

At the time of Apollo 11 there was no doubt that the Saturn V was the most powerful operational rocket on earth, although there was some doubt about how long that would continue to be the case.[3] Its massive size had dictated the location of major construction and testing facilities near waterways; only stage 3, the S-IVB, was small enough to travel by air in the Super Guppy. Stage 1, the S-IC, would take up four lanes of a superhighway and even lying down would be thirty-three feet high, ripping out telephone and power lines if it were taken across countryside. It was assembled at Boeing's Michoud plant in New Orleans, then put on a barge and moved through seven and one-half miles of man-made canal to NASA's Mississippi Test Facility, nudged through the last few hundred yards on the site by the only jet-powered tugboat in the world. (The MTF was put together in a little less than three years, between May 1963 and April 1966, on thirteen thousand acres of government-purchased land, surrounded by another 128,000 acres leased to create an acoustical buffer zone. Some seven hundred people had to be moved, including quite a few moonshiners.)[4] The engines of stage I for Apollo 11 were tested on August 6, 1968, then shipped by barge through the Gulf of Mexico, around the Florida Keys and up the east coast of Florida to Cape Kennedy.

Stage II, 81 feet 6 inches long, had an even longer trip — a sixteen-day voyage on a covered vessel named *Point Barrow* from the Seal Beach, California, plant of North American Rockwell past the Pacific coast of Mexico, through the Panama Canal, across the Gulf to the MTF for test-firing (October 3, 1968), then on across the Gulf and around the Keys to Cape Kennedy. Stage 3, the S-IVB, was flown directly to Cape Kennedy after McDonnell Douglas gave it a static test at Sacramento, California, on July 17, 1968. Each stage of the Saturn V had been built by a different aerospace giant, but that was not the whole story: the subcontracting involved twelve thousand companies and some 325,000 people. Batteries were made in Joplin, Missouri; tunnel assemblies in Traverse City, Michigan; loading systems and components for cutoff sensors in Davenport, Iowa; cathode ray tubes in Waltham, Massachusetts; spools, harnesses and ducts in St. Louis; metal hose in Waterbury, Connecticut; temperature controls in Minneapolis; castings in Kokomo, Indiana; flex lines in Batavia, Illinois; communications antenna in Falls Church, Virginia; and component parts for the rocket's instrument unit, for which International Business Machines was the prime contractor, at Huntsville, Alabama. Neil Armstrong always seemed particularly conscious of the numbers and the quality of the people involved on an Apollo mission . . .

« In our preparation we weren't concerned primarily about safety; it was a matter of mission success. The nation was depending on the NASA-industry team to do the job and the NASA-industry team was staking its reputation on Apollo 11.

I knew about our crew assignment a couple of days before the news was announced in January 1969. It wasn't a shock, but it was very pleasant news to hear. It is true that most of the people available for flight status were already assigned to crews and were doing jobs from which they couldn't be removed. That left a limited number of us available for this mission, and I certainly had strong hopes that we would get it. Yet when we did get it, I suspected that it was highly unlikely that Apollo 11 would in the final analysis be the first lunar landing flight. The lunar module had not flown, and there were a lot of things about the lunar surface we didn't know. We didn't know if Mission Control could communicate with the lunar module and the command module simultaneously and successfully. Then, with the magnificent successes of Apollo 9 and Apollo 10, it became increasingly evident that we were in fact going to get a crack at the lunar landing. As this conclusion was unfolding, the preparations just became more and more relentless

— not just for the crew but for all the many people that were involved in the planning and execution of Apollo 11. »

This is Apollo Control at 2 hours 8 minutes into the mission. Apollo 11 about to be acquired at the Tananarive [Malagasy Republic] station. As expected this orbit is changing slightly as the S-IVB third stage vents. . . . We have acquired Tananarive now. . . . About forty minutes to go now to translunar injection. . . .

HOUSTON (McCandless): Apollo 11, through Tananarive. How do you read? . . . This is Houston standing by through Tananarive. Over.

ALDRIN: Houston, Apollo 11. We have the pyros armed.

HOUSTON (McCandless): Tananarive has loss of signal. The Carnarvon station will acquire at two hours, twenty-five and one-half minutes and during the Carnarvon pass the go/no go decision will be made for the translunar injection maneuver. . . . The spacecraft is near the Gilbert Islands, about halfway between Australia and Hawaii.

HOUSTON (McCandless): This is Houston through Carnarvon. Radio check, over.

COLLINS: Roger, Houston through Carnarvon, Apollo 11. Loud and clear.

Ten minutes away from ignition on translunar injection. Add 10,435 feet per second to the spacecraft's velocity. Look for a total velocity at the end of this burn of about 35,575 feet per second — about twenty-five thousand statute mph.

HOUSTON (McCandless): Apollo 11, this is Houston. Slightly less than one minute to ignition and everything is go. . . . Slightly less than one minute to ignition.

ARMSTRONG: Roger. . . . Ignition.

Coming up on 27,000 feet per second. Telemetry and radar tracking both solid. Velocity 27,800 feet per second. . . . Twenty-nine thousand feet per second building up to 30,000 feet per second. . . . Altitude 115 nautical miles. . . . 31,200 per second now. Altitude 125 nautical miles. Velocity 32,000 per second. Altitude 130 miles. . . . 34,000 feet per second now. Altitude 152 . . . 35,000 feet per second . . . Cut out. Velocity 35,570 feet per second. Altitude 177 nautical miles. . . .

ARMSTRONG: Hey Houston, Apollo 11. That Saturn gave us a magnificent ride.

HOUSTON (McCandless): Roger, 11, we'll pass that on, and it looks like you are well on your way now.

ARMSTRONG: We have no complaints with any of the three stages on that ride. . . . It was beautiful. . . . It was all a . . . all a good ride.

Considering the fact that it had been only a little more than ten years since early-model rockets fired at the Cape frequently aborted and dived into the Atlantic like fatally wounded geese, the men of Apollo 11 showed an odd indifference toward the Saturn V. Earlier flight crews had been more impressed; to Wally Schirra of Apollo 7, the booster was unbelievable. "It was a big *maumoo,*" he said. "A *maumoo* is an all-encompassing word to denote something that's bigger than almost anything else." And Schirra's crew was powered only by an uprated Saturn I, much less powerful than the Saturn V stack used for succeeding manned missions. Schirra liked to quote Wernher von Braun as saying, "If you make a booster that's very powerful I will cluster it and make it bigger." In effect von Braun had clustered eight V-2 type rockets and strapped them together; that was the genesis of the Saturn which, true to his word, von Braun kept making bigger and bigger. Finally it became something just to be depended on. Neil Armstrong once explained . . .

« It either does its job or not. And formerly there was very little you could do with the Saturn. You either rode it into orbit and it did its part, or you got off it and then you got somewhat less out of the mission. By the time we lifted off that was not necessarily true. You could not fly the earlier models from the spacecraft; had there been a failure on the Saturn's inertial system on Apollo 9, for example, McDivitt, Scott and Schweickart would have had to splash down into the Atlantic. Or maybe land in Africa, with a high risk of physical injury.

For our flight we had added an alternate guidance system in the command module's gear so that if there were a failure of some kind on the Saturn we could switch to the alternate system and fly the rocket from the spacecraft. I never had much to do with booster development, but I was very interested in getting this guidance capability. In 1959 an engineer named Ed Holleman and I did a study that showed the feasibility of manually flying a large booster. We ran a large number of simulations at Edwards and at the Johnsville, Pennsylvania, centrifuge and confirmed that the effects of acceleration on the people who were doing the guiding task inside was not significant. They could fly, manually, the booster as you would fly an airplane, and fly it into orbit. I wanted that alternate guidance system in Apollo. It was ready for Apollo 10, and of course I was glad to have it in Apollo 11. »

As Apollo 11 shook itself free of earth gravity, five hundred miles above the Pacific Ocean, it was now down to three units: the command and service module, the lunar module and its adapter and the S-IVB with its instrument unit. Now even the S-IVB had outlived its usefulness and had to be dropped; but first the positions of the command and service module and the lunar module had to be reversed in a critical maneuver called "transposition and docking." Following the discard of the emergency escape mechanism, shortly after liftoff at the Cape, the command module had at its top a "probe," a triangularly shaped assembly with a pencil-like point, which would fit precisely into the "drogue," an obversely shaped assembly opening at the top of the lunar module. In the world of space, words like up and down are relatively meaningless, but they are the only words to be used in describing a situation for which terrestrial language is inadequate. The command module was "up" in the sense that it was the forward piece of the mechanism hurtling toward the moon; and the lunar module and the third stage were "down" in the sense that they trailed behind. The idea was to make it possible for Neil Armstrong and Buzz Aldrin to exit through the "top" escape hatch of the command module, enter the lunar module and at the proper time separate to land on the moon. Therefore the command and service module first had to separate from the lunar module, which was still hooked to the S-IVB, drift off into space, moving at the rate of one foot per second, then make a 180-degree turn and stick its probe into the drogue of the lunar module, an operation completed by spring-loaded latches which automatically snapped the two vehicles together. Separation was to take place at approximately twenty minutes following translunar injection — GET 3 hours plus 23 minutes and a few seconds. It was essentially Mike Collins's job: he was the command module pilot.

Four minutes away from separation, four minutes. Velocity 26,314 feet per second, about eighteen thousand statute mph, decreasing as the spacecraft entered the long coast toward the moon. Distance from earth, 3,140 nautical miles.

Nassau Bay, Texas

It was 11:43 A.M., C.D.T., July 16, and it was time for Patricia Collins to walk out of the house for her first press conference. She had instructed the children: "Be polite, say that you thought it was nice or whatever you thought, and don't say too much." Into a one hundred degree temperature and a hot breeze Pat walked out the door and said: "I'm thrilled, proud and pleased." Then there was a distraction: a straggly parade of neighborhood youngsters, some on foot, some pulled in small wagons, all carrying American flags. The press conference over, Pat Collins went back into the house to learn that there had been a loss of communications with the spacecraft — one of the failures most dreaded by astronaut families. The room grew quieter as everyone tried to remain calm and maintain casual conversation while straining to hear any sound from the silent squawk box. Barbara Young reminded all hands that on Gemini 10 (Mike Collins was on that flight, too) "somebody, Mike or José [as the astronauts called John Young], bumped a switch and we kept calling and calling before they discovered it." Suddenly the squawk box was making noises again:

HOUSTON (McCandless): Roger, I read you very loud and clear, Buzz. Mike is pretty weak.

ARMSTRONG: Roger, we've got the high gain locked on now, I believe. Auto tracking now.

HOUSTON (McCandless): Okay, you're coming in loud and clear but Mike is just barely readable.

COLLINS: That was Neil. How are you reading Mike?

HOUSTON (McCandless): Loud and clear, Mike, and we understand that you are docked.

COLLINS: That's affirmative.

Thirty-four minutes now from extracting the lunar module from its adapter in the S-IVB. Time to start pressurizing the lunar module. . . .

HOUSTON (McCandless): Could you give us comments on how the transposition and docking was?

COLLINS: I thought it went pretty well, Houston, although I expect I used more gas than I've been using in the simulator. . . . And during the course of that we drifted slightly further away from the S-IVB than I expected. I expected to be out about sixty-six feet. My guess would be I was around one hundred or so, and therefore I expect I used a bit more [fuel] coming back in.

ARMSTRONG: And Houston, you might be interested that at my window right now I can observe the entire continent of North America, Alaska, over the Pole, down to the Yucatan Peninsula, Cuba, northern part of South America and then I run out of window.

ALDRIN: Houston, Apollo 11. All twelve latches are locked.

The twelve docking latches had locked automatically, as they were supposed to. So far, so good. Four hours and four minutes into the flight. Velocity 17,014 feet per second, about 11,600 statute mph. Five minutes away from ejection of the lunar module, separating forever from the S-IVB. Thirty-six minutes away from the evasive maneuver to get the two spacecraft out of the way of the S-IVB. Arm the logic switches. . . .

ARMSTRONG: Houston, we're ready for LM ejection.

HOUSTON (McCandless): Roger, you're go for LM ejection.

COLLINS: Thank you.

ARMSTRONG: Houston, we have sep. We have a cryo press light.

The stowage tanks containing cryogenic [supercooled] propellant which supply the fuel cells of the service module were registering properly. Velocity 14,972 feet per second, a little over ten thousand statute mph. Five minutes away from the evasive maneuver. . . . Standing by for the burn. . . .

Now the command and service module was under its own power, supplied by a 20,500-pound thrust engine, gimbal-mounted for steering and burning the hypergolic fuel that had been put into the tanks a month earlier. This was the service propulsion system engine that Mike Collins was about to fire for three seconds, adding 19.7 feet per second velocity to get away from the S-IVB, and Collins knew that it had to work. . . .

« It has to work. There's no redundancy built into it. You just hang your neck on it. You can look at that thing as just one motor or two motors, and a lot of its important components are redundant. But when you get right down to it there's just one little engine bell sticking out the back end, and if it blows off, or if you have an explosion, or if it fails to light — you know what I mean? It has to work like a jewel. »

Ignition, shutdown. It had worked like a jewel, and it would have to work like a jewel again when it came time to escape the moon's gravity and come home.

The S-IVB still had one more job to do: commit suicide. At ground command, its residual LOX (liquid oxygen) and propellant (liquid hydrogen) would be vented through the J-2 engine bell under some pressure, thus giving the now-useless stage a small additional push out into a "slingshot" maneuver, a trajectory which would take it behind the trailing edge of the moon, the right-hand side of the moon as seen from earth. But since the moon was moving away from the path of the spent stage, it would not be captured by lunar gravity but would slide past the moon and, like the descent stage of the lunar module flown by Thomas Stafford and Eugene Cernan on Apollo 10, go into solar orbit. In a little less than seventy-two hours the Apollo 11 spacecraft would approach the leading edge of the moon, the left-hand side as seen from earth, and — if all was well — permit itself to be "captured" by lunar gravity. If all was not well — and this was one of the moments Dr. Thomas Paine had in mind when he told the crew not to hesitate to abort ("Just don't get killed") — Armstrong, Collins and Aldrin could take advantage of a different "slingshot" situation. They could ride around the moon once, then — if something was going wrong — avoid lunar gravity and be "slung" back to the earth. It would take them some time to get home that way, but they would get home on momentum plus earth gravity.

Roughly nineteen thousand miles from earth now, speed decreasing. . . .
COLLINS: Houston, Apollo 11. We've completed our maneuvers to observe

the slingshot attitude, but we don't see anything — no earth and no S-IVB.

HOUSTON (McCandless): Roger, stand by. In GET I have a LOX dump start time for you. It's supposed to start at 5 plus 03 plus 07 and stop at 5 plus 04 plus 55.

COLLINS: Roger, thank you.

HOUSTON (McCandless): We now recommend the following attitudes: roll 307.0 pitch 354.0 yaw 019.5 and the LOX dump has already been enabled so we can't hold it off any longer. . . . It doesn't look to us like you'll be able to make it around to this observation attitude in two minutes. We recommend that you save the fuel. Over.

COLLINS: Okay, Houston. You got to us just a little late. Our maneuver has already begun, so it's going to cost us about the same amount of fuel to stop it no matter where we stop it and we may as well keep going.

HOUSTON (McCandless): Roger, go ahead.

ALDRIN: We've got the — what appears to be the S-IVB in sight. Oh, it has to be a couple of miles away.

HOUSTON (McCandless): Roger. . . . We've got about twenty-three and one-half minutes or so until the APS burn [firing of small attitude thrusters to steady the spent stage in its trajectory, keeping it at the proper attitude].

At a few minutes before six hours GET the last bit of life went out of the S-IVB. There was a pop song on every jukebox at the Cape in those days which went "Goodbye, goodbye, goodbye my darling. . . ." Saturn V's third stage was not exactly a darling, but it was time to say goodbye to a faithful workhorse even though the crew of Apollo 11 — because of the attitude of the spacecraft — never did see it disappear. They did know that it was safely out of their way, and that was enough.

Translunar injection, or "TLI," was a sophisticated combination of initials. It designated the "big burn" of the S-IVB which was to send the crew of Apollo 11 on the long coast to the moon, engines shut down, using — for some two hundred thousand miles — no propellant at all. It was a little like putting a motorcar of the 1930's into overdrive. But the timing of TLI had to be quite precise, dependent on where the moon would be a few days later. In the beginning the Soviet Union did not practice TLI; on their first three unmanned moon shots it appeared that they more or less aimed from the ground and fired. But this was not a very accurate method, and the Soviets changed. According to Sir Bernard Lovell of the Jodrell Bank Observatory in Britain, "There is no doubt at all that the Soviets have used this [American] system regularly on all their lunar and planetary flights from, I think, Luna 4 onwards. We do not have the full details here but it seems well known that their lunar and planetary probes have been placed in parking orbits initially." [5] Midcourse corrections and TLI were not the same; in fact the existence of TLI in the flight plan had the net effect of eliminating some midcourse corrections. Neil Armstrong had something to say about that . . .

« We like TLI. It gives us an orbit and one-half around the earth, or two and one-half orbits if need be, to check the spacecraft, and check out a lot of other things. We don't need to take off for the moon until we are sure that everything is ready. If something is wrong on the spacecraft, we have that time to decide whether we should forget the whole thing and abort. We like it. It gives us another chance to hit a moving target. »

From TLI, through transposition and docking, then separation and evasive maneuver — the whole process had taken a little more than two hours. But behind those two hours of accomplishment, which Armstrong, Collins and Aldrin had made look relatively easy, was all the experience of the manned Gemini flights — and a ten-year-old argument about how best to go to the moon.

There had never been more than three or four ways to get to the moon, and President Kennedy was not passing judgment on "which way" when he made his fateful pronouncement on May 25, 1961. He was simply putting the proposition: land a man on the moon and return him safely to earth. But how? At the time Kennedy spoke, that question was already engaging men in high places in acrimonious argument.

One way was to build a Buck Rogers rocket, a monster bigger even than anything on the drawing boards, send it to the moon with a flight crew, land and take off again for the return flight to earth. At the beginning of the 1960's there was some suspicion, lingering in the postflight shock of the first Sputnik, that this was the road the Soviet Union had chosen — the Soviets *did* have these monstrous boosters. There was a variation to this "direct ascent" approach: an enormous rocket, again bigger than anything we had or had drawn up, which would boost a spacecraft directly to the moon, land it and go away, leaving the spacecraft to boost the crew back to earth.

The most likely approach, at the time of Kennedy's pronouncement, was the one known as earth orbit rendezvous (EOR). Wernher von Braun and most of his rocket scientists favored it: two Saturn rockets — one carrying extra fuel, one a spacecraft — would be launched into earth orbit. The rockets would rendezvous and, with the extra fuel, the spacecraft would then be launched to the moon. It would carry a sufficient fuel reserve to lift off from the surface, escape lunar gravity and return to earth. In the back of everybody's mind was the idea of a space station in earth orbit, where a crew launched from the Cape could get off the express train and pick up a rental car for the remainder of the trip, then change conveyances again on the voyage home. (Many observers, noting the Soviet Union's interest in space stations and their repeated launch and linkup of identical spacecraft in earth orbit, believed that they still favored this approach.)

Meanwhile, a forty-one-year-old NASA engineer named John C. Houbolt brought forth something he called LOR — for lunar orbit rendezvous. There was an element of accident in this. Houbolt was not involved with the manned space effort; he was associate chief of the dynamic loads division at NASA's Langley facility in Virginia. But he was also chairman of a six-man committee which was studying the problems of rendezvous in assembling and operating stations.

When Houbolt put forward his calculations on LOR, Dr. Maxime Faget, one of the chief designers of the Mercury spacecraft and later director of engineering and development at the Manned Spacecraft Center, shouted: "Your figures lie!" Wernher von Braun shook his head and said, "No, that's no good." The scientists, early in 1961, were having as much trouble making up their own minds as the President had making up his mind; and Houbolt's LOR idea initially got as pleasant a reception as a tribe of pink elephants might expect in a hangover-studded saloon.

Yet it had occurred to Houbolt that rendezvous around the moon was like being in a living room: "Why take the whole darn living room down to the surface when it is easier to go down in a little tiny craft? As soon as I saw it in that broad context, the concept looked very appealing." He had been shot down at one meeting, but he went back and did some recalculations on the back of an envelope: "Almost simultaneously, it became clear that LOR offered a chain reaction of simplifications: development, testing, manufacturing, launch and flight operations. All would be simplified. I said, 'Oh my God, this is it. This is fantastic! If there is any idea we must push, it's this one.' "

Yet when NASA administrator James Webb announced that earth orbit rendezvous appeared to be the best road to the moon and direct ascent the second best road, he did not mention lunar orbit rendezvous. Once in 1961 Houbolt was even asked to cut out of a speech, concerning rendezvous in space, a passage pleading for the logic of LOR. Finally, in November 1961, six months after Kennedy had said "go," Houbolt wrote a frustrated letter to NASA associate administrator Robert Seamans, who was to become President Nixon's Secretary of the Air Force: "Somewhat as a voice in the wilderness, I have been appalled at the thinking of individuals and committees. . . . Give us the go-ahead and we will put men on the moon in very short order — and we don't need any Houston empire to do it."

Seamans liked Houbolt's tone and spunk and had the letter circulated. But unknown to Houbolt, some people in the "Houston empire" had been having second thoughts about LOR — among them Max Faget and Wernher von Braun. Later Houbolt was to say, "When von Braun changed his mind, and I admire him greatly for it, I figured that the last hurdle had been cleared." But it was not until the spring of 1969, shortly before the liftoff of Apollo 9 (McDivitt, Scott, Schweickart) that Houbolt, now a consultant with Aeronautical Research Associates of Princeton Inc., came across the work of a self-educated Russian mechanic named Yuri Kondratyuk. Fifty years earlier, Kondratyuk had calculated that LOR was the best road to the moon. "My God," thought Houbolt. "He went through the same thing I did." True enough, but with a different outcome: Kondratyuk died in 1952 without ever having managed to get the right people to listen; Houbolt lived to get an Exceptional Scientific Achievement Award from NASA — and to see his theory tested the bite-the-bullet way.[6]

By the end of 1962, well before the last manned Mercury mission — Gordon Cooper's twenty-two earth orbits in May 1963 — it had become apparent that it was not feasible to leap straight from Mercury into the Apollo mission sequence leading to a lunar landing. First there was the matter of hardware: to implement Houbolt's LOR approach, a totally new spacecraft had to be designed, built, tested. No one knew for sure what it should look like, though it was to be the world's first true "space" craft, a vehicle designed to operate outside the drag of the earth's atmosphere in the windless vacuum of space, and then to be abandoned out there. The contract for such a vehicle was awarded to Grumman Aircraft and nobody knew how long it would take to get it ready to fly. (The first name given to the thing was the rather wistfully wonderful LEM — for lunar excursion module. Only after strenuous objections from a number of people was the "excursion" eliminated from the title, on grounds of frivolity.)

In addition to such hardware problems, however, we had no experience in joining two vehicles together in space and then separating them at specified times — the rendezvous and docking procedures. Nor did we have experience in extravehicular activity — the "space walk" — to test man's ability to operate in an airless environment. We had to have both, and therefore we had to have a sequence of missions between Mercury and Apollo. That was the sequence called Gemini.

About the time Houbolt was winning his argument for the LOR approach, NASA officials began to hear interesting reports about some original work being done at MIT by a man whose application to become an astronaut was on file. The man's name was Aldrin. He had been turned down for the 1962 class of astronauts because he had no test piloting experience (that requirement was soon to be eliminated); moreover he was having some trouble convincing his supervising professors that he knew what he was talking about. Those "original ideas" Aldrin had were a little outré at the time, even on a campus as scientifically sophisticated as MIT. . . .

« I suppose it was the usual kind of academic trouble. They kept asking me to rewrite parts of the thesis, and even after I decided to stay at MIT and not go to Edwards to get the test piloting experience, I knew I was on limited time. I'm not sure I ever did succeed in communicating successfully what I had in mind in certain aspects of rendezvous and docking, particularly the manual aspects. They finally gave me the degree with reservations, and I guess I was sensitive about that for a long time. Not anymore; the ideas worked out and that was enough. I never did want knowledge just for the sake of knowing. I wanted knowledge that could be put to work, and that was why I had wanted this additional formal education. »

When NASA planners had a look at Buzz Aldrin's work, even before he was accepted in the 1963 class of astronauts, they understood quickly enough what he was driving at. But his contribution was not widely recognized, even within NASA, for a long time. Christopher Columbus Kraft Jr. chose to set the record straight in a memorandum dated April 4, 1966, four months after Walter M. Schirra Jr. and Thomas P. Stafford had accomplished the first rendezvous in space with Frank Borman and James Lovell flying the rendezvous target vehicle, and three weeks after Neil Armstrong and Dave Scott had successfully docked their Titan-powered spacecraft with an unmanned Agena rocket in space. (The end of the mission was unsatisfying; a short circuit sent the whole package into a tumbling spin and Armstrong and Scott had to abort and land in the Pacific after six and one-half orbits, but we were on the right track.) Kraft's memo said:[7]

Mission planning for a complex operation such as rendezvous must be carried out well in advance of the actual mission. This applies particularly to the development of basic concepts since such things as guidance computer program development, both for the ground and on board the spacecraft, require a great deal of time to carry out the formulation, verification and error analysis necessary. . . . [It] would be difficult to find anyone who contributed more in the area of flight crew activity and definition of associated spacecraft guidance requirements than Major Aldrin. Even before . . . [he became] an astronaut, MSC had the benefit of his creative thinking and intimate familiarity with the dynamics of rendezvous terminal phase.

In the early stages of the development of the Gemini rendezvous mission plan, Major Aldrin almost singlehandedly conceived and pressed through certain basic concepts which were incorporated in this operation, without which the probability of mission success would unquestionably have been considerably reduced. The most striking example of this was his recognition that complete dependence on flawless operation of the basic spacecraft rendezvous guidance system was unnecessary and would be foolhardy. As a result, he proposed and played a leading part in the development of a terminal phase trajectory which was optimized for utilization of backup crew procedures for carrying out the rendezvous in spite of various spacecraft systems failures, such as loss of the radar, computer or the inertial reference system. . . .

In addition to his contributions in the basic mission planning, Major Aldrin significantly influenced the development of the spacecraft's computer program. For example, he recognized the desirability of providing flexibility in this system which would permit modification of the mission plan as results of analyses and flight experience were gained. . . . These flexibilities were introduced in the form of providing flight crew control over certain critical parameters. . . . He also introduced techniques for utilizing visual operations of the target to be used in place of the radar data if it is recognized by the flight crew that the optical data is superior. . . .

At this date, after two highly successful Gemini rendezvous operations,[8] these contributions still appear very significant. Many people have joined in the final development of these concepts and ideas; however, when it is recognized that Major Aldrin conceived and sold them some years ago, when sufficient time still remained for their incorporation, the significance of his work is more readily apparent. . . . Major Aldrin has been and is currently exerting a similar influence on the Apollo program. . . .

The first space walk was accomplished by the Soviet Union's A. Leonov in March 1965; it lasted ten minutes. In June of the same year Edward White, later to die in the pad fire, topped that with a twenty-one minute walk, linked by a twenty-five-foot gold-embossed tether to the spacecraft (Gemini 4) flown by James McDivitt. Other walks followed: by Eugene Cernan; by Michael Collins, who lost his grip on a $470 camera and saw it float off into space; by Richard Gordon and finally — Gemini 12, last of the sequence — by Buzz Aldrin himself. He did three EVAs during the 94.6-hour, 59-orbit flight, but it was more satisfying to Buzz Aldrin to know that his ideas on rendezvous and docking, tested by stages in the last seven of the ten manned Gemini missions, had proved out.

Yet only a small percentage of the nonscientific public ever comprehended the significance of that aspect of Gemini. There were several reasons for this. One, the whole LOR concept sold by Houbolt was difficult for laymen to grasp, and no wonder; even men with the knowledge of Wernher von Braun had been among the original skeptics. Two, the subtle relationship of things like translunar injection to a technique which eventually came to be called transposition and docking was equally difficult to grasp; the average human being had to have time for his mind to gain purchase on such a subject. Three, a little over two months after the last Gemini flight, the pad fire blacked out public education concerning the subject of space. The next four months were full of recriminations, and all the headlines about space were bad headlines. By the time manned Apollo flights finally began, much of the meaning of Gemini had been forgotten.

Five hours and twenty-two minutes in flight. Coming up on 22,000 miles distance from the earth. Velocity 12,914 feet per second, decreased now to less than 9,000 statute mph.

ARMSTRONG: Down in the control center you might want to join us in wishing Dr. George Mueller a happy birthday.

HOUSTON (McCandless): Roger, we are standing by for your birthday greetings.

ARMSTRONG: I think today is also the birthday of California and I believe they are two hundred years old and we send them a happy birthday. It's Dr. Mueller's birthday also but I don't think he is that old.

Six hours, sixteen minutes. Velocity 11,479 feet per second, less than 8,000 statute mph. Distance from earth 27,938 nautical miles.

HOUSTON (Duke): Hello, Apollo 11, Houston. We see your middle gimbal angle getting pretty big. Over.

COLLINS: Well, it was, Charlie [Duke], but in going from one auto maneuver to another we took over control and have gone around gimbal lock, and we're about to give control back to the DAP.

HOUSTON (Duke): Roger, Mike. We see it increasing now.

COLLINS: Hey, maybe you better call Lew [Hartzell] and tell him we might be a bit late for dinner.

Men of Apollo 11 pose with their families and a model of the moon a few weeks before the flight. At left, Edwin E. Aldrin Jr., his wife Joan, children Michael, 13, Janice, 11, and Andrew, 11. Top, Michael Collins with children Michael, 6, Kathleen, 10, and Ann, 7, and wife Patricia. At right, Neil A. Armstrong, his wife Janet, and sons Mark, 6, and Ricky, 12.

On his second evening at home after the flight and the three-week protective quarantine, Neil Armstrong discusses the whole adventure with an air of "I-don't-quite-believe-it-myself."

Janet Shearon Armstrong watches her husband's rocket roar off through a thin cloud cover at Cape Kennedy. She and the couple's two sons watched liftoff from a borrowed private sailing yacht on the Banana River.

Buzz Aldrin perspires happily in the Texas sun on his first day at home after the flight. A day he devoted to his wife, their children, the family swimming pool, and long, thoughtful silences.

Joan Archer Aldrin chose to stay home to watch the launch on television, and managed during the crucial seconds of early flight to be all alone in a room full of children, relatives, neighbors.

Mike Collins still wears his "moon moustache" at home after quarantine. A couple of days later he shaved it off because it itched. "I always grow a moustache on a long trip," he explained, "but this was the first one I ever brought home with me."

Boston-born Patricia Finnegan Collins, perched on the fireplace step in her present home near Houston, laughs at a friend's joke after the liftoff of Apollo 11 on July 16. She and the children visited Mike at Cape Kennedy before flight, but also chose to watch the launch on television at home.

Neil Armstrong and his sons Mark, 6, and Ricky, 12, share a rare peaceful Sunday afternoon's fishing in a bayou near their home in El Lago, Texas, before the flight. Neil's cap is a special one which he believes is irresistible to fish.

BOOK TWO

Where the Stars Don't Twinkle

The mass gross absence of sound in space is more than just silence.

EUGENE CERNAN

You almost wish you could turn off the COMM and just appreciate the deafening quiet.

RUSSELL SCHWEICKART

We've been given a tremendous responsibility by the twists and turns of fate.

EDWIN E. ALDRIN JR.

5

"There isn't any magic selection"

THE WEIGHT of the Apollo 11 spacecraft was now computed to be a little more than ninety-six thousand pounds, but as Armstrong, Collins and Aldrin coasted toward the moon at a decreasing rate of speed — the speed would increase again when they neared the influence of lunar gravity — they carried a few things that had not been weighed at all. Even the official cargo (the plaque, American flags to be presented to the House and the Senate, the flags of all fifty states, etc.) included late starters. When Frank Borman returned from Moscow, only a few days before Apollo 11 lifted off, he brought back two medals which the Soviets had struck for Yuri Gagarin and Vladimir Komarov, the two cosmonauts who had lost their lives — Gagarin in an air crash; Komarov when the landing parachute of his Soyuz 1 spacecraft fouled during reentry after seventeen earth orbits and he smashed straight into the ground. Borman also brought word that the Soviets would appreciate it if the Gagarin and Komarov medals could be left on the moon. (NASA had planted the idea with Borman before he left.) And of course an Apollo 1 shoulder patch, the mission Gus Grissom, Ed White and Roger Chaffee never lived to fly, would be left on the moon — an outer rim of stars and stripes, along with the profile of a light blue globe, a white Apollo spacecraft and a silver moon.

The complexity of a lunar landing mission left little room for individuality, but it did leave some. Each man was entitled to a personal preference kit (a PPK). It was a little white bag, about eight inches by four inches by two inches, with the crewman's name written on the outside, and it was closed by a drawstring. For this flight Armstrong and Aldrin had two PPKs each, one for the command module and one for the lunar module. Mike Collins had only one, since he would not fly in the LM. At the start of the flight they were all stored in a little box underneath the right-hand couch of the command module. Buzz Aldrin had some first day philatelic covers, some small flags, a few medallions and rings and some gold olive-branch pins he had had made; one of these would come back from the moon as a gift for his wife Joan. He also took along a gold bracelet which had belonged to his mother, who died in 1968. On it she had attached little discs for each of her children and recorded the dates of their marriages. She had kept adding discs for each grandchild — circles for boys, little hearts for girls. And in his PPK for the lunar module there were two other items which Buzz Aldrin had not talked about much: a small container of wine and a little silver chalice.

Mike Collins carried "mostly small flags and little medallions of our shoulder patch — a whole bunch of small things for people in the program, people we work with." He did not know what all of them were, but they seemed to be things like gold tiepins, cuff links and gold crosses. (One of his Gemini flights, Tom Stafford remembered, carried an extraordinary number of St. Christopher medals: "That Titan didn't need any fuel. It was going to go!") Collins also had his college ring and one of Buzz Aldrin's gold olive-branch pins; for his daughter Kathleen, age ten, he carried a little gold cross which had been in the family for a long time; for his daughter Ann, age seven, he carried a gold locket. For young Michael Collins, age six, he carried nothing special — "He'll get a medallion or a patch or something when he's old enough to choose."

Neil Armstrong also took along tiny American flags and medallions, plus "a lot of little packages for friends. I don't even know what was in them. They were wrapped and I stuck them into the bag, then gave them back when we got home." These were all for members of the NASA team with whom he was personally close; there was, of course, a little package for Guenter Wendt. There were small, sentimental things for Neil's wife and mother, and — without Jan Armstrong's knowledge — there was a musical tape recording which Neil intended, at some point of his own choosing, to play back to earth. He hoped that Jan would be listening; the music would mean something special to her.

Eight hours and fifty-nine minutes into the flight, distance 42,753 nautical miles, velocity 9,100 feet per second, just over six thousand statute mph, still slowing down. Time to set up the spacecraft systems for the PTC (passive thermal control) mode which would rotate Apollo 11 on an X axis like a chicken on a powered barbecue spit (an analogy that had occurred to Mike Collins) at a rate of about three revolutions an hour. The idea was to maintain a temperature balance on the outside. (It remained sixty-five to seventy degrees, inside the cabin.) If the spacecraft flew in one fixed position without using PTC, the sun would overheat one part of it and a fatal "barbecue point" would be attained long before the men got near the moon; the fuel and oxidizer tanks would explode.

Strangely enough, there had been a "musical problem" of sorts in space flights. Not during the early Mercury flights; those flights were too short and too intense. But early in the Gemini program Mission Control began broadcasting pop music on a high frequency radio band which crewmen could turn on anytime they wished. Then a NASA secretary named Geri Ann Vanderoef decided that whoever had been choosing the music had been doing a poor job. In late 1965 she began taping music at home, with more of an accent on the classics, which she turned over to the flight controllers. Some of the engineers raised their eyebrows, but they finally decided Miss Vanderoef's taste was good. Given her head on Gemini 7 (Frank Borman and Jim Lovell), Miss Vanderoef packed the tapes with Dvořak, Handel, plus Bach's Air on the G String for the benefit of Jim Lovell, who was scheduled as an experiment to spend most of the flight in his underwear. There was also some Beethoven, which bored Lovell; Mission Control came to the rescue with a number from his favorite singer, Trini Lopez, and received Lovell's thanks.

There had been little or no impromptu music from space to ground since mid-December 1965, when Tom Stafford, who with Wally Schirra had just accomplished a rendezvous high over the Mariana Islands with Frank Borman and Jim Lovell, solemnly reported sighting "an unidentified satellite in a low trajectory, in polar orbit." Before Mission Control had time to think, Schirra was playing "Jingle Bells" on a miniature harmonica and Stafford was ringing some small bells. (The story that they had smuggled the miniature instruments aboard inside their mouths was pure apocrypha; they took them along in their PPKs.) But the classicists had their onboard innings during Apollo 9; Rusty Schweickart had recorded in advance one of his favorite pieces of music, Vaughan Williams's *Hodie*: "I truly like music, and Williams is one of my favorites. He has such a breadth of composition, some based on English folk songs, plus some other elements that make him hard to place as a composer. *Hodie* is a beautiful, beautiful thing. I knew ahead of time that I wasn't going to listen to it more than once because that would have been taxing relationships a little too much. Jim [McDivitt] and Dave [Scott] appreciate Sarah Vaughn more than Vaughan Williams." Schweickart chose lunchtime on the ninth or tenth day of the flight, then could not find the tape. After searching all over the spacecraft, he located it in one of the pockets of Dave Scott's suit; Scott swore that he had not hidden it. Schweickart played the tape softly. Then he turned it off and tried to whistle the music. It was then he discovered that one cannot whistle in an atmospheric pressure of 5/psi — five pounds per square inch.

On Apollo 11 neither Mike Collins nor Buzz Aldrin had special preferences, so NASA put some contemporary music (Frank Sinatra, Nancy Wilson and parts of the sound track of the new motion picture *Romeo and Juliet*) on tapes for their onboard tape recorder — the same one which was to record technical information and flight observations for return to earth. Neil Armstrong did have two requests. One of them was unexceptionable — Dvořak's *Symphony From the New World*. The other meant more to Neil and Jan, and it was harder to find. Some years earlier Neil had fallen in love with a recording by Samuel Hoffman called *Music Out of the Moon.* One of the principal instruments used in making the recording was the theremin, named for its Russian inventor Leo Theremin; it generated tones electronically, controlled by the distance between the musician's hands and two metal rods which served as antennas. When Neil was flying experimental aircraft at Edwards, Jan used to listen to this record by the hour at their home in the California mountains. It was one of the many things which was lost when their first Texas home was destroyed by fire in 1964, and before the Apollo 11 flight Neil asked that a search be made for the record: "Somebody found it and taped it. I didn't even know who it was." [1] He did know that he wanted to play it back to earth.

So the men were individuals after all, and not quite overcome by the pressures of training which, especially in the last six months before the flight of Apollo 11, tended to dehumanize them. Yet their presence on the lunar landing mission was a vivid illustration of Buzz Aldrin's remark concerning the element of fate that tended to determine where you fitted into the puzzle. Armstrong and Aldrin had been crewmates for a year and a half, long before it was remotely evident that their mission might go all the way; Collins was added to the crew in December 1968, after his spinal operation had cleared him for restoration to flight status. Deke Slayton, director of flight crew operations, had had it in the back of his mind to hold a slot for Collins and put him on the Armstrong-Aldrin crew if Mike's surgery went well: "He got well so fast he almost could have made Apollo 8. But we didn't know that when he went to surgery, and we couldn't risk it."

It was Deke Slayton who had the most to say about such groupings; he worked them out with Al Shepard, chief of the astronaut office, then sent the three names to MSC director Robert Gilruth for endorsement; finally they went to associate administrator George Mueller in Washington. None of Slayton's recommendations had ever been turned down, and the names of Armstrong, Collins and Aldrin were approved in the most routinely possible manner. They were not picked for arcane policy reasons, or even because it was expected, at the time of choice, that Armstrong and Aldrin would land the lunar module on the moon. The widespread notion that Armstrong had been named to command the lunar landing mission because he was a civilian was totally unfounded. Indeed, had Rusty Schweickart not overcome his nausea and been able to do his space walk with the PLSS backpack on Apollo 9, or had the LM failed its own flight test on either Apollo 9 or Apollo 10, the lunar landing attempt probably would have been delayed and another crew would have drawn the assignment. Mental stability was taken for granted by Slayton: "They get a psychiatric interview during the selection process as part of the basic medical exam. But that's a minor part of the whole physical thing." Indeed Slayton always thought that there had been too much psychological emphasis and not enough technical emphasis in the selection of the original seven, of whom he was the only one never to fly: "They were interested in biological specimens."

"There isn't any big magic selection that goes on for each mission," Slayton once said. "Everybody wants to be on a crew; that's normal. I can appreciate that more than anyone. I'm not in a job where you worry much about whether people love you or not. You've got a job to do and you hope it's right. This is like handling a squadron of fighter pilots. You've got a mission to do and you've got so many flights to fly and you assign guys to fly them. Sometimes you ask the commander about the composition of his crew, but there again it goes far back. We know who gets along and who doesn't. If you've got some options, sometimes you give the commander a choice of three or four guys. Sometimes you don't have any choice. We've never pulled a man because of any personality conflict. If that happened, it would mean you had made the wrong selection in the first place. It's always been health, or . . ."

. . . Or death. C. C. Williams had been on Pete Conrad's crew, which eventually drew the Apollo 12 mission, when he was killed; Alan Bean replaced him. And Tom Stafford and Eugene Cernan had to move up on a Gemini flight because Charles Bassett and Elliot See were killed.

Nine hours and thirty-six minutes into the flight. . . . The mission continuing to go smoothly at this point. . . .

COLLINS: Hear you loud and clear, Charlie [Duke]. We pitched down some to get a better COMM attitude. . . .

HOUSTON (Duke): Roger. Stand by.

COLLINS: Gosh, the reason for delaying it [the star check], Charlie, is that [it is] difficult to find two stars that are not occulted by the LM and also are not in the midst of a man-made star field up here with dumps.

HOUSTON (Duke): Roger. We copy.

Collins had a star chart, of course; but he was now in a world where stars did not twinkle. *They just did not twinkle* — there was no atmosphere to intercept, even for a millisecond, their light waves, although the stars themselves might be sending their signals from light years away.

Was individualism dead in the world where stars did not twinkle — and where most scientists had presumed terrestrial individualism to be nonexistent? Not quite. True, the American flag was on the mission, as it should have been; neither Armstrong nor Collins, nor Aldrin, had reason to be ashamed of wearing the Stars and Stripes on his left sleeve, although it was not until Gemini 4, the first manned Gemini flight (Jim McDivitt and Ed White, June 1965) that anyone wore the American flag on his sleeve.[2] It came about then only because McDivitt and White asked permission to wear the flag, having been denied permission to name their spacecraft American Eagle; NASA was getting sensitive about nomenclature. For some time there had been an interior squabble about naming spacecraft; Al Shepard had called his Mercury capsule Freedom 7 because it was factory model No. 7, not because the original astronauts numbered seven; yet the rest of "the original seven" who flew in Mercury used the figure seven in naming their spacecraft (Friendship 7, Faith 7). The practice stopped when NASA vetoed Gus Grissom's desire to name his Gemini 3 capsule "Molly Brown" — a rather obvious reference to a very sinkable Mercury spacecraft which had splashed down (the second manned Mercury flight, July 1961), blown its hatch, sunk and almost drowned Grissom. Yet Grissom had the last word, in a way: when Gemini 3 lifted off with him and John Young aboard, Gordon Cooper erupted a whoop over the radio: "Molly Brown, you're on your way!" At tracking stations from Spain to Australia other astronauts[3] picked up the signal and — thanks to some newspaper assistance — the name Molly Brown seized a proud place in public affection. But after that the ground rules were rigid: no more names.

Yet the programs themselves had names, and some thought had gone into their naming. "Mercury" first was suggested by Dr. Abe Silverstein, a veteran space scientist who had been appointed director of space flight development. Mercury had a lot of appeal. He was a familiar pagan god, and he was a colorful fellow with his helmet and winged sandals. On the theory that most Americans were not sufficiently familiar with the pagan pantheon to know that Mercury was also associated with thievery and being usher to the dead, he of the winged foot was accepted — and announced on Wright Brothers Day, December 17, 1958, fifty-five years after the famous flights at Kitty Hawk (which in 1903 got precious little newspaper attention in Dayton, Ohio, where the Wright brothers came from). The name Apollo was what a professional football quarterback would have called an automatic — he was Mercury's colleague and the god of sunlight and prophecy. The name Gemini — for a program that had to be invented more or less suddenly — came almost as easily: Gemini would stand for the twins of the Zodiac, quite appropriate for the two-man crews of an intermediary program.

The practice of wearing individually designed mission patches, which the astronauts wore over their hearts, went back only to Gemini 5, after Gordon Cooper and Pete Conrad had been denied permission to name their spacecraft. Pete Conrad had half-jokingly suggested calling it Lady Bird. Instead of a name they came up with a patch — a covered wagon and the slogan "Eight days or bust." Even then they had to go to Washington to argue their case over dinner with NASA administrator James Webb. They finally got the design approved but not the slogan; somebody thought it would be embarrassing if this mission, by far the longest yet undertaken, had to be cut short.

Muttering bitterly that "some of the covered wagons didn't make it either, but California is *there,*" Cooper and Conrad got a parachute rigger to sew a little piece of silk over the words; once back on earth, having "busted" only one hour short of their eight-day goal, they removed the camouflage. After that every flight had its patch. For Gemini 6 Wally Schirra and Tom Stafford came up with a design including the "northern six" stars which navigators have used for centuries. There was some awe in Stafford's voice afterward as he told the story: "We were up there, aiming for the rendezvous, and when we first saw our rendezvous vehicle glittering in the reflected light of the sunset, it was right between Sirius and the twins Castor and Pollux, exactly where we had placed it on the patch." John Young's wife, Barbara, a former commercial artist, designed the handsome red Roman numeral X for the patch he and Mike Collins wore on Gemini 10; it was a distinctive patch also because the astronauts chose to omit their names. The X was revived in a different color for the Apollo 10 mission which Young flew with Tom Stafford and Eugene Cernan in May 1969.

The design of a patch for Apollo 11, when it became obvious that this was indeed the lunar landing mission, seemed to bother a lot of people, including Neil Armstrong. . . .

« We were, all of us, interested in a number of small peripheral elements which go along with a flight. Things like the patch which we wear on our suits, and the names we select for in-flight communication between the two vehicles and between our vehicles and the ground. We were very conscious of the symbolism of our exploration, and we wanted the small things to reflect our very serious approach to the business of flying a lunar mission. The patch we designed was not intended to imitate the Great Seal of the United States; it was meant simply to symbolize a peaceful American attempt at a lunar landing. All of us in the crew particularly liked the idea of the olive branch [an idea first suggested by Thomas L. Wilson, a simulator engineer at the Houston MSC], and as time went on we began to attach more and more importance to its inclusion in the patch design.

Then there was the mattter of the call signs. We got millions of suggestions, and we chose ones which had some particular significance to us. The names had to have both dignity and symbolism, and of course clarity in radio transmissions. The name Eagle was adopted subsequent to the selection of the patch design and was intended both to reflect the theme of the patch and also a degree of national pride in the enterprise.

I believed, I think we all believed, that a successful lunar landing could, might, inspire men around the world to believe that impossible goals were possible, that the hope for solutions to humanity's problems was not a joke. We were particularly pleased to deposit on the moon the medals that were struck in commemoration of Gagarin and Komarov, and we were pleased to take with us some other symbols which I was not free to discuss for some months after the flight. We had with us, from the Wright estate, a piece of fabric from the wing of the first aircraft which ever flew, the Wright brothers' 1903 airplane, and a piece of its propeller. These things, these symbols if you will, will now go to the Air Force Museum at Wright-Patterson Air Force Base in Dayton, Ohio.

We wanted all the little peripheral elements to be universal, to echo the thought on the metal plaque which we left on the moon, fastened to the descent stage of the lunar module: 'Here men from the planet earth first set foot upon the moon, July 1969, A.D. We came in peace for all mankind.' Mike, Buzz, and I all signed it, as did President Nixon, and we left it to be discovered in some future age by intelligent beings who come after us, whether from earth or elsewhere. None of us wanted our names on the patch. Adding them would have made it too crowded. So much for the department of aesthetics. More important than that, we were three individuals who had drawn, in a kind of lottery, a momentous opportunity and a momentous responsibility. »

But, as Mike Collins said, "We wanted a pretty good-looking eagle." Collins consulted one of the Audubon bird books, selected an eagle, then sketched it out — he sketches as well as cooks in his spare time — on tracing paper to take to one of the informal "patch meetings." The other two crewmen liked his eagle, but there were still hurdles to be overcome. The eagle Collins had drawn looked rather fearsome to some NASA people in Washington, particularly as it seemed to be landing in a crater field with outspread claws. It was about that time that Tom Wilson came up with the idea of the olive branch. At first it was put into the eagle's beak. Then it was moved down to be carried by the eagle's claws, although Buzz Aldrin had an airman's reservation about that decision: "I think maybe the eagle should have had the olive branch in his beak. I believe his landing gear was made for landing and not for carrying things." But, Aldrin added, "The important thing was not the location. It was the presence of the olive branch. So I had made up four olive branch pieces of jewelry, one for each member of the crew and one to be left on the moon."

HOUSTON (Duke): How's Baja California look, Buzz?

ALDRIN: Well, it's got some clouds up and down it, and it looks pretty good.

HOUSTON (Duke): We are ready at Goldstone for the TV. It'll be recorded [at the Goldstone tracking station in California] and then replayed back over here. Neil, anytime you want to turn her on, we're ready.

ARMSTRONG: Okay, it'll take us about five minutes to get rigged.

HOUSTON (Duke): We'd like you to position the antenna at pitch three zero yaw two seven zero, go to react. That will give us a narrow beam width.

ARMSTRONG: Okay. I think we've got you.

HOUSTON (Duke): Hello, Apollo 11, Houston. Goldstone says that the TV looks great. Over.

ALDRIN: And how about the F stops? Is 22 going to be accurate?

HOUSTON (Duke): Stand by . . . Hello Apollo 11, Houston. The F-22 is good; we have no real white spots. They're real pleased with it. Over.

HOUSTON (Duke): We'd like ten minutes' worth of TV. We'd like a narrative if you could give us one.

ARMSTRONG: Roger. We're seeing the center of the earth as it appears from the spacecraft in the eastern Pacific Ocean. We have not been able to visually pick up the Hawaiian island chain, but we can clearly see the western coast of North America, the United States, the San Joaquin Valley, the High Sierras, Baja California and Mexico down as far as Acapulco, and the Yucatan Peninsula; and you can see on through Central America, to the northern coast of South America, Venezuela and Colombia. I'm not sure you'll be able to see all that on your screens down there.

ALDRIN: I didn't see anything but the DSKY so far.

HOUSTON (Duke): Looks like they're hogging the windows.

ALDRIN: You're right.

The television transmission started coming into Goldstone at 10:32:40 GET, when the spacecraft was at an altitude of fifty thousand nautical miles. It lasted a little more than sixteen minutes. Two and one-half hours later, starting at 8:45 P.M. Houston time, the first Apollo 11 show was put on the American networks. But this was an unscheduled transmission; it had not been in the flight plan, and one little thing did go wrong.

Nassau Bay, Texas

Joan Aldrin had spent a fairly relaxed day; at one point, when she heard Buzz's voice on the squawk box, she smiled broadly and exclaimed: "Doesn't he say numbers beautifully?" Someone remarked on the absence of idle chatter in the spacecraft. "Not with those guys," Joan said. Someone else said that Mike Collins extemporized in an articulate manner. "Yes, but he can't very well talk to himself," Joan said. A photographer climbed the high wooden fence around the backyard to sneak some pictures; he had to be chased away. Joan was upset by that, and after her own press conference she was bothered by the fact that she had not taken the children with her so that pictures outside the home would show the family image. "I'm worried about the Spaniards," she said. Shaking salt over the potato salad, she forgot to stop and almost spoiled the children's lunch. Neighbors kept dropping in with enormous quantities of food: "What shall we do with it all? Let's freeze it!" Two orchids in a little champagne glass also arrived with a card from John Glenn's wife Annie.

Nobody had told Joan Aldrin about the unscheduled television transmission. She went to bed and missed it completely. Pat Collins also missed it. Jan Armstrong, returning from Cape Kennedy in a NASA plane, got home in time to see it ("I turned on everything" — the squawk box, the TV, the lot), but afterward she could not remember what she had seen: "Views of the earth, wasn't it?"

Nearly sixty thousand nautical miles from earth now, traveling at a speed of 7,368 feet per second – about five thousand statute mph.

Color television from space was another of the many Apollo miracles, but like all the other miracles it came about after a lot of argument. In the early days of Mercury the engineers saw no point to it. TV was tried on Gordon Cooper's Mercury flight, using a slow-scan black and white camera, but the pictures were receivable only at Cape Kennedy and the quality was poor. On the eve of the manned Apollo flights, NASA's George Mueller still resisted giving television any priority. One night Roy Neal of NBC asked Mueller over dinner, "What if the networks gave you a million dollars to expedite it?" Neal said later, "He told me he could think of better things to spend it on." A slow-scan camera did go aboard Apollo 7, but there was no monitor in the spacecraft, and Tom Stafford at Mission Control had to tell Schirra, Cunningham and Eisele where to point. This made for a lot of additional work, and the crew was not happy. Neither was Roy Neal. Then, after the first Apollo 7 transmission, Neal encountered Tom Stafford at a cocktail party thrown by the Radio Corporation of America: "He told me his only regret was the poor quality of the picture. And why not color too? Stafford thought TV would be great from an engineering standpoint; everything happening on board could be monitored visually. Also he felt the public had every freedom to watch. He knew about Apollo 10 at that time — not all about it, but he knew that he would be in charge. There was a fellow at the party from a firm in Arizona who said the camera on board could be converted to color, and Stafford began nosing around."

What Stafford soon discovered was that twenty-eight years earlier a CBS engineer named Richard Goldmark had designed a color television system which used three separate cutout discs of primary colors, mounted on a wheel at the transmission site. At the receiving end he installed a similar wheel synchronized with the first wheel. It was not the compatible system RCA later developed, but it worked. The Westinghouse Electric Corporation was already making a camera for the lunar module, but the model had such a high light sensitivity that a lighted cigarette would have blown it up.

Westinghouse pulled another camera off the shelf, a high resolution model designed for military purposes. Then, encouraged by Stafford, who by now had become a passionate advocate and was determined to have color television on Apollo 10, Westinghouse turned to CBS for Dr. Goldmark's half-forgotten device and proceeded to modify it. Roy Neal later explained the rest: "The transmitter they designed sent back three separate impulses — red, blue and green. These were stored on a video disc recorder like the one we use for sporting events. After the astronauts had shot something of known color value — the NASA patch for instance — it was the simplest thing in the world to mix the colors in the right hues and intensities. Westinghouse got the contract, put a cost of $250,000 on the camera and did it all with items off the shelf." So Apollo had its camera, and Tom Stafford and Eugene Cernan were as pleased as two small boys on Christmas morning. The distances from which they transmitted did affect the signal strength, of course, but in some ways they were favorable: there was no earth curvature to stop a TV signal from being received by the three deep receiving stations — Madrid, Goldstone in California and the Honeysuckle complex in Australia. One of these three was always in contact with the spacecraft to relay signals to receiving stations in America via a COMSAT satellite positioned over the Atlantic or the Pacific. After the Apollo 10 color spectacular, even the engineering fraternity began to come around in favor of television. "Gee, you can learn things from these pictures," one engineer told Roy Neal.[4] "It's a hard act to follow," Joan Aldrin lamented.

Fourteen hours and six minutes, 66,554 nautical miles from earth, speed 7,095 feet per second, about forty-eight hundred statute mph. All three crewmen sleeping.

Once upon a time — or so legend had it — there was a Chief Wapa and a Princess Koneta. That was supposed to account for the name Wapakoneta, Ohio — Neil Armstrong's hometown. The 1969 population of Wapakoneta was still about seven thousand, as it had been when Neil was a boy in the Depression 1930's. There was still the same long main street, along with a new residential district where Neil's parents, Viola and Stephen Armstrong, lived at 912 Neil Armstrong Drive in a house they built in the middle 1960's. A little airport seven miles from town was now called Neil Armstrong Airport, and green and white signs on every road leading into town reminded the visitor that this was the home of Neil Armstrong, "first civilian astronaut."[5] Other Ohio towns had a fractional claim on Neil Armstrong, but Wapakoneta's claim was the soundest; he was born on his maternal grandmother's farm, about six miles away, on August 5, 1930. Neil was the oldest of three children; his sister June was born three years later, and his brother Dean nineteen months after that. The grocery store, the pharmacy and the hardware store where Neil Armstrong worked as a boy were still there in 1969. So was the grassy park where Neil flew his model airplanes. The name of the county, and the river running through, was Auglaize — reflecting the penetration of the early French explorers. The ethnic makeup of the county reflected the German immigrations of the second half of the nineteenth century. Neil Armstrong's great-great-grandparents, on his mother's side, were German immigrants, and his grandmother Korspeter lived to see the flight of Apollo 11 on television. (She was then eighty-one years old.) The Armstrong branch of the family was solidly Scots-Irish; Neil's mother had traced its origins back seven generations to the early 1800's, when a Captain John Armstrong settled on land near St. Mary's, Ohio. Neil spent some time in St. Mary's, but he remembered other Ohio towns from his boyhood — Warren, Upper Sandusky, a half-dozen more. That was because his father worked for the state of Ohio as an auditor, and it took about a year to audit one county's set of books. So the family was constantly moving; in 1969 Viola Armstrong calculated that she and Mr. Armstrong had moved sixteen times in Ohio, and although Neil completed high school in Wapakoneta, he actually lived there only three years before he went to college. It was a boyhood that had a classic American stamp. . . .

« I'm afraid I did a great many ordinary things. I was a Boy Scout, although I didn't complete my Eagle requirements for a long time, about the time I had started going to Purdue. The small towns, the ones I grew up in, were slow to come out of the Depression. We were not deprived, but there was never a great deal of money around. On that score we had it no worse and no better than thousands of other families. Part of the pattern in those days was that kids got part-time jobs at an early age. I was a freshman in high school at Upper Sandusky when I worked for Neumeister's bakery. I made one hundred ten dozen doughnuts every night. I probably got the job because of my small size; I could crawl inside the mixing vats at night and clean them out. The other thing I remember about the place is that the most appreciated delicacy there was not cake or candy. The experienced bakers would take a loaf of bread, fresh and hot out of the oven, tear out the center and throw it away, put a whole quarter pound of butter inside the remaining crust and return it to the oven. The oven was always hot. After a few minutes, just enough time for the butter to melt and the crust to harden again, the loaf was excellent eating. »

Thursday morning, July 17, United States time: 19 hours and 43 minutes GET, distance from earth 87,409 nautical miles. Continuing to decelerate in velocity, now traveling at 5,930 feet per second, about forty-one hundred statute mph. Crew in deep sleep.

Despite his conventional boyhood, there always was a stubborn streak of nonconformism in Neil Armstrong. He would play football or baseball with the other boys, but rarely for very long; his mother remembered that he would soon disappear and crawl into a chair with his books. Or go back to his model airplanes. Much later, after he became an astronaut, he avoided both jogging and doing push-ups. This made him virtually unique in a group of men to whom physical fitness was a fetish. There was even a widely published bit of Neil Armstrong lore which quoted him as having said that "I believe every human has a finite number of heartbeats available to him, and I don't intend to waste mine on running and doing exercises." Like many a quote which gets printed once and therefore enshrined in the libraries of all newspapers and magazines, this particular one was erroneous. Neil recalled having heard the quote, and he even recalled having repeated it once. He did not subscribe to its thesis, however, and he only quoted it so that he could disagree with it. He didn't know how many heartbeats he had available to him, but he intended to make the best use of all of them. He is a very light drinker, but he laughed at the sense of a remark once attributed to the economist Beardsley Ruml: "If you ever hear of me dropping dead on a tennis court, you'll know I was crossing it to get a Scotch and soda." Neil never minded expending his heartbeats on music: as a boy he learned to play the piano, and in later years he remembered it well enough to play duets for fun with his wife Janet. He also played the baritone horn in the Wapakoneta school band. In a high school six-boy combo he played the bass horn, "upright, not wrap-around." This group, which called itself the Mississippi Moonshiners, played one professional engagement at a little civic festival. Neil's friend Jerre Maxson, who was on trombone, remembered that they got paid five dollars each for two nights of work: "We weren't very good. We played 'The Star-Spangled Banner' and we were so bad that only half the people stood up. Before we got done, everyone had sat down." In any event, Neil Armstrong always drew the line on getting overly involved with anything which might consume time he wanted to spend following his primary interest. . . .

« I think that interest goes back farther than I can remember. My father tells me about going out to the Cleveland airport when I was two years old to see the 1932 air races, so I must have been a staunch aviation fan before I was even conscious of it. It was my father, also, who took me for my first airplane ride. I was six years old and we flew in a Ford trimotor — the old 'tin goose' — which was carrying passengers at Warren, Ohio. We were supposed to be at church, I think, but we sneaked off and later my mother caught us, just because of the guilty, and probably excited, looks on our faces. By the time I was nine, I was building model airplanes. They had become, I suppose, almost an obsession with me. People talk a lot about the

Depression, about deprivation. They miss the point. People in the Depression — the parents, the children — lived with a lack of money. It wasn't an oppressive thing. You didn't think about it. It simply never *occurred* to you to be extravagant. I bought airplane models without engines because it never occurred to me to buy models *with* engines. I bought, and I built, rubber band-powered models. I didn't feel put upon, in any sense. I knew that was what I could afford, and I was very happy with it. Anyway, that was during the Second World War, so engines and gasoline were not available. Neither was balsa wood, so I used straw, paper, hardwood — anything I could find. Naturally I also read everything I could lay hands on concerning aviation — I still get a great deal of enjoyment out of the collected papers of the Wright brothers — and I filled notebooks with scraps of information about makes of aircraft, specifications, and performances.

By the time I was fourteen, the family had moved to Wapakoneta and settled down. I got a job at the West End Market as a stock boy, then at Bowsher's hardware store, then at Rhine and Brading's pharmacy on Main Street. I don't think I ever got paid more than forty cents an hour on any of these jobs. At the pharmacy I swept out the place before school in the morning, then went back after school and on Saturdays — to stock the shelves, clerk, and try to make myself generally useful. When I had a day off, I went out to the local airport for flying lessons. There's another bit of apocrypha here. I am supposed to have ridden my bicycle three miles out there on what is still called 'the old brewery road.' Actually I hitchhiked. I didn't go down there very often, either, because at forty cents an hour it took some time to save nine dollars for a flying lesson. I was fifteen years old when I started doing that, going out first to meet Aubrey Knudegard, who gave me my first lessons in an Aeronca Champion. He also soloed me. Then Frank Lucie and Charlie Finkenbine taught me some of the finer points of flying. I got my student pilot's license on August 5, 1946, my sixteenth birthday. I was pretty skinny then — I probably looked twelve or fourteen. I don't suppose anybody would have rented me a secondhand automobile. I didn't have a driver's license, anyway. »

If there was any parental objection to Neil's setting his sights on living a life which had an incidence of high risk, nobody has ever suggested as much. His mother thought he might give up flying after he saw one of his fellow students die in a plane crash. Neil and his father were driving home from Scout camp and came upon the scene of the accident. Neil spent most of the next two days in his room, his mother remembered, but he did not give up flying.

There still remained the problem of technical education, which was expensive. A limited number of scholarships were being offered by the Navy; recipients could use them at the university of their choice. Neil was not primarily interested in a military career, but somehow he had to get to college, so he applied to the Navy. One day in 1947 the letter arrived: he had been accepted. Neil's mother remembered the day only too well: "I was in the basement getting out quart jars of canned fruit to bake some pies. He called so loudly, 'Mom, Mom,' that he scared me to death. I dropped a jar of blackberries on my big toe. I must have broken the toe — it was black and blue for weeks. But that was such a great day."

Neil was still only sixteen years old when the news came. His first instinct was to go to MIT, but his high school instructors suggested nearby Purdue because of its strong aeronautical engineering school. That became his choice. He liked Purdue so well that his brother Dean also chose it, and followed him there a few years later. But after a year and a half at Purdue the Navy exercised its option and ordered Neil to Pensacola, Florida, for flight training; he could return to Purdue later to get his degree. He chose to train in single-engine fighters; flying multi-engine planes meant responsibility for crews, and his mother remembered Neil explaining why he had gone the way he did: "Mom, I didn't want to be responsible for anyone else." Then, on June 25, 1950, the Korean war broke out. Soon after that Neil received his wings; clearly he would not return to Purdue soon. At the age of twenty he was assigned to a West Coast jet squadron. He went to Korea in the middle of 1951; his unit, Fighter Squadron 51, was one of the earliest all-jet carrier squadrons to see action. The pilots were trained for air-to-air combat, but Armstrong recalled, "There were no enemy planes in our area. So our work was air-to-ground. Bridge breaking, train stopping, tank shooting and that sort of thing."

"That sort of thing" meant, as it had meant in the Second World War, low-level work and heavy casualties, both from antiaircraft fire and mishaps like the clipped wing which crippled Neil Armstrong's plane and forced him to eject. By the time Neil returned to the United States in the spring of 1952, a lot of faces were missing in his squadron. The official report on Squadron 51 had some somber lines in it: "3 September, 1951, Ensign Neil A. Armstrong bailed out of his F9F-2 aircraft, successfully using the ejection seat. 4 September, 1951, Lieutenant (junior grade) James J. Ashford, USN, was killed . . . while on an armed reconnaissance flight . . . cause unknown. 4 September, 1951, Lieutenant (junior grade) Ross K. Bramwell, USN, was killed. . . . 16 September, 1951, F2H-2, Bureau No. 124968, VF-172, bounced over the barriers on landing and crashed into aircraft parked forward, bursting into flame. The following men were listed missing and presumed to be dead: Neifer, Earl K.; Barfield, Wade H.; and Harrell, Charles L. . . ."

Neil Armstrong survived all this. He wanted to go back to Purdue, which he did in the fall of 1952. There he met a girl named Janet Shearon; they simply encountered each other in a chilly dawn when no other students were around. Neil had taken a job delivering the campus newspaper. From high school he had always had some kind of early morning job; that is probably how he got out of the habit of eating breakfast. (Even during his training for Apollo 11 he settled for a cup of coffee and more often than not skipped lunch.) Janet was up early for another reason. A home economics major, she had some 6 A.M. laboratory courses. At least these didn't interfere with her own passion, which was swimming. She also knew something about airplanes; her physician father in Wilmette, Illinois, had bought a plane to commute to the family's summer home in Wisconsin and both her mother and her two older sisters had taken flight instruction. It was a couple of years before Neil asked Janet for a date. "Neil," Jan Armstrong said later, "is never one to rush into anything." Nor was Neil one to overlook anything. . . .

« In watching the airplane business I had noticed that the National Advisory Committee on Aeronautics was doing some interesting things. I was out of the Navy now, and in January 1955 I was graduated from Purdue. I applied for a job at Edwards in California, but there were no openings. Then I got a call from the Lewis Laboratory in Cleveland. I worked there in a free-flight rocket group, and joined the pilot group. I can recall telling Abe Silverstein, then associate director at Lewis, that I thought space travel was going to become a reality and I thought I would like to have a part of it. In those days — 1955 — space travel was almost a dirty expression, but Edwards looked like the place to be. It was about this time that the X-15 was being contracted. Then there was an opening at Edwards and they asked me if I would like to come out. I thought about it all of fifteen seconds and agreed. »

He had proposed to Janet Shearon, and when he started driving from Cleveland to southern California he made a detour to northern Wisconsin, where Janet was working as a summer camp counselor. Years afterward Jan Armstrong still teased Neil about that visit: "He said if I would marry him and come along in the car he'd get six cents a mile for the trip. If I didn't, he'd only get four." But Jan was not to be rushed either, and Neil drove to California alone. They were not married until January 1956.

Thursday, July 17, a little past six in the morning Houston time, 21 hours and 38 minutes GET, 93,085 nautical miles out from earth. Velocity 5,638 feet per second, about thirty-nine hundred statute mph. . . . Twenty-two hours and forty-nine minutes. . . .

HOUSTON (McCandless): Apollo 11, Apollo 11, this is Houston. Over.

ALDRIN: Good morning, Houston. Apollo 11.

HOUSTON (McCandless): When you're ready to copy, eleven, I've got a couple of small flight plan updates and your consumable updates, and the morning news, I guess. . . . [*There was one item in the morning news which would be of particular interest to the crew of Apollo 11. . . .*]

On July 13, a Sunday, at 5:55 A.M. Moscow time — 10:55 P.M. Saturday, EDT, at Cape Kennedy, a little less than seventy-two hours before the launch of Apollo 11 — the Soviet Union launched an unmanned moon rocket which was called Luna 15. This was three days after Frank Borman had returned to America following a week-long visit to the Soviet Union. When he arrived at John F. Kennedy International Airport he found a message to telephone Washington. Everyone in NASA wanted to talk to him about his trip to Russia — Monday morning. Borman said, "How about eight o'clock?"

He was still in Washington Saturday night when the news came through about Luna 15. Chris Kraft called Borman: "He said we're interested in what's going on. I asked, 'How interested are you? Is it an academic interest or do you have a concern?' He said, 'We're a little concerned.' So I went and talked to the people on the White House staff, Dr. [Henry] Kissinger and so on. They all said that if we were seriously concerned about safety, my God, we'll use the hot line. If it's just a concern about the potential, then call. So I just picked up the telephone and called the director of the Soviet Institute of Science, a man named Keldysh. But it was something like two o'clock in the morning in Moscow and I couldn't reach anyone. So I got in the airplane and flew back to Houston.

"Then, at six the next morning, they called my home number, and Sue was lying in bed complaining about the telephone ringing so early. She was spooked enough to conclude that telephone calls and telegrams at that hour undoubtedly meant trouble. . . . *Moscow, Missouri? Moscow, Texas? Then it dawned on her. . . .*

"The trouble was, they had our trajectory but we didn't have theirs. I talked to the guy who had been our interpreter, but a little more than that — a very knowledgeable man. Then I talked to Keldysh's number two guy; he spoke English, and Keldysh did not speak English. They said they would give me their trajectory, and they stuck by their word. They said they would confirm, and they did."

They did indeed — with two telegrams, one addressed to Borman in care of the White House, another addressed to his home in El Lago, Texas. The first telegram read: "Informing you that on July 19 at 1608 Moscow time the orbit of Luna 15 has been changed. Elements of corrected orbit preliminary constitute: to create a revolution two hours three minutes thirty seconds. . . . Luna 15 probe will remain in this orbit for one day. You will be kept informed of further changes." A second telegram arrived within a few hours: "Inform you that on July 17 at 1300 Moscow time the probe was placed in selenocentric orbit [roughly, an equatorial lunar orbit] with the following preliminary values. . . . It is supposed in this orbit probe Luna 15 will remain for two days. In case of further change in the orbit of the probe you will receive additional information. The orbit of probe Luna 15 does not intersect the trajectory of Apollo 11 spacecraft announced by you in flight program."

Both telegrams were signed "Keldysh." Frank Borman had achieved some kind of breakthrough, but how much of a breakthrough? In the final analysis, he said, "The results of breakthroughs are human achievements. If you ever want to get anything done I've found you have to cut through the mass of things. You can generate an awful lot of enemies along the line. I've got a few." But he had liked the Soviet cosmonauts — as had other American astronauts like Jim McDivitt and Dave Scott, who had met their opposite numbers at Paris air shows. The Bormans brought back pictures of cosmonauts Gherman Titov and the late Yuri Gagarin. Someone remarked that Gagarin looked like James Dean. "Who's James Dean?" asked Frank Borman, and some embarrassed laughter followed.

HOUSTON (McCandless): Okay, from Jodrell Bank, England, via AP. Britain's big Jodrell Bank radio telescope stopped receiving signals from the Soviet Union's unmanned moon shot at 5:49 EDT today. A spokesman said that it appeared the Luna 15 spaceship "has gone beyond the moon." Another quote: "We don't think it has landed." . . . The House of Lords was assured Wednesday that an American submarine would not "damage or assault" the Loch Ness monster. Lord Kilmany was told it was impossible to say if the 1876 Cruelty to Animals Act would be violated unless and until the monster was found.

M-m-m-m-m-m-m-m-m. . . . It was the treble whine of the supersonic jet, a whine Neil Armstrong could not hear in the cockpit when he was flying faster than sound but which Jan Armstrong could hear on the ground. Those first days at Edwards, in 1955 and 1956, were wonderful days for Neil Armstrong; for the first time he was doing what he really wanted to do. . . .

« As a research pilot, I was experiencing the most fascinating time of my life. I had the opportunity to fly almost every kind of high-performance airplane and at the same time do research in aerodynamics. I first flew the X-15 in 1960, and altogether I flew it seven times. I flew it to 207,000 feet, not a record. [As of 1969 the world record for altitude was 354,200 feet, set in the X-15 by the late Joe Walker. Walker was killed in 1966 when his Starfighter jet collided in midair with the experimental XB-70 and both planes crashed.]

What was it like? That is a difficult thing to describe. You're very busy. The flight takes only about ten minutes and you want to get a lot of information while you're up there. It has taken a lot of time and money to get you there. There's very little time for wondering, but above two hundred thousand feet you have essentially the same type of view that you have from a spacecraft when you are above the atmosphere. You can't help thinking, by George, this is the real thing. Fantastic. You can see the curvature of the earth. And from that mountain house where we lived, Jan could use her binoculars to see what was going on. She could see the X-15 drop away from the B-52 mother ship, and she could see the puffs of dust down in the valley as the X-15 landed. That was quite a house; when we first moved in we had cold water but no hot, and she used to bathe our son Eric in a plastic tub in the backyard, after the sun had heated the water. We worked on that plumbing, and on other refinements to the house, for almost six years; and then it was time to move to Houston.

Meanwhile, at Edwards we had been able to cover a lot of experimental work. I worked on an advanced flight control system that was tested on the X-15 — the idea was to make the airplane handle in the same way in all positions, whether at high or low altitudes, high or low speeds. I thought, and still think, that this sort of thing will lend itself to applications far beyond aircraft. It will be useful in systems governing cameras, tape recorders, automated industry controls, many things. We were not just pilots; we worked as engineers and developers of programs. In fact we spent very little time flying. Most of the time we were organizing and planning, using airplanes as tools to get information. This was the only justifiable approach. The pilot could not be just a passenger on what somebody once called a galactic joy ride; we were using research aircraft like the X-15 to extend our investigative capabilities. If we didn't take that approach, probably we did not deserve to be there.

We even worked on the Mercury project's drogue parachute, which in the early days tended to be unstable at sonic speeds. We rigged up a bomb-like capsule which had approximately the same weight as the old Mercury capsule, and dropped it repeatedly from seventy thousand feet — the highest 'bombing,' I suppose, ever done anywhere. The idea was to test the drogue parachute at approximately the correct altitude and speed, about Mach 1.1. At the time we were doing this, the Mercury project looked like a dark horse to us. We thought we were far more involved in space flight research than the Mercury people.

I judged them wrongly. »

When the Armstrongs left Edwards they took with them one terribly painful memory. Their first son, Ricky, had been born in 1957; then, early in 1959, they had a daughter whom they named Karen. (A second son, Mark, would be born in 1963, after the family had moved to Houston.) In 1961, when some of Neil's work took him to Seattle, the family would accompany him for as much as a month at a time. One day, when they had all gone for a walk in a Seattle park, Karen took too big a step running through the grass, tripped, and fell. She had a bump on her head and a nosebleed, and that evening Neil and Jan noticed that her eyes were not focusing properly; a possible concussion, they thought. They took her immediately to a pediatrician, who advised them to take Karen to an eye doctor back home in California, which they did. In the next few days Karen got progressively worse. Jan had to cope alone, because Neil had had to go to Minneapolis on yet another working trip. It all happened in a couple of weeks, and the day before Jan took Karen to the hospital her eyes began to roll and she could no longer talk plainly. X rays showed an inoperable brain tumor. The only thing to do was try to reduce the size of the tumor with X-ray treatments. Neil took a week off and they moved into a Los Angeles motel near the hospital; for six weeks after that, while Karen was being treated as an outpatient, Jan stayed in Los Angeles, either at a motel or with a friend who lived nearby, and Neil came down when he could. The results were encouraging; Karen learned to crawl again, then to walk, then to run and play, although she had a big bald spot on her head as a result of the X rays.

Jan remembered that Eric ("Ricky") was a big help: "He seemed to understand, although he was only four. She did beautifully for a month and a half. It was Ricky who told me, in October, that something was the matter with Karen again. He had been playing with her and he noticed that she had started to regress." This time the doctors tried cobalt, but without results. Soon Karen could not walk again, although at Christmas in 1961 she could still crawl. Then, Jan recalled, "She just went downhill."

Karen died on January 28, 1962—the Armstrongs' sixth wedding anniversary. She was just short of three years old.

Twenty-three hours and forty-eight minutes. Distance from earth 100,685 nautical miles, nearly halfway to the moon. Velocity 5,356 feet per second, a little less than thirty-seven hundred statute mph. . . .

A little levity is appropriate in a dangerous trade.

WALTER M. SCHIRRA JR.

What the space program needs is more English majors.

MICHAEL COLLINS

6

Noun, Verb, Dsky and Refsmmat

THE LANGUAGE of space flight was a necessary form of shorthand. It had had to be built, like the spacecraft itself, and as missions became more sophisticated the acronyms of ground-to-air talk became more baffling. Yet there was a reason for most of them. The DSKY (pronounced "Disky") was an acronym for "display and keyboard." Mike Collins called the DSKY "our number one toy." It was a box, eight inches by eight inches by seven inches. Made by the Raytheon Company, it weighed 17.5 pounds. There were two on board the Apollo 11 command module, one mounted just above the spacecraft's center seat and the other in the lower equipment bay. There was a third in the LM. Each DSKY had a keyboard, a power supply, a decoder relay matrix, status and caution circuits and displays. Each had a twenty-one-digit character display and a sixteen-button keyboard. Two-digit numbers represented programs — VERBs and NOUNs. Five-digit numbers stood for data indicating such situations as position and velocity. An astronaut could punch data and commands into the system and these commands were displayed to him, for verification, in the DSKY's readout window. If the computer wished to sound a warning of some kind, the numbers would flash to attract attention. Mike Collins would use the DSKY most on this flight. . . .

« You know we have this crazy computer and we talk to it and it talks to us. We tell it what to do and it spits out answers and requests, and it complains quite a bit if we give it the wrong information. Our way of communicating with it is through the DSKY. It has little things like typewriter keys on it but they don't read 'a,b,c,d.' They read zero through nine. In addition we have a VERB key, a NOUN key, an 'enter' key, a 'clear' key and a couple of other little things. So you want the computer to do some general category of thing, like integrate a state vector. You push the VERB key and then give it VERB so-and-so, a number, and the computer gets ready to do something. Just *what* it is supposed to do you tell it by pushing the NOUN key and giving it the NOUN number. This computer has the equivalent of a thirty-eight-thousand-word vocabulary, and by using the DSKY you can see your own commands in the readout window, then the computer does the calculation and shows its solution. We spend a lot of time punching keys up there. »

So there was an excuse for DSKY; besides, the pronunciation had good carrying power over an open radio circuit. The same could be argued for REFSMMAT, a crucial setting in the spacecraft guidance system but a concept so arcane that those who understood it were hard put to explain it to those who didn't. Even Mike Collins, who preferred to talk in colloquial language, admitted that "It's one of my favorite explanations, but it takes about an hour and a half. You start with the mean Basilian year, which doesn't do much except establish the inertial position of a vector relative to the stars at something like .05 after midnight on January 1 . . . in our case on January 1, 1969."

REFSMMAT was an acronym for "reference to stable member matrix," the word matrix in this case referring to the mathematical process involved in determining angles with reference to navigation stars. In a spacecraft moving in transearth or translunar flight, crewmen could have no sense of right side up, upside down, forward or backward without an inertial guidance system. This system, or "platform," had three sets of axes each of which measured acceleration along its own axis. Said John Mayer, a NASA trajectory expert, "The platform gives the reference for attitude changes, the reference from which you can navigate and guide. Think of it as a sort of life raft. You want it to float but you want to hold it in a certain position, so you fasten it to gimbals. As the platform rotates the gimbals rotate, but the REFSMMAT is the one reference which stays in the same position, relative to the stars, no matter what the spacecraft does." Each of the three gimbals measured the spacecraft's attitude with respect to the angle of the *stable* member, and the stable member was aligned to the REFSMMAT.

When Apollo 11 lifted off from Cape Kennedy, this REFSMMAT was a reading of the local vertical position — that is, an imaginary line drawn from the center of the earth up through the launch pad itself and straight out. Other REFSMMATs had to be taken at certain critical points during the flight — one for translunar coast, for example, another for lunar landing, another for the transearth coast home. With a REFSMMAT determined by the ground, the crew could calculate two angles from the stars relative to their alignment; with this and other information the computer would tell them where they were, how fast they were going, and at what attitude they were flying.

COLLINS: Houston, Apollo 11. Are you copying these NOUN 49's? I've been going through them fairly swiftly.

HOUSTON (McCandless): Yes, we surely are. . . . Yeah, Mike, we show you now in — we're in 59 right now, over.

COLLINS: That's right. I — I haven't entered. I gave it back to the computer for a second. I put the mode switch from manual back to CMC while I fooled with the DSKY, and the computer drove the star off out of sight, so the delay here has been in going back to manual and finding the star again which I've finally done, and in just a second here, I'll go to enter and get a 51 and mark on it. As I say, for some reason the computer drove the star out of sight.

Necessary as it was — or was it that necessary? — this kind of exchange made the mission of Apollo 11 almost impossible for all but the most sophisticated observers to follow. Yet all agencies tend to throw up their own special clouds of rhetoric. "Graceless as this patois was, it did have a certain, if sometimes spurious, air of briskness and efficiency." So wrote Arthur M. Schlesinger Jr., but he was writing about the State Department, not about the space program. It was far worse, he added, when the State Department stopped talking and started writing. He credited President Kennedy with leading "the fight for literacy in the Department" — but without much success. Kennedy responded to one speech draft with the comment, "This is only the latest and worst of a long number of drafts sent here for Presidential signature." [1] Schlesinger added, "It was a vain fight; the plague of gobbledygook was hard to shake off. I can only testify with what interest and relief the President and the White House read cables from ambassadors who could write."

Had the language of space acquired, in ten years, its own protective patois? Neil Armstrong sometimes was inclined to think that it had . . .

« The language of engineering has always been a very precise language. Though a lot of technical words were used, a great effort was usually made to define them clearly so that the audience or the reader should be aware of precisely what had been meant by the statement or by the sentence. We used to make a lot of fun about those other professions which were less careful with their phraseology and terminology — particularly the Madison Avenue approach to speech and writing. However I guess that in recent years we have tried to out-Madison Madison Avenue. If we can't find a word to misuse properly, we'll make one up. An example of misuse is our use of the word 'nominal,' which most of the English-speaking world interprets as meaning small, minimal — and we usually use it in the sense of being average or normal, for reference value. I think that most of the English-speaking world would say that a nominal tab for dinner would be a dollar or less. To a space scientist a nominal tab would be maybe six or seven dollars. The difference is that most people think of nominal as being small, and we tend to use it as average.

We can degrade further the usefulness of a word like 'nominal' by adding modifiers — for example, 'nominalize,' which might be translated into 'make standard' or 'make normal.' And 'denominalize' might mean make abnormal, or make unusual. This kind of chicanery, when carried to the extreme, might produce such useful words in the English language as 'denominalizationmanshipwise.' We have even become a bit careless in our use of technical terms — for instance the word 'perigee,' which comes from the Greek, means the closest point to the earth in a trajectory. But it is frequently used as the lowest point in an orbit about *any* body — for example, the lowest point to the moon might be called a perigee even though it is more correctly called 'perilune.' I think the astronomers who originated these terms are perhaps turning over in their graves because of our flippant use of their very carefully determined words.

Then there are those abbreviations. Like CSM for command and service module, LM for lunar module, MCC for mission control center, RCS for reaction control system, ECS for environmental control system — and so on. NASA encourages, in fact practically demands, that such abbreviations be used throughout the system. This has led to literally thousands of phrases and groups of initials, insuring that the newcomer and the layman are going to be confused by the use of abbreviations and acronyms scattered liberally throughout the sentences spoken or written by anyone who is attempting to explain what's going on.

From a talking point of view these abbreviations probably do not help much. CMC, which means command module computer, takes up three syllables. The word computer itself also takes only three syllables, so you don't save much time in speech. It is true that you save some time, or some space, in writing and printing, particularly with respect to written notes back and forth between people at the working level. The problem is that this shorthand is used so much, and so frequently, that it becomes a crutch, and it is difficult to make any point without leaning on it. The computer people are reaching the absolute epitome of short-cut technical English. Of course they must speak in machine language when they are talking with the machines, but they carry over that kind of phraseology into their daily conversation.

And into their writing. In 1968 I received a copy of a memorandum which said in part: 'A small (but interesting) change in the interpreter makes it possible to call from interpretive, using RTB, in general any basic subroutine which may be called using BANKCALL: in particular any basic subroutine which (1) ends with a TC Q, or a TC K if it stores Q in K, (2) does not clobber BUF2 or BUF 2 + 1, (3) does not clobber interpreter temporaries LOC, BANKSET, EDOP, and of course such erasables as FIXLOC and PUSHLOC and PRIORITY with which no one should trifle. A TC Q from such a routine leads through SWRETURN to DANZIG. This amounts to a quantum jump in the sexiness of the RTB op-code; this change merges the RTB op-code with the larger set of basic subroutines callable using BANKCALL. . . . This immediately opens a large virgin territory to interpreter users; and as TCF DANZIG routines are converted to TC Q subroutines a significant area may be opened to users of basic. . . . [Some] subroutines which have required interpretive interface routines can now do without; for instance the SGNAGREE interface for TPAGREE can be dispensed with. . . . Note that . . . Q points to SWERETURN: BUF2 to a TCF DANZIG.'

Understand?

No wonder that when they put a little cupboard in the wall of the Apollo 11 spacecraft to hold between-meals snacks they wound up calling it 'the smorgasbord mode.'

Doubtless Robert Benchley could make something of all this if he were alive. So could W. C. Fields. »

Yet, for better or worse some of this language was getting into ordinary talk. "Glitch" was probably on its way to becoming a dictionary word, and many people spoke of what they called a "reentry problem." Then there was that lady at the cocktail party who was "thirty-nine and holding." Meanwhile Mike Collins had come up with his own definition of one space-age word. "I'll bet you don't know what the ocean-earth interface is," he once said. "No? It's the beach." [2]

Twenty-six hours and forty minutes into the flight, just under four minutes to the midcourse correction maneuver. Distance from earth 109,245 nautical miles, velocity 5,033 feet per second, just under thirty-five hundred statute mph.

Nassau Bay, Texas

It was Thursday, July 17, and Joan Aldrin started the day by going outside and putting up the American flag personally. She did it a bit self-consciously, but at least she had delivered her daily photograph at an early hour to the newsmen outside 18622 Carriage Court. She wanted to go shopping, despite the risk of being followed by photographers. Her daughter Jan had to choose a present for a friend whose birthday party was coming up, and Mike wanted to exchange the top of the snake cage he was building; he had bought the wrong size. Mary Campbell, a friend from the Clear Creek Theater, picked up Joan at the rear of the house and drove her to Foley's Department Store in the Almeda Mall shopping center twelve miles away. "The outside world!" Joan exclaimed with delight as the car moved through the streets of Nassau Bay.

There was a big sale on at Foley's and the other women there were too busy bargain hunting to notice Joan. She bought Jan a blue miniskirt with a bright red shirt, and they found a present for Jan's friend — pettipants edged with green lace. Mike wandered off to look at the snakes in the pet department. Over sandwiches, Joan talked with Mike about the problems of having a boa constrictor in the house — feeding it, getting a regular supply of mice. She looked at the boa herself and said, "Have you ever seen anything so ugly?" But she finally half-promised that he would get it for his birthday — "If your father agrees when he gets back." On the way out Joan looked at the dress department, stopped and pulled a long-sleeved hunting-pink dress off the rack: "I've found my dress!" It had some snagged threads, and some telephoning was required to produce another copy in the right size. She paid with a credit card, and the clerk, who obviously had recognized her and remained silent up to now, said, "Thank you, Mrs. Aldrin. We all wish you the best of luck. We'll be thinking of you all the time."

In nearby El Lago, the same storm which had destroyed a tree in the Collins yard, the night before liftoff, had made a mess of the Armstrongs' swimming pool. It was full of leaves, and Jan spent most of the afternoon in the backyard vacuuming the bottom of the pool. It was something to do while waiting for the scheduled telecast at 6:30 P.M.

In Nassau Bay, Patricia Collins remembered that the three men in Jules Verne's fictional spacecraft Columbiad had taken along a certain kind of wine. She thought it would be appropriate to serve the same wine to Mike when he got back, and there was a scramble to locate a paperback copy of *From the Earth to the Moon.* It turned out that Verne's travelers had a weight problem too; three cases of Chambertin got left behind, along with a Steinway piano. But in Verne's sequel, *All Around the Moon,* there did seem to be at least one bottle of Chambertin, plus some Château d'Yquem and Clos Vougeot. After all, there was a Frenchman named Ardan on board, and something was required to wash down reactivated tenderloin steaks, Southdown mutton and calf's-foot jelly.[3] But there was a mixup and somehow Pat Collins wound up with two bottles of Nuits St. Georges. Not so bad, though; like Chambertin, it at least was a Burgundy. Both Mike and Pat were certificated by the Confrérie des Chevaliers du Tastevin, in the French Bourgogne, and during the two months preceding the flight of Apollo 11 Mike was supposed to test one of the space meals every day for lunch and tell NASA what he thought. Making judgments on a survival rather than an aesthetic basis, he liked everything, but he confused NASA. Instead of marking an X in the "good," "fair" or "poor" boxes on the checklist of meals, he drew little knives and forks and awarded stars in the manner of Guide Michelin, the famous handbook of French gastronomy. One morning he was called into the office to interpret and be given a mild reproof; after that his checklist was marked in the appropriate boxes, with little clarifying statements added: "Rich and mellow . . . A gustatory delight . . . The perfect blend of subtle spices . . ."

It may be that Collins had a tongue in one cheek when he pronounced such extravagant praise, but the food had become better since John Young defied the system on one of his Gemini flights by eating a forbidden corned beef sandwich on board (crumbs could be a menace). By the time Apollo 9 flew there were "wet pack" meals which could be eaten with a spoon, and John Young thought the turkey and gravy especially good. However, Young said, "Deke Slayton ate the same food we did throughout the flight and I guess he almost starved to death. But in our zero-G situation we were always full." There had been some complaints during the Gemini flights about the food tasting like mush, and a good deal of research had been done to improve the quality. An Army laboratory in Natick, Massachusetts, figured out a way to keep bread from going stale. At the Swift and Company Research Center in Oak Brook, Illinois, Dr. Robert L. Pavey, who held a Ph.D. in animal nutrition from Cornell, expanded the freeze-dried (and "rehydratable") menu to twenty-six meat and fish items. (U.S. choice beef was better for rehydratable meat cubes than prime, which had too much fat.) [4] The Apollo 11 spacecraft had, for eight days of meals, about seventy items of food on board, including dried peaches, pears, apricots and coffee, with or without sugar and cream, in the snack pantry — the "smorgasbord mode." It was now possible to make allowances for individual taste; Armstrong, Collins and Aldrin elected to start the same meal, on the second day of the flight, with (respectively) spaghetti, potato soup and chicken salad. Probably none of it would have brought three knives and forks and a Michelin star to a French chef at Chambley, but it all represented enormous improvement.

It was at Chambley, a NATO air force base in eastern France thirty miles from the ancient fortress city of Metz, that Mike first met Pat, early in 1956, and first impressed her with his sophistication about food and wine. Though he never volunteered much information about his background, he had been born into a distinguished military family and had lived the typical kaleidoscopic military life at home and abroad, acquiring widening tastes in each post-to-post move.

He actually had been born in an apartment just off Rome's Borghese Gardens, on October 31, 1930, the youngest of four children of General and Mrs. James L. Collins. General Collins had been with General John J. Pershing in the Philippines and in the 1916 Mexican campaign against Pancho Villa; he had won the Silver Star in combat in France during the First World War and later had been aide to Pershing. By the time Mike was born, General Collins was military attaché in Rome. Mike had two sisters, Virginia and Agnes, who were six and ten years older than he, and a brother, James L. Collins Jr., thirteen years older. The family left Italy when Mike was only a year and a half old, though proud Italians still claim him as their own,[5] and moved around from Oklahoma to Governor's Island in New York Bay, to a post outside Baltimore, to Ohio, to Texas, to Washington — the lot. Mike had one reason to remember Fort Hayes, Ohio, near Columbus, with some distaste: "I was thirteen or fourteen, and had a job in a bakery. Did you ever wash one thousand pie plates in one day?" But he liked San Antonio, Texas: "That city has something extra going for it, a bit of the cosmopolitan." And he liked Fort Hoyle near Chesapeake Bay . . .

« There were horses to ride, and woods to tramp through and, best of all, fishing. I had a little closed ecological cycle going there. During the summer I'd catch crabs for bait to catch fish. Then I'd use the fish in a trap to catch crabs. It was a sort of endless chain of converting crabs to fish and fish back to crabs again. I never missed having a real 'hometown.' On balance I think all that moving around was an advantage. You had to learn some things twice in school, and other things you never did learn, and you had to leave a lot of friends when you moved. But it was an interesting life for a kid. »

Best of all, Mike remembered San Juan, Puerto Rico. Early in the Second World War his father was commanding general of the Puerto Rican Department, which included the island, some of the Virgin Islands and bits of the West Indies. The family lived in a four-hundred-year-old building which was reputed to be the second oldest house in the western hemisphere. The original Spanish settlers had built it with walls thick enough to withstand the fire of sixteenth-century cannon. . . .

« It was an immense place with a ballroom and formal drawing rooms and gunports and tunnels, extraordinary play places for a boy ten years old. There was a strip of beach, and when a shark appeared a guard would run out and hose it down with a tommy gun. The military area was surrounded by a high stone wall and right up against the back of it was a poor section of town. There was a house there, the kind of house that isn't a home. It did a thriving business and I used to sit on the wall and watch everyone come and go. I can remember — oh, I was a rat fink — the girls used to toss me money if I would come down and talk to them but I never would. I was scared to death to go, but I would sit there and exchange pleasantries and I'm afraid a lot of their profits got converted into ice cream cones for me. »

By the time Mike was twelve the family was living in Washington, D.C., and his parents sent him off to St. Albans, an Episcopalian preparatory school next to Washington's National Cathedral. Mike's father had an ancestry which was straight County Cork, but he was not, personally, a Roman Catholic. His mother, Virginia Stewart Collins, was an Episcopalian whose Scottish-English family went back to pre-Revolutionary American times. Mike was a tall, skinny boy, athletic enough to captain the St. Albans wrestling team and play guard on the football team. He was a popular if indifferent student with a reputation for being the prime suspect in whatever school mischief went on.

« I was just a normal, active, troublesome kid. I liked airplanes, and kites, and climbing trees and falling out of them. I didn't like school much. I think St. Albans let me in sort of to be nice. I usually looked sleepy in the classroom, and the school yearbook poked fun at me about that. I was an altar boy at the Cathedral and I had to get up for the 6:30 A.M. service; I never did seem to catch up on sleep. Reading was something else, though. I discovered books I guess when I was about eleven, and it was just like somebody had turned on the light or opened the door. I became a voracious reader. »

He also seemed to have a gift for leadership. He was a prefect in his last year at St. Albans, and one of his teachers, John Davis, who later became an assistant headmaster, noted a quality of "intellectual precision." When it came time for college, Mike had no strong preferences. His father, who died in 1963, had gone to West Point and so had his uncle J. Lawton "Lightning Joe" Collins, who became Army Chief of Staff in 1949. So, in fact, had his older brother James L. Collins Jr., who was graduated in 1939 and was a field artillery battalion commander during the Second World War. James Jr. won the Purple Heart, Bronze Star, Silver Star, Distinguished Service Medal, Legion of Merit, and finally a brigadier general's star in 1965. He retired in 1969. But family tradition had nothing to do with Mike's decision to accept the appointment to West Point offered by a Louisiana congressman, F. Edward Hébert . . .

« If my father exerted any pressure one way or the other, it was probably away from going into the military. He was a very independent cuss himself and he raised his kids to think independently. He didn't want me to do something just because it seemed the thing to do. My mother I believe liked the field of diplomacy as a profession better than the military: I think she was trying to steer me subtly toward foreign service with the State Department. I can recall thinking I might want to be a doctor. In the end I chose West Point because I wanted to go to college and I knew West Point was a first class education. Also it was free. You don't get rich in the service, and at that time my father was retiring from the Army, and buying a house. . . . »

Collins was graduated in 1952, in the same class as his close friend Ed White, two years behind Frank Borman, one year behind Buzz Aldrin. He and Aldrin were not acquainted at school. Mike took a balanced course ("I do about as well in liberal arts as I do in the sciences and I don't do particularly well in any of them. I'd rather do something than study about it or talk about it"). He finished a respectable 185 in a class of 527. He had chosen the Air Force, so he had to spend another year getting his wings. By that time there was an armistice in the Korean War and he wound up in Chambley, France, as a first lieutenant flying F-86's. He wasn't at the base a lot because its runways weren't quite ready and his squadron spent most of its time flying out of another base in the south of France. But one night Mike walked into the Chambley officers' mess, which had in it a little bar the men had pooled their own money to build. There he saw a slim, dark, attractive girl sitting on a barstool. He walked over and said "Hi, I'm Mike Collins. Do you live here?" The girl's name was Patricia Finnegan, and it occurred to her that nobody would have introduced himself that way in her home city, Boston, because there she would never have been sitting alone at a bar — not even one which, like this one, was "more like a living room." Mike Collins was very polite, and after a decent interval he made a suggestion Pat could not resist: how would she like, he asked, to drive with him to nearby Metz and have a really good dinner at a really good French restaurant?

COLLINS: My compliments to the chef. That salad salmon is outstanding.
HOUSTON (McCandless): Roger, understand that's the salad salmon, over.
COLLINS: Something like that, salmon salad.
HOUSTON (McCandless): Is that music I hear in the background?
COLLINS: Buzz is singing.
ALDRIN: Pass me the sausage, man.
Twenty-nine hours into the mission, distance from earth 115,837 nautical miles, velocity 4,788 feet per second, a little less than thirty-three hundred statute mph.

Mike Collins was, to Patricia Finnegan, "a whole new world, from the beginning." She had grown up in suburban Boston, on Train Street in middle-class Dorchester, the oldest of eight children of Joseph and Julia Finnegan. Later she remembered that "Everybody I knew was either a lawyer or a priest. My father was a lawyer and a politician, my brothers became lawyers, and everybody I dated was a lawyer." Pat majored in English at Emmanuel, a women's Roman Catholic college on the Fenway, but to help pay her way she had done social work with Aid to Dependent Children during summer vacations. When she was graduated she kept on with that work: "It was mostly with unwed mothers. You went out and visited them every month and, hopefully, steered them on how to deal with health problems and truancy problems, with schools and clinics and sometimes psychiatrists. When I think back on it, it staggers me, particularly now that I have children of my own. I was only twenty-one years old, right out of college, and they took me seriously."

After two and a half years of this, Pat Finnegan was ready to take the grand European trip for which she had been saving. Then: "I realized that in three weeks I couldn't do half the things I wanted to do. A friend of mine knew a girl who was recruiting civilians for various jobs in Europe, so I went to see her. Ten days later I was hired as a recreation worker for the Air Force service club and was getting my arms full of shots. I went to Europe on a troopship and wound up in the mud at Chambley Air Force Base. That was in 1955. The runways weren't finished; there wasn't a club, a program, or anything — just a librarian, a court stenographer — what for, I can't imagine! — a girl who was to be my boss, and me. There were about forty people in the little village, and four thousand chickens, and we got very well acquainted with them. My high school French was pretty rickety, but it improved because it had to.

"Little by little the base began to grow and the men began to come, bringing their families. They lived in aluminum trailers which they put out the way you plant trees, on the highest elevation around. We called it Spam Can Row. They finally did build a service club which we operated pretty much as a community center. We had ballet lessons and piano lessons for the children, French classes, Bingo. Bingo was very big at the time!"

In those early days at Chambley, Pat had been dating a doctor and a dentist. The fighter pilots who flew in and out seemed to her "so young, so completely frivolous." Then she met Mike: "He looked just as dashing, as white-scarfy, as the others, but he was different. He loved to eat and he understood French food and introduced me to a lot of things I wouldn't have tried by myself. Some of the Americans were joining wine study clubs but Mike didn't have to. He had learned a lot about wines from his father, and he had studied up on vineyards so he could tell wonderful stories about them. In fact he could talk about anything. He knew books; he knew poetry; he was interested in the theater. He was bright about technical things, and he was lots and lots of fun. I couldn't get over all this combination in one man!

"One of the first things I remember about him was his approach to life. It was, and is, that everything will be okay; that everything will work out. He was always ready to go and see what's new, and yet he was serious and concerned and sentimental and kind of tender. He is never irresponsible, yet he is always sure that everything will turn out right."

Mike Collins flew away again only a couple of days after they met, but he and Pat saw each other each time he came back to Chambley. As their mutual affection grew, so did a nagging question in both their minds. It was Mike who brought it into the open: "I hate even to ask, but I guess you must be a Catholic, huh?" Pat remembered this years later as a moment she had been dreading: "Mike wasn't a Catholic, and it was a major concern for both of us. I couldn't budge from my faith no matter how much I might have wanted to. I began to consider asking for a transfer to somewhere else, to get away. In the end, Mike solved it. He must have done some pretty deep soul-searching. He was off on a trip as usual, and when he came back there simply was no problem anymore. It just wasn't there."

So they were engaged, at least informally, and Mike Collins wrote a letter to Joseph Finnegan in Boston. It was about six sentences long, and Pat never forgot her father's response. He telephoned her from Boston and, in the days before satellite communications and transatlantic telephone cables, kept her waiting for seven hours in a cubicle for the call to come through. Joseph Finnegan told Pat that Mike had said all there was to say in six sentences. Never, he added, had he seen such a good lawyer's brief in six sentences. "Marry him," concluded Joseph Finnegan. "Marry that young man before he gets away."

That was the autumn of 1956, and Pat and Mike started making plans for a wedding the following June in Boston. One October morning Pat woke up conscious that she had heard a lot of air activity during the night. The Hungarian revolution had erupted; the situation and the immediate future were frighteningly uncertain, and all American military personnel were being redeployed. It was five days before Mike was able to telephone her, from Germany, and it was early 1957 before they saw each other again.

ARMSTRONG: Houston, we're just looking at you out our windows here [they were then 100,000 miles away]. Looks like there's a circulation of clouds that just moved east of Houston over the Gulf and Florida area. Did that have any rain in it this morning?

[*It did. At the moment Neil was giving his report, Jan was cleaning leaves out of the drain so the garage wouldn't flood.*]

HOUSTON (McCandless): Roger. Our report from outside says that it's raining out here, and looks like you've got a pretty good eye for the weather there.

ARMSTRONG: Yes — well, it looks like it ought to clear up pretty soon from our viewpoint. The western edge of the weather isn't very far west of you.

[*The scheduled television transmission was still nearly three hours away, but the crew wanted to test the reception, using the Goldstone receiving station in California. . . .*]

COLLINS: Hey, you got any medics down there watching heart rate? I'm trying to do some running in place down here. I'm wondering just out of curiosity whether it brings my heart rate up.

HOUSTON (Charles Duke): They will spring into action here momentarily. Stand by. . . . Hello, eleven. We see your heart beating. [The rate was ninety-six.]

COLLINS: Okay, well, look at the CDRs and the CMPs and see if they go up any. We're all running in place up here. You wouldn't believe it.

HOUSTON (Duke): I'd like to see that sight. Why don't you give us a TV picture of that one?

COLLINS: I think Buzz is trying.

ALDRIN: He got it.

HOUSTON (Duke): Okay. It's coming in at Goldstone, Buzz . . . Rog, Rog. Goldstoners say they see you running there, Mike.

ALDRIN: Ask him what he's running from.

[*Thirty-one hours into the mission, 121,158 nautical miles from earth, speed 4,613 feet per second, just over thirty-one hundred statute mph. . . .*]

COLLINS: You sure do get a different perspective in this thing in zero G. Right now Neil's got his feet on the forward hatch and he can with his arms reach all five windows.

HOUSTON (Duke): Sounds like plastic man to me.

COLLINS: I'm hiding under the left-hand couch trying to stay out of his way.

[*Charlie Duke put in a call; there had been a drift of a couple of degrees in both pitch and yaw. . . .*]

HOUSTON (Duke): We can prevent that by just immediately prior to going to rate command on the manual attitude switches, if you cycle the spacecraft control switch to SCS, then back CMC, over.

COLLINS: I'm not going to let these LM guys play with my DSKY anymore.

HOUSTON (Duke): You sound like you'd better protect it.

[*Thirty-three hours, thirty-four minutes. . . . Standing by for television. . . .*]

HOUSTON (Duke): Latest on Luna 15. Tass reported this morning that the spacecraft was placed in orbit close to the lunar surface, and everything seems to be functioning normally on the vehicle. Sir Bernard Lovell said the craft appears to be in an orbit of about sixty-two nautical miles. Over.

COLLINS: Okay. Thank you, Charlie.

HOUSTON (Duke): And also, President Nixon has reported — has declared a day of participation on Monday for all federal employees to enable everybody to follow your activities on the surface. . . . It looks like you're going to have a pretty big audience for this EVA.

El Lago, Texas

In the Armstrong home Jan settled down on the floor in front of the television set: "Marky, hurry up. We're going to see Daddy." Mark came in as the screen was showing views of the earth; the voice was Armstrong's: "We're looking at the eastern Pacific Ocean. . . . We can see North America, Alaska, United States, Canada, Mexico and Central America . . ." Ricky Armstrong said to Mark, "Watch TV!" Mark said, "Why?"

COLLINS: Hold onto your hat, world; I'm going to turn you upside down.
HOUSTON (Duke): Eleven, that's a pretty good roll there . . . [but] you'll never beat out the Thunderbirds.

Mark had become tired of the show and was tumbling on the sofa behind his mother. Jan Armstrong coaxed him: "Come over here and get off the table." Mark wanted to know, "Where's Daddy?" Jan said, "That's Mr. Collins."

ALDRIN: That's big Mike Collins there. You got a little bit of —
COLLINS: Yeah, hello there, sport fans. You got a little bit of me, plus Neil in the center couch, and Buzz is doing the camera work just now.

Somebody's arm came on the screen, and Jan Armstrong said, "That must be Daddy right there!" Mark said, "I can't see his face." Jan said, "There he is, there he is!" Mark ran up to the screen, pressed his nose against it and said, "Hi, Daddy!" Jan was delighted by the standard spaceflight trick — rotating an object in weightlessness. In this case it was a flashlight. "Looks like a rotating beacon! Watch out, Daddy, there's a police car!" She studied the picture of the Apollo 11 patch and concluded: "You can't really tell it's an eagle unless you know what you're looking at." At that point Mark, who had wandered away, ran into the room crying, "The bird! You know that dresser thing in the garage? He's behind it!" Not the Apollo 11 eagle; this was a young bird found that same morning in the backyard, fully feathered. It presumably had fallen from a nest, and Mark had put it in a box with some grass. "We'll work on it after a while," Jan said; she was watching Mike Collins's head begin to appear at the bottom of the screen, and she did not notice Mark wander away. Then he came back: "I know what's the matter. He wants some food." He was covered with dirt and Jan Armstrong asked, "*Where* have you been?" Mark said, "I was behind the dresser thing in the garage. I had to get him. Do you have some bird food?"

In Nassau Bay Joan Aldrin sat on the couch, wearing the same pink linen two-piece dress she had worn during the afternoon shopping expedition. Buzz's Uncle Bob Moon was trying to monitor and record the squawk box and the television commentary simultaneously. Dee O'Hara was there, wearing her nurse's white uniform. Jeannie Bassett was there too, and everyone laughed at Mike Collins's "Hold onto your hat" line. When the camera started to bring pictures from inside the spacecraft, Joan leaned forward, smiling, watching intently and laughing a lot. There was a gabble of conversation: "Who's that?" Joan said. "That's Mike." Then, "Isn't it clear?" "Where's Buzz?" Joan said, "He's taking the pictures." Somebody's head appeared on the screen, upside down, and Andy Aldrin said, "That's Daddy!" Turning herself half upside down, Joan said, "No, it's Armstrong." Where was Buzz? The camera panned around on an unidentified hand. Throwing her own hands wide, Joan said, "Come *on!*" At last there was Buzz, performing gyrations, and Andy wondered what he was doing. "I think he's doing push-ups," Joan said. "How can he when it's weightless?" Finally there was a good head-on view of Buzz. "Finally!" Joan said triumphantly. Then her daughter Jan said, insistently, "I *must* eat" — and Aunt Audrey Moon went into the kitchen to fix something. "How long have they been on?" Joan asked. "About twenty-five minutes? You'd better watch yourselves, boys — you're going to run out of material." Somebody added, "Especially those three!" The comment was not quite fair, but there was some sensitivity about what Joan herself had called "a hard act to compete with" — the Apollo 10 television show. Mike Collins himself, when asked if Buzz Aldrin was holding cue cards during the July 17 transmission, answered, "Cue cards have a no. We have no intention of competing with the professionals."

In the Collins home young Michael, playing with his pet rabbit Snowball, was unimpressed by the telecast; he reluctantly joined his mother and his two sisters for two minutes, then went away. He had been much more excited the night before, at the end of liftoff day, when he saw a light in the sky, bigger than a star, flickering on and off; he had, he announced, spotted the spaceship.[6] The two daughters, Kate and Ann, did wave to their father when he appeared on the screen, but Kate soon left the room and Ann was more interested in mixing baby powder, salt and water to make "perfume and hair lotion." Pat Collins was soon complaining about seeing too much of the earth: "Let's get on with it. I've seen all this." As if on cue, the camera switched to the interior of the spacecraft, and Pat exclaimed, "They look great, don't they? Look, Mike's growing a moustache! There's Buzz explaining the star chart — that's his thing . . ."

HOUSTON (Duke): We see the sun shining in through [the window] behind him [Buzz] and blotting out the equatorial — correction, ecliptic — plane, and the stars that you're using for the navigation.

ALDRIN: You're right.

COLLINS: He doesn't really need the charts. He's got them memorized. This is just for show.

Pat Collins said, "I can't understand the lunar strip maps. I have a block about maps. When I ask Mike for directions, I expect him to say, 'Take the freeway south and get off at exit ten.' Instead he gets out a map and I'm sunk."

Early in 1957, after Mike had returned to Chambley, it seemed for a time that their marriage might sink in a sea of red tape. They had decided not to wait until June in Boston; they would be married in France in the spring. Then they discovered that all the regulations were set up to accommodate American military personnel who wanted to marry French girls. There was no easy routine for two Americans who wanted to marry each other.

"It took months, literally, to get all the regulations sorted out and the documents drawn up and signed," Pat remembered. "And Mike wasn't there. Every time he would fly in, it would be a Saturday and all the offices would be closed. So I ran around from office to office and it seemed that nineteen people had to sign everything. We had to do a lot of it through the French liaison office, and all the papers had to be translated into French. The man in the liaison office kept saying, 'It's a shame. It's a disgrace that they would keep you apart like this.' Little did he know we were apart anyway because Mike by this time was in North Africa for gunnery practice.

"I think Mike got back two days before the wedding. According to French law we had to be married first in a civil ceremony by the mayor of Chambley — on April 28, 1957. We had to have witnesses of course and ours were an American boy and a French girl he had married. They understood the words of the marriage contract, which was fortunate because we never knew when the mayor was talking to us because of the way he pronounced our names. Every once in a while he would look up and say *'Dîtes oui'* and we would croak *'Oui.'* When it was over we turned around and saw that we had two other witnesses we hadn't known about. They were the custodians of the city hall, and there they were in their hip boots and coveralls with little mashed caps on their heads. As soon as we signed the book they came running over with *félicitations.*"

Back at the base chapel, they went through another ceremony. The Roman Catholic priest married them all over again while the Baptist minister's wife played the organ. But even that was not the end of it. In 1967, when Mike and Pat went to Paris for an international air show, their French hosts learned that they had been married in Chambley and invited them back for a "hometown" celebration. "They took us back up to the same little courtroom," Pat recalled. "We thought it was just for speeches, but oh no. They had decided that we had not been properly married there in the first place. The mayor had died, but the deputy read the entire thing over again and then said that the trouble with the first wedding was that there had been no procession in the streets. There were champagne toasts in the courtroom, and then we came out to find a mass of schoolchildren. I don't know where they got them because that many people don't live in Chambley.[7] Then there had to be a wedding breakfast, more toasts and dancing in the garden. The NASA man with us was overwhelmed, because this was just after General de Gaulle had ordered all NATO bases in France closed down, and there was a lot of touchiness in the political climate."

By then the Gemini program had been completed, Mike Collins was fairly famous and the French (like the Italians) were claiming him for their own, transatlantic politics notwithstanding. But there had been some frustrations between 1957 and 1967, starting with Mike's assignment to the Chanute, Illinois, Air Force Maintenance Officers' School following his return to the United States a few months after he and Pat were married. . . .

« I said I never did like school, and I guess I meant any kind of school. I wanted to fly airplanes, and here we were going back over electrical engineering and the guts of airplanes and repeating some training I already had. As things turned out I think the experience was probably good administrative training, but I couldn't wait for graduation so I could get back to flying jets. When graduation came, I got an assignment as an instructor in the same school. I had an awful couple of weeks. Finally I went over and said, 'Look, I'm just not instructor material. I just don't like it.' So they said, 'Okay, we've got you in training command, so if you don't want to be an instructor you can go into one of the mobile training units.' Those units had to go from one base to another, all over the country, when the base was due to get a certain kind of airplane, and get a cadre of personnel ready to handle the plane when it arrived. It was a weird sort of existence, because my permanent station was Chanute. But I was always in Columbus, Ohio, or Mount Clemens, Michigan, or Montgomery, Alabama. But at least I was flying again. . . . »

Pat more or less kept up with him in a station wagon loaded with footlockers and pots and pans until they got to Montgomery, Alabama; by then Pat was seven months pregnant, so she went back to Boston to have Kate. Then came six months in North Carolina, and another transfer, to Las Vegas. By then Mike Collins had his sights set on something else: test pilot school at Edwards. "He had always talked about it," Pat said, "but he had to have all those hours of flying time, it must have been fifteen hundred then, even to apply, so I sort of forgot about it. But he kept adding up the hours, and the moment he had enough, off went the application." He was accepted. Men named Frank Borman, Neil Armstrong, Tom Stafford and Ed

White were already at Edwards, and they would all be heard from later. Pat settled down there and rapidly had two more babies — happily, because Mike was obviously happy in flight operations. Then it dawned on her that, although he had never talked about it, Mike was pointing for something beyond flight ops. He had put in to become one of the second group of astronauts — the class of 1962. "I couldn't quite get it through my head for a while," Pat said. "I'm the gal who used to sit there when all the Mercury shots went up and listen on the radio to make sure they got off the pad. Then I'd turn on the TV. I wasn't going to watch one of those things go to pieces right in front of me."

Neil Armstrong, Ed White, Frank Borman and six others made the class of 1962, but Mike Collins did not. Mike said, "I've got the second best job in the Air Force anyway." But his disappointment showed, and when he applied for the class of 1963 Pat was pulling for him: "I wouldn't have picked it, but I wanted him to have it, even though I knew it would change our whole way of living. The second time around Mike was beginning to sweeten the pill: 'Of course it will be the first chance we've had to have our own home, buy a house where we will live for a long time. . . .' " Then, in the fall of 1963, Mike went off on another one of his cross-country flights. It was two o'clock in the morning when Pat got a call from Mike: "Thumbs up. You say nothing, you know nothing. I'll be home Saturday." The next thing Pat knew, Mike was pushing airline tickets into her hand, telling her to go to Houston immediately, stay with the Bormans and pick out a lot. They were going to build a house.

But why plunge into this at all? Mike Collins had his own answer to that question. . . .

« I think man has always gone where he has been able to go, and I think that when man stops going where he can go he will have lost a lot. Man has always been an explorer. To me there's a fascination in thrusting out and going to new places. It's like going through a door because you find the door in front of you.

I can't call myself a natural-born explorer. I think I'll be a good explorer and do a good job, but so would a lot of other people. It's a peculiar twist of circumstances that got me here. I guess I'm just a normal human being with a normal curiosity. I feel very definitely that man should go to the moon and should go to the planets. I don't do a good job of explaining why. It's been one of the failings of our space program, I think, that we have been unable to delineate clearly all the reasons why we should go to the moon. I think the key to it is that man loses something if he has the option to go and does not take it.

As for the worry part of it, in this business you have to come to grips with worrying. There are so many pieces that can fail, so many unpleasant possibilities, that you'll get a horrible case of ulcers if you worry about them all. There are some things that you had better devote your attention to, but you also have to draw a line somewhere and decide it's going to work all right, then take it on faith that it will work. I'm not a perennial optimist; I think I'm more the fatalist — what's written is written and I don't know what it is. On the other hand, there's nothing wrong in acting as if things *will* work out. I mean, if I tell my wife I believe in the Easter bunny — well, why not? Either he exists or he doesn't, and I choose to believe. I think that is much more pleasant. But if you really cornered me, I'd have to admit reluctantly that there is no Easter bunny. »

Forty hours and fifty-eight minutes into the flight, 146,300 nautical miles from earth, velocity 3,917 feet per second, a little less than twenty-seven hundred statute mph. . . . Six hours remaining in the ten-hour sleep period. . . . Lunar landing 61 hours and 48 minutes away by the clock in Mission Control.

The real friends of the space voyager are the stars.

JAMES LOVELL

Have news. This sextant is fantastic.

EDWIN E. ALDRIN JR.

7

The Ecliptic

IF IT WERE POSSIBLE to get "high enough" and "look down," our section of the universe would appear to be rather flat. This table-top effect is due to the fact that the orbits of the planets of our solar system lie approximately in a single plane — the plane of the ecliptic, which is the name given to the apparent annual path of the sun as it travels through the heavens. As viewed from the earth, the sun appears to change position each day. This is because of the rotation and revolution of the earth as it spins on its tipped axis around the sun; and the earth's canted attitude to the sun creates the variations in terrestrial climate which we call seasons. If one were to chart the position of the sun as seen from earth each day by points and — at the end of 365 days, five hours, 48 minutes and 46 seconds — join these points together, one would have a reasonably correct picture of the ecliptic orbit. The line drawn in such a manner resembles a snake with a couple of humps in its back, and the plane that this orbit creates is the "stable reference" which distinguishes three-dimensional space navigation from two-dimensional terrestrial navigation. The point is that the earth is not precisely "right side up" as it swings through the skies; it is tilted at the equator. Terrestrial navigational aids, like the North Star, indicate north only from our own, slightly tipped, viewpoint. They are quite sufficient for getting a man from one point to another on land and sea, or even for earth-orbital flights such as Mercury, Gemini, and Apollo 7 and 9. But for travel to the moon they no longer apply. It makes more sense to ignore the familiar earthbound references and line up on the ecliptic plane, through which the sun moves and in which earth, the moon and all the planets lie.

Plotting along this plane was, until after Apollo 8, a navigational theory but an unknown in practice, because men had never traveled that kind of road before. Uncertainty about this factor was one of many reasons why NASA executives (and, to some extent, the Apollo 8 crew) looked back on the success of Apollo 8 with a sense of wonder; on the eve of Apollo 11 Dr. Thomas Paine still thought that the decision to "go" on the December 1968 circumlunar mission had been the most difficult of all to take.[1] Paradoxically, this mission was easier to fly than Apollo 9 in March 1969, which got much less public attention.) There was, to begin with, no star chart specifically applicable to lunar missions; conventional charts related the positions of the stars to the earth itself, and this was the first time flight planners had been forced to think seriously about celestial road-mapping. They tried to do a chart by hand, at first relating the stars to the lunar equator, but — in Ted Guillory's words — "It was a mess!" It was then that they went to the Air Force's Aeronautical Chart and Information Center in St. Louis, which had done some map work for Mercury and Gemini. A new program had to be devised; the stellar positions were plugged into a computer's memory and worked into a chart from there. Once the computer had the information, it was not too difficult to throw the stars up into a new position when Buzz Aldrin suggested that Apollo 8 switch to the ecliptic reference.

Although chance had thrown Armstrong, Collins and Aldrin together on Apollo 11, the choice of the three men did illuminate a fundamental principle particular to the space age if not peculiar to it; *balance*. Each man on the crew of Apollo 11 had a job to do concerning which he was an expert. In Armstrong's case it was flying, flying the lunar module safely to the surface of the moon. In Aldrin's case it was rendezvous and systems; he was the man who knew the machines, whose job was to interpret, update, support. Collins's job was flying the command and service module, and handling the celestial navigation. He navigated Apollo 11 out there, and he navigated it home again. Buzz Aldrin also had a few thoughts on that subject. . . .

« The stars are, as Jim Lovell said, your best friends. They sit there and you line up on them and you know where you are. We used the Morehead Planetarium in Chapel Hill, North Carolina, and the Griffith Planetarium in Los Angeles to study the positions of the stars as they would be seen from the moon when we got there. But it turned out that the earth would be in our field of view and the chance of spotting bright stars from the moon was not good.

For spacecraft navigation — getting there and back — we use thirty-seven stars — plus the earth, the sun and the moon. We don't use Venus, but we do use Polaris, Rigel, Capella, Canopus, Sirius, Antares, Vega, Arcturus, Altair and a big one called Fomalhaut which is less widely known. Each star has a computer number. You sight a pair of stars and mark on each one.

Then the computer will calculate spacecraft attitude. The velocity and position can be calculated once the ground has sent up the vector. The mileage signposting on a high-speed turnpike will tell the motorist the same thing if his watch is keeping correct time. The moon is a moving target, but so is Chicago if that city is the motorist's destination. This is no problem for the motorist because he is moving also as the earth rotates, and Chicago and the motorist have a stable relationship to one another. The sun also moves — not very rapidly, to be sure, but it takes the rest of our solar system with it when it does move. So if we use the plane of the ecliptic as our stable reference, we should not get lost, and we should, if we do our calculations correctly, arrive at the space equivalent of Chicago more or less on time.

There are actually three elements in the guidance system, all with heavy-sounding names. One is the inertial guidance subsystem, which measures changes in attitude and acceleration and assists in generating steering commands. Someone once described inertial guidance in this fashion: if you are walking down a street carrying a briefcase with inertial guidance and try to turn a corner, you can't do it. You keep going straight ahead. Then there is the optical subsystem, which means you have your eyes and your instruments, and you use them to take star sightings and measure angles. Then the computer subsystem takes all this data and gives you the answers you want: where you are, your attitude, and what it will take to get where you want to go. »

HOUSTON (McCandless): We have been working under the assumption that we'd take about an hour for the interference from a waste water dump to dissipate to the point where you can reasonably take star sightings for platform alignment navigation or something of this sort. If you have a spare minute or two, could you comment on the observation conditions now? Over.

COLLINS: My guess would be that the telescope is probably pretty useless, but you can differentiate in the sextant between water droplets and stars by the difference in their motion. . . . Buzz is looking through it now. Just a second.

ALDRIN: Houston, Apollo 11. It looks like at this time the sextant would be quite usable for any alignment. There are actually very few small particles drifting by.

HOUSTON (McCandless): Roger, Buzz. How about the telescope?

ALDRIN: Well, it's not quite as useful. It doesn't seem to be. Depending on the position of the sun it's got that band that seems to go across the center. I don't think it's because of the waste water particles that it would lack its effectiveness. Over.

HOUSTON (McCandless): Roger. What is this band — something that's deposited on the outside of the optics? Over.

ALDRIN: No, it's a reflection from the sun.

COLLINS: The sun bounces off the LM structure. With the LM attached the telescope is just about useless. Those star charts that MPAD [Mission Planning and Analysis Division] has provided, I think would be most useful . . . if for some reason we had to burn through the telescope we could use those as a guide for what we're looking at and say, well that bright blob over there has got to be that star because that's the position we're in, but so far we've not been able to pick out any decent star patterns while docked with the LM using the telescope.

HOUSTON (McCandless): This is Houston. We copy.

Forty-nine hours and seven minutes into the mission, 163,040 miles from earth, velocity 3,476 feet per second, about twenty-three hundred seventy statute mph.

All the navigation instruments aboard Apollo 11 had historical antecedents; the instruments were among the many refinements of the computer age, the same age which made it possible to reposition the stars — which also move in space — to where they were at the time of Jesus Christ's birth and come to the conclusion that Christ was probably born in the spring. (The Biblical reference to lambing time had been lying there as a clue for nearly two thousand years.) But the basic tools used by Mike Collins had a long history. . . .

« The space sextant we use is a natural descendant of something the Greeks called the astrolabe when they invented it two thousand years or so ago. This represented the beginning of man's ability to navigate in the unknown, because the astrolabe made it possible to get a fix on the angle of the North Star to the horizon, and then it was possible to calculate distance north or south of the equator — latitude. Our space sextant has a small field of view — 1.8 degrees — and a magnification of twenty-eight for precision sightings and light gathering. Usually we use it with our telescope, which has a sixty-degree field of view and no magnification; as a one-power telescope it is good for general or 'coarse' sightings of space landmarks, but you have to have the sextant to get precision. »

The telescope, of course, was nothing new. Three Dutch opticians named Jensen, Metius and Lippershey apparently were working on this invention by the end of the sixteenth century, although Lippershey seems to get credit for a crude model he built in 1608. He was refused a patent, but Italy's Galileo Galilei heard about it and a year later made a model of his own, which eventually got him into serious theological trouble.

There is no space equivalent to earth's longitude, but in space navigation

— as on one of the earth's oceans — the principle of time is involved. When Columbus's first voyage inaugurated the Age of Discovery in 1492 there was no instrument available to measure east-west distances, and for two hundred years there was frantic competition among rival maritime nations to invent one. The race was won in 1761 with the successful demonstration of the chronometer, the English invention of a Yorkshire carpenter named John Harrison. It was a kind of seafaring clock. It would be useless in space, of course; yet the space navigator had to have something which would keep time in relation to position in flight. Even "slowed down" to two thousand statute mph, the spacecraft would be moving at three thousand feet per second, and a small error in timing would result in a significant and possibly fatal error in position. The spacecraft's computer subsystem was, in a sense, a descendant of the first chronometer; the circuits in the computer itself did for the crew of Apollo 11 essentially what the chronometer did for eighteenth-century navigators.

The old mariner's compass also had a descendant aboard Apollo 11. The compass itself is an invention attributed variously to the Chinese, the Arabs, the Greeks, the Etruscans, the Finns and the Italians, but in any event it was being used by Europeans as long ago as the twelfth century. A navigational device dependent on a magnetic needle, oriented to a point in the far north of earth, would do little good in a spacecraft a couple of hundred thousand miles away. That is where the inertial guidance subsystem came in — a stable platform, suspended on gimbals and inclining freely regardless of spacecraft position, aligned to star direction references and retaining its alignment regardless of rotational movement or whatever else might be going on. Three gyroscopes kept the stable platform fixed with respect to some point in space, and three accelerometers noted any change in the speed of the vehicle. If Aldrin, or Mike Collins, who during the period of the actual lunar landing would have to navigate the command module on his own, should get away from a predetermined flight path, these devices would generate electronic signals to tell the computer — and tell the navigator. All together, they did the job a pocket compass does for a schoolboy on an overnight hike.

HOUSTON (McCandless): I've got the morning news here if you are interested. . . . The interest in the flight of Apollo 11 continues at a high level but a competing interest in the Houston area is the easing of watering rules. . . . The Senate Finance Committee has approved extension of the income tax surtax. . . . In sports, the Houston [football] Oilers are showing plenty of enthusiasm in their early preseason workouts. . . . National League baseball . . . American League baseball . . . and in Corby, England, an Irishman, John Coyle, has won the world's porridge eating championship by consuming twenty-three bowls of instant oatmeal in a ten-minute time limit from a field of thirty-five other competitors. Over.

COLLINS: I'd like to enter Aldrin in the oatmeal eating contest next time.

HOUSTON (McCandless): Is he pretty good at that?

COLLINS: He's doing his share up here.

HOUSTON (McCandless): You all just finished a meal not long ago, didn't you?

ALDRIN: I'm still eating.

HOUSTON (McCandless): Okay, is that . . .

COLLINS: He's on his nineteenth bowl.

HOUSTON (McCandless): Roger. Are you having any difficulties with gas in the food bags?

COLLINS: Well, that's intermittently affirmative, Bruce. We have these two hydrogen filters which work fine as long as you don't hook them up to a food bag. . . . Their efficiency ranges anywhere from darn near perfect to terrible, just depending on the individual characteristics of the food bags we're putting through them. Some of the food bags are so crimped near the entryway that there's no way we can work them loose to prevent back pressure.

HOUSTON (McCandless): Roger.

Fifty-four hours and six minutes into the mission, distance from earth 172,748 nautical miles, velocity 3,260 feet per second, about twenty-two hundred statute mph.

Nassau Bay, Texas

It was Friday, and Joan Aldrin had had the idea of a swimming pool party for all three Apollo 11 wives. Pat Collins arrived at noon, having dropped her children off at day camp; her sister, Ellie Golden, was with her. Jan Armstrong arrived about a quarter of one, accompanied by her sister Carolyn Trude and the Armstrongs' two sons, Eric and Mark. She also brought gifts of "moon cheese" — made in Wapakoneta, Ohio, with Neil Armstrong's picture on top of the rounds. She had spent the morning studying the flight plan, trying to work out for herself the schedule of events leading to the lunar orbit insertion burn. She wanted to know what she didn't know, and then find out about it. Dave Scott devoted three evenings, after he had finished his work at Mission Control, going over the flight plan and the fine points with her. Earlier Mary Campbell had brought Joan Aldrin a good luck candle, and it was burning on the coffee table in the family room. Michael Aldrin was taking pictures of Makita the Siamese cat while his mother went away to change into a blue swimsuit with green polka dots, topped with a blue and green shift. Andy Aldrin was wearing a rather ragged blue and yellow striped shirt which Joan thought scruffy; she sent him off to change into a white sweat shirt and khaki shorts. Joan and Pat Collins discussed the candle. "This poor little candle," Joan said, "has assumed international proportions." Pat said, "Yes, I read about it in the papers." The three wives went outside to give the assembled photographers their picture for the day, then got out of their dresses in the family room and went out to the pool in their swimsuits. The party had given Joan's houseguest Jeannie Bassett an idea: she went off that morning and bought a new swimsuit, a black-and-white patterned one-piece.

The Armstrong and Aldrin children were all water babies. They spent more time in the pool than their mothers did, especially after someone brought around a copy of the afternoon paper. While the children frisked in the water with Kurt Henize, the son of astronaut Karl Henize and his wife Caroline, Jan Armstrong, Pat Collins and Joan Aldrin had a private huddle around the swimming pool to discuss the press conference situation. They were all, on the third day of the flight, beginning to feel the strain; Pat Collins had become confused when foreign press representatives had stood behind her to tape — presumably — a simultaneous interpretation of her answers to questions, and she had found herself unable to distinguish the questions from the interpretation of her answers.[2] At one point Joan spoke rather sharply to Michael: "Don't make all these plans about the snake while you still haven't been told definitely that you can have it." She was trying to defer the decision on buying a boa constrictor until Buzz got home. Bob Moon suggested that a vodka and tonic might be in order, since Audrey Moon was busy getting together a cold lunch, mostly leftovers. There was no problem about dinner; Mary Campbell arrived again, carrying a huge casserole of lasagna and a great stick of French bread wrapped in foil for reheating. Caroline Henize produced a copy of a Tennessee paper which had carried a colored drawing of the Apollo 11 astronauts on the front page, and Joan exclaimed: "That's Buzz, but it doesn't look like him! They've got his eyes all wrong."

Bill Anders's wife Valerie came, and with her another friend, Milburn Stone, "Doc" on the television series *Gunsmoke*. For the first time in three days Michael Aldrin seemed truly excited; he went around showing off a sheet of paper with a scrawl on it: "To Mike Aldrin, good luck from his old friend 'Doc.' " Michael's father was going to the moon, but for the moment he considered that he had a more valuable autograph.

Fifty-five hours and ten minutes, receiving live television at Goldstone, presumably a test of the system. Stand by for a call from the crew. . . .

HOUSTON (Duke): We're getting ready to flight configure here at Houston for the transmission. We'll be up in a couple of minutes. Over.

ALDRIN: Rog, this is just for free. This isn't what we had in mind.

[*It turned out that they had quite a lot in mind. . . . It was time to remove the probe and drogue assembly linking the two spacecraft. Then Buzz Aldrin would go "up" and through a tunnel to enter the lunar module for the first time and check it out; the lunar landing was almost exactly forty-eight hours away. . . .*]

HOUSTON (Duke): It's a pretty good show here. It looks like you almost got the probe out.

ALDRIN: Yeah, it's loose now. Can you see that? . . . Coming down.

HOUSTON (Duke): Looks like it's a little bit easier than doing that in the chamber. Over.

ARMSTRONG: You bet.

ALDRIN: It's pretty massive but it goes where you direct it.

ARMSTRONG: Mike must have done a smooth job in that docking. There isn't a dent or a mark on the probe.

HOUSTON (Duke): With a twelve-foot cable, we estimate you should have about five to six feet excess when you get the camera into the LM. . . . We can see the probe now. A correction, the drogue.

COLLINS: Okay, drogue removal is coming next.

[*Once that was out of the way, they would have access to the lunar module hatch. . . .*]

HOUSTON (Duke): Looks like it's pretty crowded in there with that drogue. Over.

ARMSTRONG: Oh, it's not really bad.

ALDRIN: This TV cable is getting in the way.

HOUSTON (Duke): We see lots of arms.

COLLINS: The only problem, Charlie, is these TV stagehands don't know where they stand. . . . well, the dock latches look good today, just like they did yesterday.

HOUSTON (Duke): Eleven, Houston. We can even read the decals up there on the LM hatch.

ALDRIN: Well, let me move [the camera] up and see how much you can read.

HOUSTON (Duke): We can see the LM umbilical connection quite well, Buzz. We see you zooming in on one of the decals now. It's — to reset, unlatch handle, latch behind grip and pull back two full strokes. That's about all we can make out.

ALDRIN: You get an A plus.

HOUSTON (Duke): Thank you very much, sir. At least I passed my eye test.

COLLINS: I'm standing six feet from it, Charlie, and you can read it better than I can. There's something wrong with the system.

HOUSTON (Duke): Buzz, the view by your left shoulder there is so good we can see the ascent engine cover. . . . Now we see the helmet stowage bag.

ARMSTRONG: We don't see anything loose up there.

HOUSTON (Duke): Eleven, Houston. Buzz, are you already in? Over.

ALDRIN: Rog. I'm halfway in, halfway out. I'll start turning around, I guess.

HOUSTON (Duke): Hey, that's a great shot right there. Guess that's Neil and Mike. Better be, anyway.

[*Aldrin had taken the camera into the lunar module, then used it to photograph Armstrong and Collins in the command module. Now Neil Armstrong was also in the lunar module, and holding the camera. . . .*]

ALDRIN: I can see the hatch and all the EVA handrails. First time we've seen the silvery outside of the command module. Charlie, is there any concern about the duration that we ought to have the window shades open? We don't have any circulation in here and it might get a little on the warm side.

COLLINS: We'll put up a couple of hoses in the command module here and get a little circulation going.

HOUSTON (Duke): It sounds like a good idea. . . . Apollo 11, Houston, as far as the window shades go . . . I don't think we've got any systems problem. Be sure to put them back up when you egress. Over.

ALDRIN: That we will do.

ARMSTRONG: There wasn't very much debris in the command module or the lunar module. We found very few loose particles . . . bolts, nuts and screws and lint. Very few in each spacecraft. They were very clean.

HOUSTON (Duke): On the LM cameras, we'd like you to do it on LOI day with the LM power. Over.

ALDRIN: Checklist stowage packet. It's got a 16-millimeter camera in it, and it's got this little cylinder and I guess — I don't understand what it is. . . . It's got an arrow on the back and it says "turn," but I'm afraid to turn it.

HOUSTON (Duke): Eleven, your friendly geologist [Jack Schmitt] says it's the camera crank. Excuse me, for the 16 [millimeter] sequence camera if it jams.

[*Nearly five o'clock in the afternoon in Houston. . . .*]

HOUSTON (Duke): Eleven, Houston. Your show is going out to the U.S. now. . . . It'll be transmitted to some other countries after that.

At the Aldrin home in Nassau Bay the children continued to splash in the pool and the Apollo 11 wives relaxed with small talk. They were under the impression that the telecast was due to start at 6:30 P.M., and the TV set was not even turned on. Around five o'clock Jan Armstrong and Pat Collins left; driving home, Jan turned on her automobile radio and was startled to learn that the telecast was already in progress. She and Carolyn Trude got home to find the house full of more family, including their mother, Mrs. C. G. Shearon of Pasadena, California, and their sister Nan Theissen and her husband Scotty, an IBM district manager for Cleveland and Pittsburgh. Jan was not too upset by being late, and she watched the checkout of the lunar module with interest and authority: "That stick there is the steering handle. It drives like a trolley car."

Pat Collins got home, forty minutes after the start of the transmission, and saw a hand on a switch in the lunar module. "That must be Buzz," she said. "See his ring?" The commentator muffed it and identified the hand as Armstrong's. "I ought to call up those commentators and tell them who's talking," Pat said.

In the Aldrin home someone remembered that Valerie Anders, before she left, had said something about the telecast being moved ahead. Five minutes of excited talk followed; finally someone thought to turn on the television set. Joan was visibly distressed to discover that the transmission had been going on for some time. Without looking from the screen, she reminded Michael to "go take one of your pills," then said dejectedly, as the camera continued to rove around, "I'm tired of all these buttons." Buzz's back appeared on the screen, and still she did not react. But when the commentator said, "There, the Aldrin bare spot [on his head] comes through beautifully," Joan got up, smacked her hands together and said, "I can say that — no one else can. Now the whole world knows it!"

HOUSTON (Duke): Eleven, you got a pretty big audience. It's live in the U.S., it's going live to Japan, Western Europe and much of South America. Everybody reports very good color and appreciates the great show.

ALDRIN: Roger. I understand. Thank you.

HOUSTON (Duke): Well, that's a good view of Mr. Collins down there. We finally see him again.

COLLINS: Hello there, earthlings.

HOUSTON (Duke): Hello there.

ALDRIN: It's like old home week, Charlie, to get back into the LM again. The traverse from the bottom of the LM to the aft bulkhead of the command module must be about — oh, sixteen, twenty feet. It's not a disorienting one at all but it's most interesting to contemplate just pushing off from one and bounding on into the other vehicle all the way through the tunnel.

HOUSTON (Duke): Roger. Must be some experience. Is Collins going to go in and look around?

ARMSTRONG: We're willing to let him go, but he hasn't come up with the price of the ticket yet.

HOUSTON (Duke): Roger. I'd advise him to keep his hands off the switches.

COLLINS: If I can get them to keep their hands off my DSKY, it'd be a fair swap.

ALDRIN: That's why I've been eating so much today. I haven't had anything to do. He won't let me touch it anymore.

COLLINS: How far out are we now, Charlie? . . . Did you notice the difference between yesterday and today? This is as large an image as we can give you.

HOUSTON (Duke): Roger. If you think we're smaller, you are now 177,000 [nautical] miles out.

ARMSTRONG: Charlie. I'd like to say hello to all my fellow Scouts and Scouters at Farragut State Park in Idaho having a national jamboree there this week and . . . send them best wishes.

Jan Armstrong, with her house full of family, sent out for some pizzas. Pat Collins sneaked out through the garage, to shake off the newsmen, and went to dine at the Rendezvous, a restaurant which she and Mike Collins had always liked: "It has a fine wine cellar. Poor Mike is probably suffering from the cuisine more than anything else on his flight." Joan Aldrin, when a picture of Buzz in his helmet came on, looked up and announced: "He's gotten more TV time in the last two days than I ever did in a year of *trying!*" Then she sat in the corner of the couch, looking tired, drawn and drained.

During the period of the television transmission, one hour and thirty-six minutes, the spacecraft had traveled something over two thousand nautical miles. Buzz Aldrin had been in the lunar module a little more than one hour and thirteen minutes, Neil Armstrong fifteen or twenty minutes less. . . . Fifty-seven hours and four minutes into the flight, 178,236 nautical miles from earth, velocity down to 3,146 feet per second, down to a little more than two thousand statute mph.

In 1952 Joan Archer's mother invited to dinner, at the Archer home in Ho-Ho-Kus, New Jersey, a twenty-two-year-old Air Force lieutenant whom she had met a few days earlier at a party. Joan's father's parents had emigrated from Italy, and the family name had been Auricchio. It was changed to Archer when Joan's father and her English mother were married; Auricchio was too difficult to spell. Joan suspected her mother of matchmaking the night Buzz Aldrin came to dinner; she had been going around with actors of whom her father, Michael Archer, an oil company executive, largely disapproved. But, according to Joan, "Nothing happened. No spark. He went off to the Korean War, I went back to banging on theater doors and we didn't even correspond. Those were the golden years of television in New York, when all the shows were live, and I was beginning to get parts and some walk-ons." Then Joan's mother was killed in the plane crash, and when Buzz returned from Korea he decided to look up "the Archer girl." Soon Joan was wearing Buzz's "A" pin and not going out much, since her father insisted that she was now off limits to other dates while Buzz was an Air Force instructor in Nevada and continuing to court by mail. Joan said much later: "I persuaded my father to take his vacation in Las Vegas for two weeks, and the night before we left there Buzz proposed. We were together only five times in our courtship, and the fifth time was the day we got married."

Joan's father strongly approved of the match; he was a business acquaintance of Buzz's father. The son of Swedish immigrants, Gene Aldrin was born in Worcester, Massachusetts, and as a teen-ager he was something of a prodigy. Edwin E. Aldrin Jr. admired his father profoundly. . . .

« I think of him as an extremely intense person. They called him Shrimp in college, and I think that he got out of high school when he was fifteen years old. He wanted to go to MIT, but they wouldn't admit him because he was too young. So he went to Clark University in Worcester, Massachusetts, where he encountered Robert Goddard. And he took a mathematics course, a six-hour weekly classroom course, that was about fifty years ahead of its time. It applied 'systems analysis' to the whole subject a long time before that expression came into general usage. He came to know a lot about the theory of aerodynamics before he had learned to fly himself, and before a lot of pilots and base commanders in France during the First World War had learned anything about the subject.

I guess there isn't anybody left from the early aviation days that Dad doesn't know or who doesn't know Dad. When I was a boy he had an address book that was just packed with all the names of early aviation people. He was always going into New York to some aviation club meeting. But he didn't really sit down and describe what he was doing, any more I guess — than I would sit down and try to describe to my children what I am doing. However I think my children are more aware of what we are doing. They ask questions. I asked a few questions, but I think the kids of today are able to identify with what we are doing a lot more, perhaps because there is so much more in the newspapers and on television. Our children are living in the world of what we are doing. One of my boys asked me to bring home some maps of the moon so that he could use them in a display for a school project. They are much more familiar with the detailed efforts of what I am doing than I think I was with what my father was doing.

What Dad was doing was more obscure. He was dealing with the business of commercial aviation, and in those days — this was during the Depression — only the big oil companies had money to spend on the promotion of aviation. He was "aviation manager" for Standard Oil of New Jersey, and I didn't really fully understand what he was doing. Standard Oil: that meant lights and tires and batteries, and they were the people who supplied gas for your car. His job was promoting their aviation interests here and abroad. He flew over to Germany on the dirigible *Hindenburg;* that must have been in 1936. I would have been only six years old at the time. . . . »

HOUSTON (Charles Duke): Eleven, Houston. We were wondering who's on horns.

ARMSTRONG: Say again, Houston?

HOUSTON (Duke): We just had a little music there.

ARMSTRONG: Just trying to keep you entertained.

HOUSTON (Duke): Roger. That was good. You can keep it coming down, eleven.

ARMSTRONG: Okay.

COLLINS: Because it's a special occasion today, Houston. This is the third anniversary of Gemini 10 [flown by Collins with John Young, July 18–21, 1966].

Sixty and one-half hours into the mission, time for Mission Control to say goodnight to the crew. Distance from earth 184,600 nautical miles, spacecraft velocity leveling off to about three thousand feet per second, just over two thousand statute mph. . . . Sixty-two hours and twenty-nine minutes, crew asleep, spacecraft thirty-two thousand miles from the moon, spacecraft under lunar influence and picking up speed again — 3,782 feet per second, about twenty-six hundred statute mph. . . .

The home Buzz Aldrin remembered was a three-story house of white stucco on a corner where Princeton Place and Park Side intersected in Montclair, New Jersey. His father had resigned from the Army Air Corps in November 1928, and his resignation was accepted only after an argument: "I had just got my doctor's degree the year before and they tried to tell me I had three more years to serve." Aldrin *père* sold everything he owned and bought the Montclair house in August 1929, at the top of the bull market which crashed in October to trigger the Depression, and he sold it thirty years later for about what he paid for it. Had he not put his capital into a home, he would have lost it in the crash, and the Aldrins would have had a considerably different life. Montclair was an affluent residential city of forty-two thousand people; the population ebbed and flowed in the next thirty years, with a net gain of about one thousand, but it remained basically residential. It was an old community; the site of the city was purchased from Indian fishermen and hunters called Lenni Lenape ("original people") by English and Dutch settlers in 1678, although Montclair did not become a town with its own name until 1868, when an act of the New Jersey legislature separated it from Bloomfield.[3] The Aldrin house was in a shaded, modest-looking area, where Buzz and his two sisters played as children and where Buzz put a crossbar on a tree to build up his shoulder muscles for football and pole vaulting. It was a big house — four bedrooms and two baths on the second floor, three more bedrooms and one bath on the third floor, which in the years immediately following his birth, in Mountainside Hospital on January 20, 1930, Buzz had all to himself. . . .

« It was in that house that I started learning about the effects of weight in an airplane. What I tried to do was stretch a long string out perhaps one hundred feet from the third story, then I'd hand out model planes and let them glide down. They were recognition type models, painted black for training in profile recognition. They were too heavy to fly on their own, so I suspended them on the string. Occasionally they crashed, and there never was much left of what had been their landing gear.

I don't believe at this point that I was specifically motivated to be a pilot. In our home, flying was rather taken for granted. I made eight or ten flights with my father, and I remember working on a school project which involved building a model airport. My uncle helped me with it, and we made an idealistic airport with four sets of parallel runways on either side of a central administration building. I don't think you would ever find a spot in the country that would require four sets of parallel runways going off in different directions. That model, however, had a lot of simple beauty which appealed to a youngster's sense of the aesthetic. In the light of practical realities, I suppose I overdid it.

I overdid the aesthetic again when I was making airplane models. I took great pains to see that every surface was as smooth as I could make it. I thought that was the thing to do. During the war, when my father flew in for a visit in a P-38 fighter, I was amazed to see, after a close look, that the panels didn't really fit together as perfectly as I thought they should. There has been an evolution in aircraft, and a similar one in spacecraft. Each generation is smoother than the one before: Gemini craft look like boilerplate by comparison with Apollo. One of these days we may yet get a nice, smooth, highly polished surface. »

Montclair was affected by the Depression less than most communities of that size, and none of the Aldrins remembers the 1930's as a period of privation. There was enough money for a nurse and a cook, and a beach house in nearby Manasquan, New Jersey. Buzz developed tastes in food outrageous even for a small boy — sandwiches of peanut butter, sliced banana and a topping of powdered chocolate; whole cans of tuna eaten straight out of the can; packages of Jell-O eaten dry. His sister Fay Ann, now Mrs. Fay Ann Potter of Cincinnati, remembers the peanut butter Dagwoods with special distaste: "I don't know how he got his mouth apart."

Fay Ann was responsible for Buzz's nickname. At first he was called "Baby Gene," but his sisters called him "Baby Brother." Fay Ann pronounced it "Buzzer." It was Buzzer for a while, then just plain Buzz.[4]

Fay Ann also recalled that she had to take him to school, and make sure he stayed there. His other sister Madeline, now Mrs. Charles P. W. Crowell Jr. of Tulsa, Oklahoma, remembers that he was more interested in neighborhood football than in grades. He got mostly B's and C's, which was all right, and his ambition was to go to Notre Dame to play football. Buzz himself says he got over that by the time he was ten or eleven.

Sixty-five hours and twenty-eight minutes into the mission, countdown clock for lunar landing now showing 37 hours and 18 minutes. Apollo 11 now 25,280 nautical miles out from the moon, velocity up to 3,832 feet per second, about twenty-six hundred statute mph. Crew still asleep.

It would have been biologically unnatural if Buzz Aldrin had not decided, by the time he was ready to enter high school, that aviation was to be his way of life. The big house in Montclair, when his father was home, was full of talk about aviation: about the DeHavilland "crates" Gene Aldrin had flown ("no brakes, no radio") and even put together, when he was Billy Mitchell's aide in the Philippines, out of fuselages and spare parts left over from the First World War: "The washers on the bolts going through the fuselage and all the plywood gear had deteriorated from the tropical heat and moisture. Bacteria had eaten up all the glue. I think we got five airplanes out of two hundred that were serviceable." Then there had been that flight over the Alps, with Mrs. Aldrin and a mechanic, in the Lockheed Vega: "It was cold as the dickens. I remember particularly when I'd blow my breath it was blue. I stopped at Bologna because the clouds were down on the Apennines and all I had was a Rand McNally map of Italy which didn't show the mountain altitudes. The field I picked out was hardly a field. It was a military setup surrounded by a stone wall, and I just barely got out of the place." Gene Aldrin was partial to that Vega: "It would be good today if they'd put flaps on it, because I could land that thing on a dime." It would fly faster than contemporary American military pursuit planes; so would the Italian Savoia-Marchetti seaplanes which General Italo Balbo flew from Italy to Chicago for the 1933 World's Fair. Gene Aldrin, who was made an Italian *commendatore* for handling some American logistics on that flight, remembers that he had to give the slow American planes at Selfridge Field in Michigan some lead time to get into the air so that the Italians, taking off from Montreal, would have a proper escort into Chicago. (After the Second World War the Italians revoked Aldrin's *commendatore* status along with a lot of other honors conferred by the Mussolini government. Eventually they gave it back.)

Then there were the air shows at the airport in Newark which the Aldrins, father and son, saw together, and the parachute jump at the 1939 New York World's Fair which left nine-year-old Buzz unimpressed. "It was a controlled jump. Not a free fall." (A girl named Joan Archer, visiting the same

World's Fair, was sufficiently impressed by the parachute jump to decide she wouldn't go.)

In 1938 Gene Aldrin left Standard Oil to become an aviation consultant on his own; then came Pearl Harbor and the American participation in the Second World War, and by 1942 he was back in the Air Force as a full colonel — with the 13th Air Force in Guadalcanal, then as chief of the All Weather Flying Center at Wright Field in Ohio. This was about the time Buzz was due to enter high school — and about the time that his attitude to the classroom changed completely. If he wanted to be an aviator, he needed an appointment to West Point or Annapolis, and at a conference with his parents and the school counselors it was made frighteningly clear to him that unless his grades improved he would never see the inside of a service academy unless he elected to take a job sweeping floors. Buzz got the message; and after that it was all A's and B's, mostly A's. He made it look almost too easy. His sister Fay Ann was a year ahead of him in high school: "If I had a math problem or a science problem I'd go in and work it out with him even though he hadn't taken the subject yet. He did seem to be able to explain it to me."

He also knew quite a lot about ice cream. He had a job. . . .

« It was at Bond's ice cream parlor. George C. Bond's ice cream parlor. At the time there was only one such shop; now there are at least eight. I was a dishwasher, not a soda jerk. But this place did serve a thick milkshake called the Awful Awful — 'awful big, awful good.' The Awful Awful even got into a comic strip, *Harold Teen.* It had three or four scoops of ice cream in it. If you could drink three of them at a sitting, you got another one free. But you weren't allowed to drink the fourth one on the premises; they wouldn't take responsibility for the cost of summoning a stomach pump. Fay Ann worked at Bond's too, and she qualified for the special list of those who had been able to down three Awful Awfuls at one sitting. I don't recall whether I ever did or not. [Neither did Mrs. Bond, who in 1969 had retired to Townsend, Vermont.]

I made the phenomenal wage of thirty-five cents an hour and I thought that was great. I worked there for about two years in the afternoons, from about four o'clock until seven or eight in the evening. Then, one day, I had a little pain in my side. I tried to keep going with the dishes. Finally I couldn't take it anymore and told somebody about the pain. They stretched me out on a little bed they had in the back. Then Mrs. Bond came in, and before long I was in an ambulance headed for the hospital. My appendix came out two days later.

Then I became a counselor at a summer camp — Trout Lake Camp, about sixty miles north of Portland, Maine. I had been going to this camp since I was nine years old, and I continued to go there until the summer before I had to take the examinations for West Point. I look back on my experiences there as being quite instrumental in leading me toward what I call competitive appreciation for associating with other people: having standards set for you, set by other people, or standards you would set for yourself.

When I first went up there, before I became a counselor, I think it was the first close exposure that I had had to small groups of boys my own age. We lived together and were given challenges in swimming, track and baseball. But I hadn't given up on football. When I was a sophomore in high school I fancied myself a pretty fair broken field runner, but I was still smaller than some of the other boys. So they made me a quarterback. In my junior year I was working hard trying to get into the Point, and I did not go out for the team. In my senior year they had all the running backs they wanted, but they needed a center. So I played center. We won the state high school championship that year — 1946. »

That autumn Buzz was all set for West Point. He had spent the summer of 1946 at a Maryland prep school, buying some academic insurance. Buzz's father was under the misapprehension that Buzz preferred Annapolis, probably because Gene Aldrin preferred Annapolis for Buzz: "I advocated the Naval Academy because they had more technical studies, engineering studies, and also because in my experience the Navy took care of its people better." Then, one day in the fall of Buzz's senior high school year, Senator Albert W. Hawkes called Gene Aldrin to say that Buzz could have a principal appointment to West Point. When Gene reported this to Buzz's mother, she said, "As a matter of fact, I think Buzz prefers West Point." He did, and that was that.

The Montclair high school yearbook forecast that Buzz would become a general (in the Lower Slobbovian Army); [5] in any event there were to be no scholarship problems. Having entered the Point at the precocious age of seventeen, Buzz ranked number one in his class at the end of his plebe year. He was too small for varsity football,[6] but he played the intramural version and he used the shoulder muscles he had developed on that crossbar at Princeton Place in Montclair to become a thirteen-foot, nine-inch pole vaulter with a Swedish steel pole, before the new whiplash poles made old vaulting records obsolete.

Saturday, July 19, 1969, 5:58 A.M. *Houston time . . .*

HOUSTON (CAPCOM Ron Evans): Apollo 11, Houston. Good morning.

ALDRIN: Good morning, are you planning a course correction for us this morning?

HOUSTON (Evans): That's negative. Midcourse No. 4 is not required. We were going to let you sleep in until about 71 hours, if you'd like to turn over.

ALDRIN: Okay. I'll see you at 71 hours.

Seventy hours into the flight, 13,638 nautical miles from the moon, velocity up to 4,047 feet per second, over twenty-seven hundred statute mph. . . .

Eventually Buzz Aldrin came to think that it was just as well he had not won a Rhodes scholarship, but it seemed a big thing when he applied. . . .

« I tried two different times, first when I was a junior at the Point and then in 1952, when I was completing my pilot training before going to Korea. Originally I applied for political science, philosophy, politics and economics — the general category was what you asked for if you came to the selection board with a West Point background. But I got a little smarter and decided I really didn't want to do this; I was more interested in the general area of physics. And I guess the fighter pilot about to go over to Korea wasn't what they were looking for. In any event, for me, this was just a way of getting further education, another rung on the ladder which appeared to be an attractive one. I remember coming back and telling my parents that I had finally met my Waterloo, but looking back on it now, I don't think I was really cut out to do this sort of thing. »

Then came Korea and marriage to Joan, who was upset when she learned that a wedding band would have to share the ring finger of Buzz's left hand with his West Point ring. She got around that by having a ring made up and inscribed "JOAN PLUS BUZZ" for him to wear on his right hand. Joan found adjustment to marriage difficult: "In my whole life I had never been alone, and all of a sudden I found myself at the Squadron Officers' School in Montgomery, Alabama, with a husband who was either studying or off flying in his off-duty time. I was always alone. It was the same when we went to Bitburg, Germany. Buzz was always flying and I was always alone. I was naïve; I had been brought up as an only child; probably I was spoiled. Men don't really chatter as women do, and Buzz is not a man who talks a lot. I am a talker, and I am very direct. It was hard for me, not to have him there to talk to."

Joan thought in those days that Buzz's greatest ambition was to command a squadron of F-100s. Up until 1959, that may have been true. But that year both Buzz and Joan were flattened simultaneously by hepatitis and they had six months of hospitalization to think about things. Buzz thought about how many pilots there were at Bitburg, how many of them could probably fly F-100s as well as he could. He wanted to excel, and that desire made him apply for MIT. The school interested him because his father had earned his Doctor of Science degree there. It also interested him because he thought it offered the kind of challenge he could accept. Both Buzz and MIT accepted the challenge: he was accepted as a student in late 1959 and proceeded to a doctorate, which was achieved with the customary anguish.[7]

Then, down at San Antonio in 1963, Buzz was undergoing physical and psychological tests to become an astronaut. . . .

« I thought many of the psychological tests were rather silly and naive, and none of us really could understand what they were driving at. They would show us a blank piece of paper and ask, 'What do you see in that?' You don't see anything there, so you make up some simpleminded answer. They had an idiot box that would test your reflexes, and it had lights that would flash up, and you would have to respond with a certain order of punching buttons. Then they had a code; a square pattern. A certain light would come on, and in order to extinguish the light you would have to maneuver certain controls. But that wasn't very realistic, because it was a completely untrained-for situation, and I didn't see where that had any particular application to the recall of anything you had been trained for. Most of it was physical and medical — put your hands in ice water, that sort of thing. . . . »

He passed. In October 1963 he became one of the third class of astronauts — "the fourteen," of whom four would not survive to see the flight of Apollo 11.[8]

COLLINS: Houston, Apollo 11. The earthshine coming through the window is so bright you can read a book by it.

HOUSTON (McCandless): Oh, very good.

ARMSTRONG: And Houston. I'd guess that along the ecliptic line we can see corona light out to two lunar diameters from this location. The bright light only extends out about — about an eighth to a quarter of the lunar radius. . . . That's two lunar — two lunar diameters along the ecliptic in the bright part, right, a quarter to an eighth of lunar radius out, and that's perpendicular to the ecliptic line on the south pole.

HOUSTON (Duke): It looks like it's going to be impossible to get away from the fact that you guys are dominating all the news back here on earth. Even *Pravda* in Russia is headlining the mission and calls Neil "The Czar of the Ship." I think maybe they got the wrong mission.

HOUSTON (McCandless): West Germany has declared Monday [July 21] to be "Apollo Day."

HOUSTON (Duke): Even the kids at camp got into the news when Mike [Collins] Jr. was quoted as replying "Yeah" when somebody asked him if his daddy was going to be in history; then after a short pause he asked, "What is history?"

HOUSTON (McCandless): You might be interested in knowing, since you are already on the way, that a Houston astrologer, Ruby Graham, says that all the signs are right for your trip to the moon. She says that Neil is clever, Mike has good judgment, and Buzz can work out intricate problems.

. . . Buzz is said to be very sociable and cannot bear to be alone, in addition to having excellent critical ability. . . .

COLLINS: Who said all that? [*Laughter from the spacecraft.*]

HOUSTON (McCandless): Ruby Graham, an astrologer here in Houston. Now that we've got a check with flight operations for all the signs of the mission . . . and then we, of course, had to make sure that everything was really all set.

Seventy-two hours and forty-five minutes into the mission, Apollo 11 was 8,188 nautical miles from the moon, velocity up to 4,324 feet per second, nearly three thousand statute mph. . . .

Joan Aldrin never really thought of her husband as "sociable." His dedication she had come to understand: "If Buzz were a trash man and collected trash, he would be the best trash collector in the United States." She thought the man she had married to be "a curious mixture of magnificent confidence, bordering on conceit, and humility." She wondered, when he was training for Gemini 12, whether their marriage would ever be the same after Buzz came back from his space walk; it would be, she thought, "so much more magical and meaningful and magnificent because he had done this wonderful thing, whatever it was that he was going to do." And then: "Maybe six months later I realized that our marriage was exactly what it had been before, that if we ever had an argument, we argued still over the same things, but we still shared the same ideals and principles."

There was, in fact, a special depth to the "curious mixture" called Buzz Aldrin. On Gemini 12 his personal preference kit had included a prayer in the handwriting of his boyhood nurse Alice, now Mrs. Robert Howard of Montclair, New Jersey:

The light of God surrounds me;
The love of God enfolds me;
The power of God protects me;
The presence of God watches over me;
Wherever I am, God is.[9]

Yet it was more than a year following the Gemini 12 mission, when Dr. Martin Luther King Jr. was assassinated in Memphis, that Joan Aldrin could still be stunned by the emotional dimension of the man she had married. "It was Palm Sunday," Bruce Catton wrote, "and they would all live to see Easter, and with the guns quieted it might be easier to comprehend the mystery and the promise of that day." [10] On Palm Sunday in 1968 there were to be marches for Dr. King in many places, including Houston, and Buzz Aldrin did something that seemed to Joan out of character. He went to his minister, the Reverend Dean Woodruff of Webster Presbyterian Church, and they agreed to march that Sunday afternoon in downtown Houston. His picture appeared on the front page of one of the Houston newspapers, but the manned Apollo flights had not been resumed and he was not even identified in the caption. He was still just another astronaut. Joan Aldrin was surprised by what he did, but proud: "He cared, and I don't really know to this day what it was that made him do it. Someone in the office remarked that he understood that some astronaut had marched in 'that peace parade yesterday.' Buzz never said anything. It was simply something that he had to do."

This was one incident Buzz Aldrin did not want to talk about: "It was something I wanted to do." Joan once said of him: "There are times when I sit down and try to predict how Buzz will react to a situation. I always come up with three possibilities, and the third possibility is always that I know that he is not going to react as I figured."

One night, his eyes gleaming with the intensity of small volcanoes, Buzz talked about some of his other motivations. . . .

« I wouldn't classify myself as a fatalist or anything like that. I just think that when I am engaged in one of these things I'm in no danger at all. It may be a question of faith, a belief that I wasn't brought here to meet with some untimely occurrence. I've wondered often why I feel the way I do about these things, but I really don't have a concern. It's more a question of being prepared to meet situations as they come up. I think a person creates his own faith somehow in order to respond to a situation, to look for something bigger and better to do.

I think we have obviously accepted a challenge to undertake this task of going to the moon. I think the challenge would have been there anyway, and there is no doubt in my mind that whether the target date had been the 1960's or later, we would have gone about this particular task sooner or later, just because it is a challenge. As man develops the tools and the capabilities to extend his reach further and further, there is no doubt we shall feel compelled to go as far as we are capable of going. »

Seventy-three hours and six minutes into the mission, distance from the moon 7,331 nautical miles, velocity 4,399 feet per second, three thousand statute mph, lunar orbit insertion burn less than three hours away, lunar landing about thirty-six hours away. . . .

I swung the earth a trinket at my wrist.

FRANCIS THOMPSON

The long ride out to the moon was, frankly, a little bit of a drag.

WILLIAM ANDERS

8

Earthshine—and "Over the hill"

ONE OF THE THINGS that surprised Bill Anders during the flight of Apollo 8, in December 1968, was that he traveled nearly all the way to the moon and did not see it once en route. The five windows in the Apollo spacecraft did not add up to a particularly good observation platform, but the chief impediment to visual sighting of the moon was the trajectory of flight, which aligned spacecraft, moon and sun in such a way that the moon was very close to the sun and vanished in its glare. Eventually Anders became aware that some great dark body was building up "behind" the spacecraft: "I knew it was there because I could see stars up to a point, past which they were obscured by a big black disc of nothing." This was the first manned space flight to enter the gravitational sphere of another solar body — that is, the moon itself; and the Apollo 8 astronauts were also the first to cut off their ties with the earth by allowing themselves to be "captured" by the moon's gravity. Their ticket home, as in the cases of Apollo 10 and Apollo 11 in the following months, was the continued perfect performance of their single service propulsion system engine, which had to fire properly to get them back out of lunar orbit. On the Apollo 8 flight Anders felt the same surging sense of exploration which later was in the minds of Armstrong, Collins and Aldrin; Anders thought that Lewis and Clark, Admiral Richard E. Byrd and Sir Edmund Hillary must have felt the same way as their own goals came into view: "To me, and I think to many Americans, there has always been a sense of exploration and a sense of the frontier. The Appalachian Trail, the wide Missouri, Antarctica — they were there, and men came to conquer them and to benefit from them. Now space was our frontier, and here I was in the lead wagon." The Apollo 8 flight had several surprises, and one of them was the visibility on the moon provided by something which, until the time of Apollo 8, had never been seen by man — the phenomenon of earthshine. Anders analyzed it later: "From there, the earth is about eight times brighter than the moon is from the earth. The earth reflects light from the sun to the moon; it's a sort of triple bounce. You really can see things on the moon by earthshine, and you can certainly recognize features. We flew over the crater Copernicus in the dark, but it was sufficiently illuminated by earthshine so that I could identify it." Yet earthshine could not be called brilliant, and making a lunar landing in that light was quickly ruled out. Some visual assistance from the sun was needed; hence the timing of the Apollo 10 and Apollo 11 shots to take advantage of the "lunar window."

Obviously the moon had the capacity to reflect light; otherwise nobody ever would have seen it, or perhaps even known it existed until the invention of the laser. So did the earth, as evidenced by the spectacular photographs taken during earlier orbital missions. But the moon and the earth reflected light differently. The difference was between "back scatter," which the moon has, and "spectral" reflection, which the earth has. Bill Anders tried to explain it after he, Frank Borman and Jim Lovell had completed their circumlunar mission: "Compare it for a moment with a polished ball bearing. On the ball bearing, a viewer will see a bright spot where the light hits it and bounces to the eye — like looking into a mirror. The edges of it, away from the light source, will be dimmer. The earth is like that ball bearing. It's very bright where the sun hits it, and it gets dimmer toward the edges. The moon isn't like that at all. It's like a reflectorized highway warning sign; it lights up all over its area when a light source hits it, no matter from what angle you look at it."

In the three years preceding the flight of Apollo 8, thanks to photographs sent back to earth by unmanned lunar probes, we had learned more about the moon than we had learned in the past three thousand years. There was, in fact, a scientific school of thought which argued that a manned lunar landing was unnecessarily dangerous and unnecessarily expensive; why not rely on man-made instruments? [1] After Apollo 8 Anders sought to rebut that argument, and not just on exploration grounds: "A camera can 'remember' a scene much better than the mind, but the eye can handle a much greater range of lighting conditions and often can detect important detail not recorded in a photograph. I peered into every valley and crack my cloudy windows would allow me to see, looking for lava flows, volcanoes and other signs of igneous activity." Most features on the moon's back side, which man had never seen before except in photographs which were not entirely satisfactory for map-making purposes, seemed to be of meteorite origin, but there were other features which could be more easily attributed to volcanism. The idea, of course, was to combine the camera *and* the human eye in a search for clues as to the age and origin of the moon, concerning which there were conflicting theories, and as to whether the moon was a truly "dead" solar body.[2] To Borman, Anders and Lovell, the first men to see the moon with the naked eye from sixty miles off the surface, it appeared to be a vast and forbidding place — in Borman's words, "a great expanse of nothing that looks rather like clouds and clouds of pumice stone"; in Anders's words, "a dirty beach," grayish-white and churned up like sand: "It looked much like a battlefield, hole upon hole, crater upon crater. There was a total lack of sharp definition. It was completely bashed." They all noted one important point: the large dark areas of the moon, which astronomers had been calling "seas" (*maria*) for about three hundred years, appeared to be relatively smooth and therefore promising places for a lunar landing.

ARMSTRONG: The view of the moon that we've been having recently is really spectacular. It fills about three-quarters of the hatch window, and of course we can see the entire circumference even though part of it is in complete shadow and part of it is in earthshine. It's a view worth the price of the trip.

HOUSTON (McCandless): Well, there's a lot of us down here that would be willing to come along.

ARMSTRONG: One of these days we could bring the whole MOCR [mission operations control room] along, and then that'll save a lot of antenna switching.

HOUSTON (McCandless): Apollo 11, this is Houston. Over.

COLLINS: The Czar is brushing his teeth, so I'm filling in for him.

HOUSTON (McCandless): Roger. If you don't get in the way of the Czar while he's brushing his teeth, we'd like you to bring up the primary accumulator quantity a little bit. . . . Seems to have gone down a bit since you've gone into the shadow.

Seventy-three hours and twenty-five minutes into the mission. . . . Seventy-five hours. . . . Only 2,241 nautical miles away from the moon, velocity back up to 5,512 feet per second, over thirty-seven hundred statute mph, forty-one minutes away from loss of signal as Apollo 11 goes behind the moon. . . .

El Lago, Texas

Tom Stafford was crawling around on the floor of the Armstrong home, explaining to Jan how the lunar module crewmen would get their bearings when it came time to land: "Right about Boot Hill here they roll over. You can't miss Diamondback here. At Faye Ridge they begin to pitch up more — at three minutes after ignition. Faye Ridge sticks up here about seventy-five hundred feet. And at the start of powered descent they'll see Weatherford Crater, right here. I told Jack Schmitt to pick me out a good crater and name it Weatherford [Oklahoma]. I sure didn't know it would turn out this way."

How could he have known? Naming features of the moon had been an inexact science, the hobby of many men, for more than three centuries. Galileo Galilei (1564–1642) unwittingly started this at the same time he got into trouble with the Roman Catholic Church, which was disturbed by the thought that the earth might not be the center of the universe. All Galileo did was gaze at the moon through a primitive thirty-power telescope and make a map of it. This was the first telescopic map of the moon; it showed specific physical formations which shrieked for names. People were willing to oblige. Five years after Galileo died, a German named Johannes Hevelius published *Selenographica*, a lunar map on which he included names. He limited himself to the use of prosaic earthly designations — Alps, Caucasus and the Urals; he feared jealousy if he used the names of scientists and philosophers. Then, in 1651, came the publication of *Almagestum Novum* by an Italian Jesuit priest named Giovanni Riccioli. Predictably, he named a large number of the craters on the visible side of the moon after prominent Jesuits, but he also named two craters in the moon's northern hemisphere Plato and Archimedes. (He allotted the great Galileo only an island in the Sea of Storms.) It was Riccioli who gave us the names Tycho, Sea of Rains and the Sea of Tranquility — in which lay two (out of five, narrowed down from thirty) of the primary landing sites for Apollo, including site No. 2, where the Apollo 11 crew actually hoped to land. Other names were added, one by one — e.g., Crater Hell, named for a seventeenth-century astronomer-priest, Maximilian Hell. But there was no coordination in the name calling, and by the beginning of the twentieth century things had become so confused that two astronomers could discuss the moon together only if both were using the same map. The International Association of Academies in England undertook to simplify everything and actually published two books, the second of which, *Named Lunar Formations*, published during the First World War, identified about six thousand place names on the moon with coordinates for their locations, descriptions of their physical characteristics and the authority for identifying places with names. But it was not until 1961, the year in which Yuri Gagarin orbited the earth and Alan Shepard made the first American suborbital space flight, that the fifteen hundred members of the International Astronomical Union laid down some basic rules: name craters and isolated peaks for deceased astronomers and scientists; call mountains the equivalents of some earthly counterparts; and since the word *mare* (sea) had been used illogically for such a long time (on the naïve assumption that there might be water in the *maria*), stick with it — or *oceanus, lacus, palus* or *sinus*. *Mare* became the most popular designation, and it was thought suitable to link the Latin with a state of mind — Sea of Tranquility, Sea of Crises. So matters stood until August 1967, when the International Astronomical Union met in Prague, Czechoslovakia. One of its commissions, No. 17, dealt with "lunar nomenclature." The Soviet Union had already photographed the back side of the moon, and the Soviet delegation arrived with a lunar atlas which named about one hundred sixty of the largest craters, seas, promontories, mountains and basins on the far side. Not surprisingly, most of them were named for Russian cosmonauts, scientists, engineers and mathematicians; but it was also noted that the Greek delegation's list included the name of every prominent Greek from Aristotle onward, and that the French list included a great many Frenchmen. The American delegation had a list also; it originally included at least twenty Soviet names. It never was read to commission No. 17.

It never had to be read. Just a few days before the meeting convened, the United States had launched the enormously successful Orbiter 5 and completed the lunar Orbiter photographic mapping project. The success of Orbiter 5 was not lost upon the delegates. The United States, for the time being, preferred that no names at all be awarded at this particular meeting; the Russians yielded with grace, and the congress voted that the identification and naming of the wealth of new topographical detail be studied methodically and spaced over the next three years — to August 1970.[3]

Then came Apollo 8 and Apollo 10, and a good deal of unofficial naming on the part of the American astronauts. Apollo 8 produced "Mount Marilyn," named for Jim Lovell's wife; craters were named for Robert Gilruth, Deke Slayton and Wernher von Braun; three small craters in a row were called Grissom, White and Chaffee. A giant crater on the far side of the moon was called America, and lesser features were called Florida and Texas. Bill Anders's wife hoped that something would be called "Valerie's Valley," but this did not come about. Apollo 10 added Boot Hill, U.S. One, Weatherford Crater and Sidewinder Rill, along with Faye Ridge — which Faye Stafford had rather expected her husband Tom to call Faye's Fault. The names were all unofficial, merely call signs for convenience, but pending agreement on more official-sounding designations they would do very well for the flight of Apollo 11.

Seventy-five hours and fifteen minutes into the mission, distance from the moon 1,516 nautical miles, velocity 5,981 feet per second, over four thousand statute mph, less than half an hour from loss of signal. . . .

It had lasted about seventy-two hours, but the long coast was over and it was nearly time for action. When the spacecraft went behind the moon, about noon Houston time on Saturday, July 19, it would be out of radio communication with the earth for an hour, more or less. The "more or less" depended on a decision that Neil Armstrong, as commander of the mission, had to take while the spacecraft was out of touch with earth: burn or no burn? If everything seemed to be functioning properly, Mike Collins would fire the SPS engine to slow the spacecraft and permit Apollo 11 to be "captured" by lunar gravity. If a serious malfunction were detected, they did not need to fire the engine; they could come around the moon, not go into lunar orbit, take the "slingshot effect," coast safely home and let another mission attempt the lunar landing. Armstrong, Collins and Aldrin had worked too hard, trained too long and traveled too far to want to do that; yet Dr. Thomas Paine's warning nine days earlier had to be in their minds. . . .

Jan Armstrong went outside the house in El Lago Saturday morning only to water the ivy in the Japanese garden — and give the photographers their daily picture. Jan had never been to Japan, but Neil had become fond of the country when he took his leaves there during the Korean war, and for a time after his return to the United States he liked to lecture his parents and friends on the many virtues of the Japanese. Back at Purdue University, he wrote the words and music for a campus variety show and insisted on a tremendous production number involving a series of Japanese torii arches onstage and a giant Buddha. When it came time to design a home in Texas he wanted a modified Oriental motif; one visitor, noting a large green Buddha on a table in the living room, asked Neil if he were by chance a Buddhist. "No," he replied soberly. "I believe you would find in a Buddhist house that the central beam up there, the one running lengthwise, would be painted red to ward off evil spirits."

Nassau Bay, Texas

Pat Collins started her Saturday with a nine-thirty appointment for a hair shampoo and set. Three women journalists had somehow managed to get simultaneous appointments at the same establishment; one kept her notebook out and asked the hairdresser to "comb me out slowly." If Pat was bothered, she did not show it; by now a number of people had remarked on her knack of appearing to say a great deal when she was actually choosing her words to say very little. Back home, there were roses from people she did not even know, plus lemon yellow orchids from Annie Glenn: *May God watch over you and your family. Fondly.* . . .

Joan Aldrin spent the morning visiting with her father and stepmother, Michael and Rosalind Archer, who had arrived the preceding evening after a hard drive from Pensacola, Florida, through rain and thunderstorms. Joan also had a hair appointment; she took along a bundle of Mission Control transcripts to read under the dryer, along with a small radio. After she had gone a neighbor brought in a large coffee cake. Bob Moon killed some time mending a pottery vase which the kitten had knocked off the dining-room table and smashed. Rusty Schweickart staggered in with an enormous casserole containing a cold roast turkey. As the minute hand of the clock edged toward noon, everyone gathered around the squawk box — everyone but Joan, who was under the hairdresser's dryer with her radio.

Seventy-five hours and twenty-six minutes, fifteen minutes away from loss of signal, 966 nautical miles from the moon, velocity 6,511 feet per second, more than forty-four hundred statute mph, 23 minutes away from the lunar orbit insertion burn. . . .

These men had uncommon skills and commensurate professional pride. They considered themselves fortunate to be chosen for this mission, and the trip home would be rather dismal if they had to abort and pass up the lunar landing. Their training had been fantastically intense; it had little to do with the public image of an astronaut jogging on the beach. And the training had changed a lot since the Mercury days, when the astronauts were subjected to a weird mixture of uncoordinated physical torture — hours on the centrifuge to become accustomed to the punishment of four or five G's, minutes in a device called MASTIF (for Multiple Axis Space Test Inertia Facility), which spun an astronaut in three different directions at once. The late Gus Grissom had described the MASTIF: "A mission began like a carnival ride. You tumbled slowly, twisted and rolled as your body lurched against the tight harness that strapped you to the couch. You rotated faster and faster until you were spinning violently in three different directions at once — head over heels, round and round as if you were on a merry-go-round, sideways as if your arms and legs were tied to the spokes of a wheel. Your vision blurred. Your forehead broke out into a clammy sweat. And unless you could stop all of this with your stick, you could get sick enough to vomit." [4] Later Deke Slayton said, "We didn't know how to train an astronaut in those days. We would use any training device or method that had even a remote chance of being useful, and we would make the training as difficult as possible so that we would be overtrained rather than undertrained. And we decided to conduct our training on an informal basis."

Those "informal" days were over after Mercury missions proved that astronauts could tolerate G forces, that weightlessness did not disorient them, that they could eat and sleep and perform necessary physical functions in space. During Gemini the training emphasis began to shift away from the specifically physical and toward the technological; and there certainly was nothing informal about the training of Apollo crews. It was an exercise in diversification and specialization, for the simple reason that Apollo was too complicated for any one man to master its whole. More than two thousand hours of *formal* training were plotted in enormous detail for each man on Apollo 11, not including exercise, home study, miscellaneous briefings, reviews and mission support activities. The major spacecraft systems had to be learned and constantly reviewed; for the command module alone there were ten hours of briefing on electrical power and ten on communications; six on docking; eighteen on sequential events control subsystems, fifty-nine on guidance and control. There was the effect of weightlessness to be learned and mastered. A KC-135 aircraft flying parabolas could diminish the effect of gravity on men riding inside; and there was that water tank at the Houston MSC, where the astronauts spent hours in weighted suits to simulate one-sixth gravity. It was not like a Sunday afternoon swim in the family pool; Clare Schweickart could never forget the evening her husband Rusty came home so exhausted he could barely walk to supper. "I got in and out of the LM *three times* this afternoon," he said. One of the few astronauts who ever sat down and calculated just how he had spent his time in a calendar year was Dave Scott, command module pilot of Apollo 9. He did it because his wife had to make a speech to a woman's club — "I didn't want those women to think all we do is push-ups and then go out and fly." He made a detailed list for 1966, the year he flew with Neil Armstrong on Gemini 8: systems briefings and lectures, 300 hours; spacecraft testing, 370 hours; simulations on lunar modules, lunar landing research vehicles, mockups and command and service modules, 680 hours; operational briefings, 270 hours; mission equipment development, 350 hours; flying check rides and personal travel in T-33 and T-38 jets, 200 hours; flying helicopters, 40 hours. He was unable to assign figures to some other things: planetarium visits, egress training in the Gulf of Mexico, procedures development, survival training, physical training, interviews, administrative duties, weightless training, experiments, lectures, suit fittings, photographic training and building a brick wall in his backyard, which he did in the same year.

The most critical part of the training for an Apollo mission was also the most odious to the crew. This involved seemingly endless hours in simulators, first at the Houston MSC and then at Cape Kennedy; hour after hour they "flew" missions in which the simulators never moved but in which a complex system of lights, computers and motion pictures made it seem that they were flying to the moon and back. A "catastrophe team" at Cape Kennedy kept thinking up hypothetical crises to spring on the Apollo 11 astronauts as unpleasant surprises: suppose Collins, in the command module, had to swoop closer to the moon's surface than the flight plan called for, say to fifty thousand feet, to make an emergency pickup of Armstrong and Aldrin in the lunar module? (This technique had been rehearsed in the training for Apollo 10 by Stafford, Young and Cernan; the nature of the Apollo program was such that each flight crew, after a successful mission, was able to eliminate some pick-and-shovel work from the training of the next crew.) Or suppose the ascent engine of the lunar module faltered after leaving the surface of the moon? (This problem was tossed at the Apollo 11 crew one Friday without warning, less than a month before the July 16 launch day. Using the guidance thrusters of the lunar module, they got just enough extra thrust to get up to fifty thousand feet; in the simulator this appeared to work, but they hoped they would not have to prove out this alternative in flight.) There had been that jungle training in Panama, learning how to survive if they had to land a spacecraft in Africa or New Guinea, learning to recognize which snakes were edible and which were not. Jungle training was primarily characteristic of Gemini; Armstrong, Collins and Aldrin all had flown Gemini missions. The training for Apollo 11 had dimensions which went beyond a snake diet. Deke Slayton said of it: "The toughest job we had to train for on Apollo 11 was landing on the moon. All the rest of the stuff — rendezvous, zero-G, extravehicular activity — had been done and done again. Each mission which had gone before contributed knowledge to the follow-on mission. We just handed the package from the mission before to a new group of guys and said 'Here you go.' Apollo 7 shook out the command module; the Apollo 9 guys had to bite the bullet on the lunar module — and that was a helluva job. Apollo 8 went to the moon and back, so we knew how to do that. Apollo 10 checked out lunar orbit rendezvous as opposed to rendezvous in earth orbit. The Apollo 11 guys took everything everybody else had proved out and learned, and went on from there. The thing we hadn't done was land anything, manned, anyplace on the moon. Training for that was something else."

Something else. . . . Both Armstrong and Aldrin had trained to land something, somewhere, on the moon; but Armstrong had had the most specialized training. He had done many hours in helicopters; he had, several times, flown the lunar landing research vehicle (and ejected when one such vehicle crashed); in the weeks before the launch of Apollo 11, he had flown the lunar landing training vehicle (LLTV) more than thirty times at Ellington Air Force Base, near Houston. Armstrong and Aldrin, additionally, had been importuned by eminent men of science to learn as much as they could about lunar geology before they landed on the moon. Which solar body was older than which? And why? The astronaut Harrison ("Jack") Schmitt, a civilian who held a Ph.D. in geology from Harvard, thought diamonds found on the moon would not be worth the cost of transporting back to earth: "We would bring back only some fantastically useful and rare thing. If we found something which would cure heart disease, or cancer. . . ." *Or water. . . .* "There won't be any surface water. But there may be subsurface water, or water existing in minerals. Using solar energy, we could make oxygen and hydrogen from it, to refuel our rockets. . . ." Mike Collins managed to avoid most of this. "I hate geology," he admitted cheerfully. "Maybe that's why they won't let me get out on the moon." But Jack Schmitt had to teach Armstrong and Aldrin all he could in a relatively short time about blocky rim craters ("That's just a crater with rocks around the rim of it") and encourage them to take whatever field trips they had time to take. They had been to volcanic Hawaii, El Paso and the Big Bend country of Texas; and Schmitt, shortly before he became an astronaut himself in 1965, accompanied them to the famous meteor crater in Arizona ("About thirty thousand years old, young to a geologist"). But if Mike Collins was able to avoid geology, he was not able to avoid four hundred hours in the command module simulator between January and July 1969; and that was hard work. Always there were those simulators; Neil Armstrong had once called simulation both an art and a science. . . .

« You'd like to hope eventually to take the art out. There's a lot of ingenuity involved, a creative aspect. The free-flight simulator was conceived back in the days before there was a lunar module on the drawing boards, when I was at Edwards. Some people at Edwards and some others at Bell Aerosystems got their heads together. Long before we became committed to the lunar effort, it was obvious that landing, maneuvering over the lunar surface and taking off from it was quite a problem, and that even without knowing the nature of the configuration of going to the lunar surface, this was a valid engineering problem. We were thinking about this as early as 1958. The primary problem was that dynamics were considerably different over, and on, the moon than they were over, and on, the earth. Since there was no atmosphere on the moon, you couldn't use the techniques of airplanes or lighter-than-air vehicles. You had to have propulsion, a lifting force, to keep any machine flying, and this lifting force had to be equal to the weight of the machine or greater than its weight. It was decided to 'simulate' the lunar gravity on earth by using a jet engine which would support five-sixths of the weight of the vehicle, plus propulsion rockets to lift the remaining one-sixth, the lunar weight.

The problem of walking on the lunar surface was also anticipated. Man's weight on the moon would be one-sixth of his terrestrial weight, but his body would have the same mass. So in order to start walking or stop walking, you had to acquire force at a steeper angle — lean farther to start or stop. If you stood erect, as on earth, you would have little acceleration or deceleration ability. This would be something like trying to walk on the bottom of a pool with some weights. You couldn't get enough traction because of the low weight force you could apply to the bottom of the pool.

Two ways proved somewhat successful in simulating a walk on the lunar surface. One was to use an aircraft flying a trajectory very similar to zero gravity; for thirty-second periods you could walk under one-sixth gravity. Another technique was to counterbalance a man — stand him erect and support him by a cable and pulley system counterweighted with five-sixths of the man's weight. It was a Peter Pan rig. They had one at Grumman, where they were building the lunar module. [There are now several of them, improved versions, including one at MSC, Houston.] You had the feeling of being able to jump very high — a very light feeling. You also had a feeling that things were happening slowly, which indeed they were. It was a sort of floating sensation. But the lunar setting will become a very easy place to work, I think, once we have mastered the problems of balance and starting and stopping. We'll adapt to it. We'll be able to lift large loads — one hundred pounds with one hand, for instance — very easily. It's our feeling that the first time we step out on the lunar surface will not be the time to try to develop a technique concerning how much area we can cover or how far we can reasonably go. It will be a kind of dress rehearsal. There's no way we can simulate all the aspects together. There's no way to do that until we get to the moon. But we can take all the different parts and do them one piece at a time.

Then, mentally, we can put them all together and comprehend what the actual problems are going to be. It's like fitting together a jigsaw puzzle. »

HOUSTON (McCandless): Eleven, this is Houston. You are go for LOI. Over.

ALDRIN: Roger, go for LOI.

HOUSTON (McCandless): And we're showing about ten minutes and thirty seconds to LOS. I would like to remind you to enable the BD roll on the auto RCS switches.

ALDRIN: Roger, and confirm you want PCM low going over the hill. Over.

HOUSTON (McCandless): That's affirmative, eleven. . . . I'll give you a MARK at thirteen minutes and thirty seconds to ignition.

ARMSTRONG: Okay, and then a GET, please.

HOUSTON (McCandless): Stand by a minute. . . . I'll give you a time hack on the GET at 75 hours 37 minutes and I'll show you a bias at about a second and a half to allow for the time of flight. . . . Stand by. MARK, 75 hours 37 minutes GET. . . . Stand by for a MARK at TIG minus twelve. MARK TIG minus twelve.

COLLINS: You were right on, Bruce. Thank you.

Three minutes away from loss of signal, 425 nautical miles from the moon, velocity 7,368 feet per second, just over five thousand statute mph. . . .

In the history of Apollo, as in the third chapter of Ecclesiastes, there had been "A time to weep, and a time to laugh; a time to mourn, and a time to dance. . . ." The fifth verse had special relevance to the mission of Apollo 11: "A time to cast away stones, and a time to gather stones together. . . ." Geologists saw the mission of Apollo 11 as "a time to gather stones together" — one hundred twenty-seven scientific laboratories from all over the world had been selected to receive samples of the "moon rocks" which Armstrong and Aldrin were to scoop off the lunar surface and bring back to earth.[5] The rocks could tell a lot or could tell nothing; but the betting was that the rocks could tell the geologists *something.* A Minnesota-born Chippewa Indian named Arnold Brokaw was convinced that they would. Out at Flagstaff, Arizona, during the eighteen months preceding the flight of Apollo 11, he had been chief of the surface planetary exploration branch of the United States Geological Survey's Center of Astrogeology. There might or might not be latent water on the moon, but rocks — with their meaningful "bathtub rings" — were certainly there.

By 1968 Brokaw and his colleagues at Flagstaff had a good deal to go on. Five lunar Orbiters had circled the moon and photographed it, and four Surveyors had successfully landed on the moon. There were literally thousands of photographs to examine. The geologists' favorite, one they often called "the picture of the century," was a low-angle photograph of the lunar crater Copernicus, taken by lunar Orbiter 2. For the first time men had a good, clear look *across* the lunar surface, not up toward it, and the geologists calculated and noted that the crater walls of Copernicus rose nearly ten thousand feet — almost twice the height of the walls of the Grand Canyon. There was also a good photograph of a prime landing site in the southwestern part of the Sea of Tranquility (*Mare Tranquillitatis*), the contours of which bore a striking resemblance to an enormous area near Flagstaff known as the San Francisco volcanic field. Almost as flat as the moon's *maria,* covered with cinder cones, lava flows and volcanic ash, the field was formed in the eleventh century by one of the last violent volcanic eruptions on the North American continent. A mountain in what is now called the San Francisco Peaks group, in north central Arizona, blew up and emitted fiery globs of lava, cinders, ashes and smoking stone, most of which — because of the prevailing westerlies — fell to the northeast. When it was all over a cinder cone one thousand feet high had been formed, surrounded by miles of volcanic trash; the place came to be called Sunset Crater because the cone had a reddish complexion at its peak. As an American Indian, Arnold Brokaw felt a special sympathy for the Sinagua ("Without Water") Indians who had been farming this barren area for about five hundred years when the mountain exploded, and who then fled — those who had time.[6] Brokaw also felt that this volcanic area had a practical relevance to the Apollo mission: it was impossible to identify and select rocks on the lunar surface without going there, but men could practice for it if they could create a truly realistic moonscape on earth.

It had to be done with explosive charges. "We began with the Orbiter photo," Brokaw said, "and scaled off all its craters and features and put them all onto a map. We went out into Cinder Lake, which forms part of the volcanic field, and staked out the center point of all the craters, in the exact relationship they were to each other in the Orbiter photo. Then we calculated the amount of explosive it would take to get the moonscape we wanted. There were three distinct ages of craters shown on Orbiter. It is fairly easy to determine this even from a distant photograph because of the overlapping of ejecta that comes out of the craters. It's as if you threw a shovelful of dirt on this piece of ground; it spreads out a little bit. Then you throw another shovelful on top of this, and you get the age relationships." Furthermore, there was plenty of evidence from the Surveyor pictures that one did not just sink into the moon's fine-grained *maria* — at least not very far. Brokaw calculated, "There's bedrock down there. So on our cinder field we went through the upper layer of cinders, down to a layer of lakebed material, and finally down to a hard basalt layer. We dug all our holes and that was quite a job, getting them to exact depth to create a properly sized crater. We used carbon nitrate as an explosive, planted all three 'generations' of craters, hooked up the 'fuse' — miles of a sort of yellow tape, really — and blew."

When the three sets of charges were fired, one after the other, Cinder Lake shook the way Cape Kennedy seems to shake at the time of a Saturn V launch. When the echoes had stopped ricocheting off the far sides of the San Francisco Peaks, even the men who had organized the project looked at each other with some awe. "The resemblance to moon craters was remarkable," Brokaw said. "We measured our diameters against the diameters on the Orbiter photo, and we had an error of less than two percent. We had a simulated prime lunar landing area with three generations of craters, rays coming out from them, double-ring craters, blocky rim craters, everything."

Alas, the Apollo 11 astronauts never walked on Flagstaff's lunar surface.[7] This was not because Mike Collins hated rocks; the relentless pressure of flight simulations, meetings and flight plan changes kept the crew from traveling to Arizona. It was not a critical omission, because Brokaw had crews of trained geologists in space suits who roamed over the man-made crater field, trying out the complicated tools the astronauts would have to use on the moon and practicing "traverses," or walks across the field, describing features as they went. They even built a flimsy "lunar module" out of steel tubing and canvas (for a paltry six hundred dollars) and parked it on the moonscape so men could practice observation and geological identification from the limited viewing field and particular height of the LM windows; they experimented with cameras to determine which types would be both practical for astronauts wearing heavy gloves and most productive of photographs which would be used to reconstruct the geological history of the actual lunar landing site. All this information was relayed to the Apollo 11 astronauts by way of the team's principal investigator, Dr. Eugene Shoemaker,[8] and astronaut-geologist Jack Schmitt. Brokaw had come to think of his branch as a kind of fifty-man backup team: "The astronaut himself is, to us, like a sensor. He's *there*, and we're down here. He can't be a fully trained geologist; he's got five thousand things to do. But when it comes time to help him with his

EVA we want to be able to feed him any information he needs, offer any advice, support. We have to be ready."

By the time Apollo 11 was due to fly Brokaw had come to think that photographs taken on the moon would be at least as important as the rocks Armstrong and Aldrin brought back: "What is important to us is how the rock got where it was, how and where it lay, how it relates to other things in the same region. We can determine a lot about its mineralogy just from photographs." He spent an hour with Neil Armstrong emphasizing the importance of photography. And as he left the Armstrong home, he made a seemingly irrelevant but strangely prophetic remark: "It doesn't matter if the guy knows where he landed on the moon; we can figure that out later. . . ."

HOUSTON (McCandless): Two minutes to LOS. . . . Apollo 11, this is Houston. All your systems are looking good going around the corner and we'll see you on the other side. Over.

ARMSTRONG: Roger. Everything looks okay up here.

HOUSTON (McCandless): Roger, out.

HOUSTON (PAO Jack Riley): And we've had loss of signal as Apollo 11 goes behind the moon. We were showing a distance to the moon of 309 nautical miles at LOS, velocity 7,664 feet per second [5,225.3 statute mph]. Weight was 96,012 pounds. We're 7 minutes 45 seconds away from the LOI No. 1 burn, which will take place behind the moon out of communications. Here in the control center two members of the backup crew, Bill Anders and Jim Lovell, have joined Bruce McCandless at the CAPCOM console. Fred Haise, the third member of the backup crew, has just come in, too, and Deke Slayton, director of Flight Crew Operations, is at that console. The viewing room is filling up. Among those we noticed on the front row in the viewing room are astronauts Tom Stafford, John Glenn, Gene Cernan, Dave Scott, Al Worden and Jack Swigert. With a good lunar orbit insertion burn the Madrid station should acquire Apollo 11 at 76 hours 15 minutes 29 seconds. Acquisition time for no burn, 76 hours 5 minutes 30 seconds.

They were around the corner now, and over the hill. . . .

. . . And out of sight and out of communication. Within the next seven minutes the three men on Apollo 11 had to decide whether to allow themselves to be "captured" by lunar gravity — or take the slingshot and come home. This was an important decision for each man on the crew. Each had had his frustrations and his disappointments; each, in his own way, had a kind of private score to settle with space. Neil Armstrong perhaps had the most to settle; despite the apparent ordinariness of his Ohio upbringing, his subsequent life had been a mix of high risk and mishap, even tragedy (with the death of thc Armstrongs' daughter Karen). *There was that night in 1964 when the house had burned down in Texas* . . . and Jan Armstrong remembered that night better than Neil did: "He went out the door to see what it was, and he came back and said the house was on fire. So I turned on the light, fumbled for the light — and Neil told me to call the fire department. I couldn't get the operator on the telephone at 3 A.M. It was a long distance call to the fire department; I couldn't call them directly. I tried dialing 116, because I had had a first aid course in California. Then I realized that number was local only for the Los Angeles area. So I put the phone down, and Neil had gone in for Marky. I ran to the back of the house, and I was banging on the fence calling for Pat and Ed White, who lived next door. Our air conditioning wasn't working, and it was a warm night. I had closed the doors and opened the windows. Otherwise Mark probably would have been asphyxiated by the time we got to him. And there I was, banging on the fence. It was a six-foot fence. The Whites' air conditioning wasn't working either, and they heard me calling. Ed came bolting over the fence. I don't know how he did it, but he took one leap and he was over. He got the hoses out immediately, and by this time I had run around to the front of the house for Neil to hand Marky out the window. But no: Neil didn't do that. They were little windows, and Neil would have had to break one of them. He brought Mark back down the hall, back to our bedroom and out. He was standing there calling for somebody to come and get Mark because he was — what, ten months old? — and he couldn't put him down because he was afraid Mark would crawl into the swimming pool and drown. By this time I could hear the fire engines on the way — Pat White had turned in the alarm. This whole wall was red, and the glass was cracking on the windows. I can remember Ed White calling me. He was saying: here, you hold the hose; I'll get Mark. Neil had gone in for Ricky, who was just awakening at the time. And I was standing with the hose, the concrete was burning my feet, and we had to keep watering the concrete so we could stand there." Neil Armstrong came back with Ricky, and he had a wet towel over Ricky's face. *It was the longest journey Neil had ever taken; that forty feet with Ricky was longer than the journey from the earth to the moon. . . .*

. . . . And longer, perhaps, than the journey from space in March 1966 when he had to abort and come down prematurely, dropping into the Pacific Ocean after 6.5 earth orbits and 10.7 hours of flight, 37.5 orbits and 60.3 hours short of the goals set by the mission's flight plan. It was March 16, 1966, when Neil Armstrong's voice was heard; the tone was calm, but the words were urgent: "We consider this problem serious. We're toppling end over end, but we are disengaged from the Agena." Armstrong had more to say later, again in a calm voice. . . .

« After our first orbit, I began the maneuvers required to bring us to a rendezvous with the previously launched Agena rocket. [Armstrong and Dave Scott had been launched by a Titan.] These maneuvers required a series of burns on our thrusters that would put us into precisely the same orbit with the Agena. On our third revolution around the earth we picked it up on our radar. Even though we knew it would be an hour or so before we actually saw our target, we started straining for the first sight. And finally: there it was, a pinpoint of light seventy-six miles ahead. When we had closed the distance to about four miles, we could see it glowing in the sunlight like a fluorescent tube.

I made the closing maneuvers, and Dave handled the computer calculations which told us the exact amount and direction of thrust needed. As we closed the final few inches and latched the two vehicles, we felt a firm contact. Outside in airless space there was only silence, but in the cockpit we heard a slight thud. We relaxed for the first time.

We flew the Agena for the next thirty minutes. But discussions with the ground made us suspicious of its performance. When both vehicles started to spin, slowly at first and then rapidly picking up speed, we shut down the Agena and tried to stabilize with our regular attitude control system. That seemed to work for a while. Then the tumbling started again. We felt something that Dave was to describe later as 'constructive alarm.' We were aware of a serious emergency. A test pilot's job is identifying problems and getting the answers. We never once doubted we would find an answer — but we had to find it fast.

Although we had no way of knowing for sure, we were concerned that the stresses might be getting dangerously high — that the two spacecraft might break apart. We discussed undocking, but we had to be sure that the tumbling rate at the instant of separation would be low enough to keep us from colliding moments later.

As we unlatched, we still hoped to rejoin the Agena. At this point we figured that the trouble was in the Agena, but it wasn't. After separation the Gemini spacecraft stopped responding to the controls and rotated more rapidly than ever — the sun flashed through the window about once a second. The sensations were much like those you would feel during an aircraft spin. Neither Dave nor I felt the approach of loss of consciousness, but if the rates continued to increase we knew that an intolerable level would be reached. The only way to stabilize the spacecraft would be to shut down the regular control system and turn on the thrusters in the reentry control system.

I made that decision reluctantly — reluctantly, because once the decision was made, the mission had to be terminated. That excluded Dave Scott's EVA, the two-hour walk in space scheduled for later in the flight — and that hurt.

After a check of all the electrical circuits, we finally pinpointed the problem: the No. 8 thruster had been firing on its own. »

Neil Armstrong had been wearing a wristwatch worn by a pioneer aviator named Jimmy Mattern when Mattern attempted the first solo around-the-world flight in 1933. Jimmy Mattern did not achieve his goal; his plane cracked up in Siberia with a frozen oil line. *We had done a lot of thinking,* Neil Armstrong reminisced, *about those old pilots and how they had to improvise so often. Little did we know that we would have to do the same that day.* At least Jimmy Mattern's wristwatch had finally gone around the world; but for Neil Armstrong, a mission aborted was a job left undone. For Buzz Aldrin, a job left undone was — well, a job that had to be done later; and as Apollo 11 flew "over the hill," behind the back side of the moon, Armstrong, Collins and Aldrin were of one mind on one point; they wanted to complete *this* mission.

By comparison with the abort of Gemini 8, Buzz Aldrin's frustration with space was minor, but to his way of thinking the flight of Gemini 12 left a job undone just the same. He did three EVAs during that four-day flight and spent 5 hours and 37 minutes outside the spacecraft — a world's record by far. (The first space walk, by the Soviet Union's A. Leonov in 1965, lasted ten minutes. Ed White's EVA lasted 21 minutes, Eugene Cernan's 125 minutes, Mike Collins's 89 minutes and Dick Gordon's 161 minutes.) But Buzz had hoped to do his space walk with something called AMU ("Astronaut Maneuvering Unit"). This was not the complete "moon suit" — that would come later. But, as developed by the Air Force, the AMU did have the oxygen backpack, a precursor of the PLSS used on Apollo 11, and it had tiny jets for propulsion and stabilization. The Air Force was working on the assumption that AMU would make the spacecraft tether ("the umbilical") unnecessary; a man could simply step outside the hatch and, since speed is relative to where you are, float alongside the spacecraft at the rate of 17,500 mph. On earlier EVAs the astronauts had received life-sustaining oxygen by way of the umbilical; the backpack would eliminate the need for that. The idea was discussed before the Stafford-Cernan flight (Gemini 9A); then it was dropped. It came up again in the planning for Gemini 12. The first thing that went out was the idea of no tether; the tether, it was agreed, was safety insurance. Then the whole idea went out the window. This was the last mission of the Gemini series, and Buzz's AMU was dropped from the flight plan entirely. . . .

« I think they took it off because the benefits that were to be gained from a onetime AMU flight at this point were a little questionable. It was an experiment, and it was felt that there would have to be a chain of many events, all of them successful, before you got to the point where a man could float out with this thing strapped on his back. There were lots of little pitfalls. And on the last flight in the Gemini program we just couldn't afford not to have a success, particularly in the EVA. We had to take a couple of steps backward and start reviewing what we had learned from previous EVAs. In most problem areas, when something hangs up you start back over again. Well, in EVA you can't. Usually you can't try again, so if you are going to stop you have to stop before you start. This is an area, as in flying airplanes or driving automobiles, where the more time you put in on something the better you know how to handle it.

There wasn't a whole lot that could hang up on Ed White's EVA. It was simple and straightforward, and he handled himself admirably doing it. They did have trouble closing the hatch, but everything had worked out well on a short EVA, and I think this created a false state of mind. The next flight was planned to include a very comprehensive EVA — Dave Scott's. Of course, due to the thruster problem on Gemini 8 he and Neil didn't get to carry that one out. It may have been just as well, because hindsight would indicate that we were not yet ready for something that ambitious. We had better equipment available for the flight on which Gene Cernan did his EVA, but some of his tasks were still difficult. There just wasn't enough anchoring capability to force your body to stay in one place while you devoted your attention to hand activity. Those little foot stirrups just weren't adequate to keep Gene's feet down, and he found himself continually coming out and floating on up.

It was Mike Collins who actually propelled himself over to another vehicle, and that was a first. His job was to retrieve a meteoroid package from an Agena rocket after he and John Young had rendezvoused with it in their Gemini. But he had trouble. The first time he just slid off the Agena. So we put handrails on that rocket for subsequent missions.

I remember that after Jim Lovell and I had splashed down at the conclusion of Gemini 12, one of the doctors aboard the carrier asked me, 'What about that EVA?' And I said that the one thing that had surprised me about it was that it was just what I expected. The reason it was a surprise was that we had had all those little difficulties on earlier EVAs. This time we had just tried to come up with a flight plan that couldn't be hung up anywhere. I didn't have that little rocket gun to propel myself in space, but I was able to move around by hand power — manual labor. The umbilical was always free and floating behind, and you paid very close attention to where it was so it wouldn't get wrapped around your feet. You'd use your feet for touching something every so often, but you used your hands to get your body moving. I managed to avoid the fatigue problem that some of the others had had, but that too was something that came out of their experience. Of course I would have liked to have flown the AMU, but the reasons for taking it off the flight were probably sound. Rusty Schweickart increased our knowledge of the life supporting backpack on Apollo 9; in fact, he went beyond the AMU as we were talking about it back in 1966. He demonstrated the PLSS — the backpack Neil and I were to wear when we stepped out on the moon. That had to be done if the goals of Apollo 11 were to be considered possible. »

Mike Collins had done his own very successful space walk; but two years after he did it he was almost washed out of the Apollo program entirely, and for a reason that had nothing to do with his competence as a pilot. There was that loss of feeling in one of his legs, and then the medical diagnosis: bone spur, loose disc. There were two possible operations. One would relieve the problem; the second, more dangerous and complex, would result in a complete cure, if the operation succeeded. Mike's position on Apollo 8 was now gone anyway; so he opted for the more dangerous choice. . . .

« It seems to me that trauma comes from having to make decisions that are very nearly fifty-fifty — 50.1 vs. 49.9. Those are the things that really bug people and worry them. But in this case there was no doubt. I had a creeping malady coming up the left side of my body, getting worse and worse, and the salient fact was that this had to be stopped. The rest of the considerations were only background noise. I had no choice. I had a condition that had to be corrected, surgery was the only way to correct it, and *voilà!* — one hoped the operation would work out. I didn't have a sense of horror about it. You get a sense of horror about things that are not clearly understood and seem very insidious. For example, you get a cancer and they remove part of your liver, and then you find it is in the lymph stream and then it shows up somewhere else — those things are insidious. The mechanisms are not well understood. But if you have a pressure on your spinal cord and it affects one leg — that is a relatively 'clean' thing. The mechanism is well understood.

I'm not the kid who takes the model all apart and then, when the thing breaks, takes it to the garage. I'm not a mechanical engineer. I'm not in love with machines, and I'm not a machine-oriented guy. I understand the theory and one hopes the machine will work, but it is not my secret love to take things apart and put them together again and fiddle with old cars. I can tell you the theory of the internal combustion engine and hopefully I can analyze what's wrong with it, but when it gets down to fixing it, getting all the rusty bolts off it — that's not my bag.

When I had to retrieve that package from an Agena rocket on Gemini 10, John Young brought me right up to the end of the Agena. I gave a little push with my fingertips and sailed out the open hatch. The Agena's docking cone was very smooth and tapered. I grabbed the leading edge of the cone and held on with both hands; there was nothing else to grab. Then I got over to where the package was stored, and when I tried to stop my movement I found that my body just kept on going. My hands came loose and I sailed nose over teakettle into space. This was where I had to use the little space gun. It was stuck onto my chest pack, and when I reached down to get it the gun was gone. Of course this gave me a moment of uneasiness; then I realized that it had probably come loose during these wild gyrations. But it was attached to a cord. I reached down my side to where the cord was attached to my parachute harness, grabbed the cord and kept pulling on it until I got to the end of it — the gun was still there. I fired the gun, first to stabilize myself, then to thrust back toward the spacecraft. Then we regrouped the two spacecraft and I tried again, this time using the gun to propel me towards the Agena. I almost skimmed over the top of it; I was barely able to snag it as it was passing below me. I used the blunt leading edge of the docking cone for a handhold, and I found that I could get one hand inside the cone and get hold of some wires. Then I reached down with the other hand; luckily the package was retrievable with a one-handed operation. Holding the package in my left hand, I pulled myself back slowly, hand over hand, on the umbilical and got back into the cockpit area. Then I had to get the fifty feet of umbilical back into that crowded cockpit, which was one heck of a job. When the hatch closed over my head I could barely see John for all that hose. That was when we told the ground that the cockpit looked like the snake house at the zoo. The only fatigue I felt was in my fingers, and that was because of all the manipulations that I had had to do wearing those gloves.

John and I had a little wager — the last of a whole series, actually — about whether or not I'd be able to get that package off the Agena. When I did, he owed me a hundred and forty martinis. The whole thing started during training when we played darts in St. Louis, on visits to the McDonnell plant where they built the Gemini spacecraft. The game got way out of hand at the Cape. There we had a little training control box with a knob on it, and we sent a series of commands to the Agena — long successions of ones and zeros which were easy to foul up. John started betting me more martinis that somewhere along the line I was going to send the wrong commands to this stupid robot, the Agena. I must have been spurred on by thirst, because I got to rattling this little box and I ended up ripping the control right off the top. It was a good thing that happened at the Cape and not during the mission; in fact they modified the design of the box before we flew to make it less susceptible to damage. Because I seemed to be going through a phase in which everything I touched broke, the backup crew finally presented me with a huge chromium hammer. As for the martinis, I think John and I wrote off that debt as uncollectible and unconsumable.

Then, in July 1968, I had to have that operation; and the risk that I might never fly again just had to be accepted. During the recovery period I took Pat and the kids down to Padre Island on the Gulf coast for a couple of weeks on the beach; Pat had to do the driving, of course. It was on the way home that I picked up a morning paper in a motel and learned that Apollo 8 was going to be a circumlunar flight. My first reaction was that this was great news; all the planning we had put into the mission was going to pay off. Then in October I got back on limited flying status, although my neck was not at full strength and I was still restricted from flying airplanes with ejection seats. That was when the frustration began, and it got worse a few weeks later when I was restored to flying status without restriction. Here was my old flight going around the moon, and without me. That was a very unhappy time. When they made me a CAPCOM on Apollo 8 I thought — well if I can't go, there is at least something useful I can do to help them. It was a beautiful flight, but then I found myself still wondering: would I get on Apollo 11? Apollo umpty-ump? Any Apollo mission?

Then, in January, the Apollo 11 assignment came through and I was overjoyed. Aside from the fact that I would be flying with Neil Armstrong and Buzz Aldrin, for both of whom I had great admiration, there was now at least a chance that Apollo 11 might be the first lunar landing mission. I would be either a liar or a fool if I said that I think I have the best of the three seats on Apollo 11. On the other hand all three seats are necessary seats. I would very much like to see the lunar surface — who wouldn't? — but I'm an integral part of the operation and happy to be going in any capacity. I'm going 99.9 percent of the way, and I don't feel frustrated at all. When Neil and Buzz are tramping around on the lunar surface I am going to be their number one fan and I'm going to be listening as intently on my radio as anybody on earth. I don't think our flight will end in disaster. There is obviously some finite possibility that it may, but there are other hazards of life, like being on the freeway. Then there are those little cancer cells — and so on. It would be a terrible thing to turn down the space flight because it's too dangerous and drown the next week at Cape Cod. Statistically the chances are on our side, and the odds are good enough for me to accept the risk. »

El Lago, Texas

When the spacecraft lost communication with earth, Barbara Young, who understood a great deal about the technical details of space flight, was with Jan in the Armstrong home. They spent some time converting Ground Elapsed Time to Houston time. At twenty-four minutes from AOS ("acquisition of signal") they synchronized watches. Barbara Young had been through this before; when Apollo 10, of which John Young was the command module pilot, came around the moon for the first time, acquisition of signal was two minutes late, and a good many nerves were showing. Yet Barbara felt more nervous now than she had felt on "her" flight; Jan Armstrong agreed that other people's flights were worse than your own. Jan had another worry; the spacecraft's signal might be heard ten minutes *early* —

about 12:40 P.M. Houston time. That would mean that Apollo 11, for one reason or another, was on the way home without accomplishing its mission. Fifty seconds away from the time for "no burn" acquisition, Jan said, "Don't you dare come around. Don't you dare!"

Twelve forty-five p.m. . . .

HOUSTON (Jack Riley): We are past the no-burn acquisition now and we have received no signal. . . . It's very quiet here. . . . Most of the controllers are seated at their consoles, a few standing up, but very quiet. . . .

It was not quiet in the Armstrong home. Jan let out a single shriek: "Yippee!" The men of Apollo 11 were going to get a chance to do what they wanted to do — *get* to the moon, see it, complete the mission. *A minute and one-half away from acquisition time. . . . Thirty seconds. . . .*

HOUSTON (McCandless): Apollo 11, this is Houston. Are you in the process of acquiring high gain antenna? Over. . . . Apollo 11, Apollo 11, this is Houston. How do you read?

COLLINS: Read you loud and clear, Houston.

HOUSTON (McCandless): Roger. Reading you the same now [the signal came via Madrid]. Could you repeat your burn status report?

COLLINS: It was like — it was like perfect.

Jan Armstrong had had her ear pressed against the squawk box. When Madrid announced AOS, she had another joyous reaction: "Hot dog!" When Neil's voice came through clearly, Barbara Young happily clapped Jan on the shoulder.

Joan Aldrin was listening to this at the hairdresser's. In the Aldrin home the center of attention was Rusty Schweickart, who was annotating the flight plan. "Just off one-tenth of a mile in apogee," Rusty said. Clare Schweickart was visibly relieved: "Everything is real close? Good. Great. Now we don't have to worry until next time."

ARMSTRONG: It [the moon] looks very much like the pictures, but like the difference between watching a real football game and one on TV — it's no substitute for actually being here. . . . We're going over Mount Marilyn [Lovell] at the present time. . . .

HOUSTON (McCandless): Roger. Thank you. And our preliminary tracking data for the first few minutes shows you in a 61.6 by 169.5 orbit. . . . And Jim [Lovell] is smiling.

Armstrong leads crewmen Aldrin and Collins out of elevator at base of launch pad 39A at Cape Kennedy after men had spent hours inside the spacecraft during the Countdown Demonstration Test before flight. Each man is carrying his flight helmet and wearing light headgear astronauts call the "Snoopy helmet," which contains communications headset.

Astronaut Aldrin practices using a scoop for lunar rocks and dust during preflight training exercises at the Manned Spacecraft Center. Bag on his left hip is for temporary storage of material.

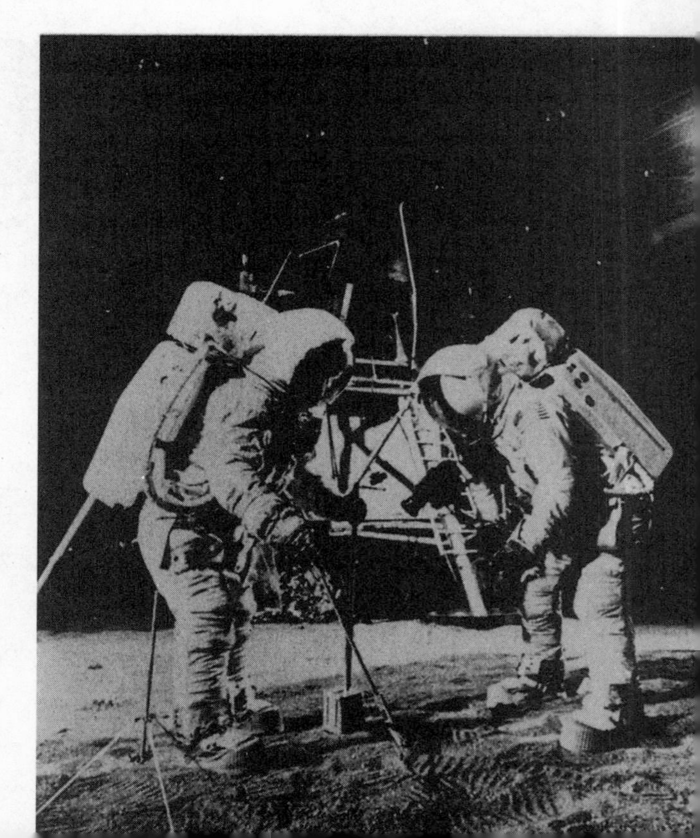

Wearing spacesuits and portable life support systems on their backs, Aldrin, left, and Armstrong practice lunar surface activities in front of a dummy LM at the Manned Spacecraft Center near Houston, Texas.

Balanced on its own flame like a broomstick on a finger, all 363 feet of the Saturn V rocket quiver into life as Apollo 11 lifts off from Pad 39A at Cape Kennedy the morning of July 16. Torpedo-like assembly on nose is the launch escape tower, designed to pull the bell-shaped command module free from the rest of the rocket in case of trouble at launch.

Fifteen seconds after ignition of the five clustered F-1 engines of the booster's first stage, the roar drowns the sounds of typewriter and camera, even the babble of languages of the thousands of journalists from around the world, who watched from a press site almost three miles away.

From 112,000 miles away, en route to the moon, the earth looks like a library globe spattered with whipped cream. The African continent is clearly visible, along with the Mediterranean Sea, the Near East, and part of the Asian continent at upper right.

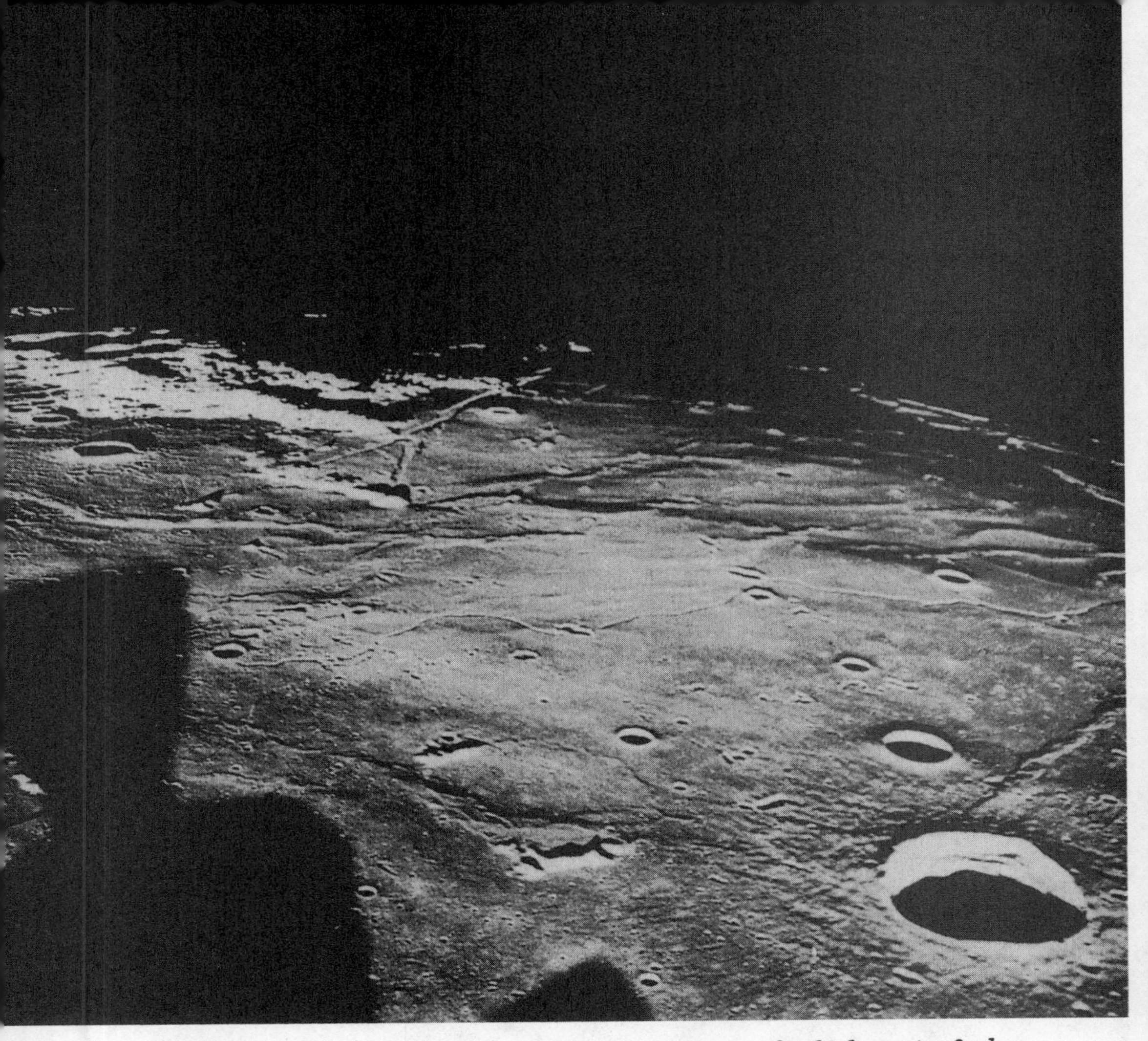

The lunar surface as seen by the LM from sixty-three miles high, on its final pass before swooping down to land. Landing zone shows roughly center at top of picture. Dark shape left is one of Eagle's control thrusters.

The back side of the moon, never visible from the earth, appears even more rugged and forbidding than the front in this picture taken by Mike Collins in Columbia. Future Apollo flights are scheduled to land in highlands, but not in the highlands of the far side, where astronauts would be out of radio communication with earth.

Eagle came down on the lunar surface about 100 yards from the medium-sized crater in background, which is eighty feet across. Structure at left is the portable stereo camera crewmen used to photograph rock specimens.

A mass of footprints, crisply defined, forms backdrop for elongated shadow of the lunar module's struts in photo taken from inside Eagle. Lunar soil, in the spot where Eagle landed, makes far better matrix for footprints than does earth dust because it compacts and stays put, does not fly up in clouds.

BOOK THREE

Eagle and Columbia

The Yankees, the first mechanicians in the world, are engineers — just as the Italians are musicians and the Germans metaphysicians — by right of birth.

JULES VERNE

I have been in high-risk businesses all my life.

NEIL ARMSTRONG

Man looks aloft, and with erected eyes
Beholds his own hereditary skies.

OVID

As soon as we go to face up in the LM we'll get a good view of the earth, but that won't be very helpful.

EDWIN E. ALDRIN JR.

9

"See that ridge? That can be five hundred feet high!"

IT WAS ONLY the first revolution, the first of twelve around the moon, before the command and service module was due to separate from the lunar module during the thirteenth pass, on the back side of the moon. But a big decision had been taken: it was go. The first lunar orbit insertion "burn," lasting about six minutes with a retrograde velocity change of about two thousand statute mph, had permitted the spacecraft — still combining the CSM and the LM — to be "captured" by lunar gravity. This maneuver took place about eighty nautical miles distant from the lunar surface and put the spacecraft into an elliptical orbit of 61 by 169 nautical miles. The command and service module's service and propulsion system engine still worked like a jewel. A second LOI burn would follow two revolutions later, to put the spacecraft into a more or less circular orbit sixty miles above the surface; the intent now was to go all the way. The landmarks were already being noted. . . .

ARMSTRONG: Currently going over Maskelyne. . . . And Boot Hill, Duke Island, Sidewinder, looking at Maskelyne W — that's the yaw around checkpoint, and just coming into the terminator — at the terminator it's ashen gray. If you get further away from the terminator, it gets to be a lighter gray, and as you get closer to the subsolar point, you can definitely see browns and tans on the ground, according to the last Apollo 11 observation anyway.

HOUSTON (McCandless): Roger, eleven. We're recording your comments for posterity.

ARMSTRONG: . . . And the landing site is well into the dark here. I don't think we're going to be able to see anything of the landing site this early.

HOUSTON (McCandless): We show you in flight plan staying in orbital rate until about 79 hours 10 minutes. Do you have some particular attitude or reason for wanting to go inertial? Over.

ALDRIN: No, that's fine . . . until 79:10 then we'll breeze around here in orbit.

HOUSTON (McCandless): Roger. And we've got an observation you can make if you have some time up there. There have been some lunar transient events reported in the vicinity of [the crater] Aristarchus.

ALDRIN: Roger. We just went into spacecraft darkness. Until then — why, we couldn't see a thing down below us, but now with earthshine the visibility is — oh, pretty fair. I'm looking back behind me now. I can see the corona from where the sun has just set, and we'll get out the map and see what we can find out around Aristarchus.

HOUSTON (McCandless): Okay, Aristarchus is angle Echo 9 on your ATO chart. It's about 394 miles north of track, however, at your present altitude, which is about 167 nautical miles. It ought to be over — that is, within view on your horizon, 23 degrees north, 47 west, and take a look and see if you see anything worth noting up there. Over.

ARMSTRONG: Houston, eleven. It might help us a little bit if you could give us a time of crossing 45 west . . . and then we'll know when to start searching for Aristarchus.

HOUSTON (McCandless): Roger, you'll be crossing 45 west at 77:04:10 or about forty seconds from now. Over. Thirty seconds from now.

ALDRIN: Houston, when a star sets up here there's just no doubt about it. One instant it's there and the next instant it's just completely gone.

HOUSTON (McCandless): Roger, we copy.

ALDRIN: Seems to me since we know orbits so precisely and know where the stars are so precisely and the time setting of a star or a planet to a very fine degree — that this might be a pretty good means of measuring the altitude of the horizon.

HOUSTON (McCandless): Roger.

ARMSTRONG: Hey, Houston. I'm looking north up toward Aristarchus now, and I can't really tell at that distance whether I really am looking at Aristarchus, but there's an area there that is considerably more illuminated than the surrounding area. It just has — seems to have a slight amount of fluorescence to it.

HOUSTON (McCandless): Roger, eleven. We copy.

ALDRIN: Looking out the same area now . . . well, at least there is one wall of the crater that seems to be more illuminated than the others. . . . I am not sure that I am really identifying any phosphorescence, but that definitely is lighter than anything else in the neighborhood.

HOUSTON (McCandless): Can you discern any difference in color of the illumination and is that an inner or outer wall from the crater? Over.

ALDRIN: I judge an inner wall in the crater.

COLLINS: No, there doesn't appear to be any color involved in it, Bruce.

[*Twenty minutes to loss of signal. . . .*]

COLLINS: Houston, Apollo 11. Could you give us a time of crossing the prime meridian 150 west? Over.

HOUSTON (McCandless): Roger. Stand by about a half second here. Okay, your time of crossing the 150 west meridian will be 77:50:05 [GET]. Over.

[*Six minutes to loss of signal. . . .*]

HOUSTON (McCandless): Eleven, this is Houston. In order that we may configure our ground lines we'd like to know if you're still planning to have the TV up with the beginning of the next pass. Over.

ARMSTRONG: Roger, Houston. We'll try to have it ready.

HOUSTON (McCandless): Apollo 11, this is Houston. A little over two minutes to LOS. All your systems parameters and orbit are looking good from the ground. We have AOS [acquisition of signal] on the other side at 78:23:31. Over.

Loss of signal. . . . Around the corner and over the hill for the second time. . . .

Because of some fascinating peculiarities in the nature of what NASA's John Mayer, chief of mission planning and analysis, called "the lumpy moon," Apollo 11's lunar orbit insertion was different from that of any previous lunar vehicle, manned or unmanned. The changes were inspired by expanding knowledge of the moon's "potential" — its gravity and the force of it.

On the flights of Apollo 8 and Apollo 10, the spacecraft went into a roughly circular lunar orbit sixty nautical miles high. But the orbit itself was pulled out of shape by lunar gravity, and on both flights the final elliptical orbit was about 54 by 66 nautical miles. This effect had been anticipated and caused no problems. But on Apollo 11 the mission planners, for added safety, wanted an initial orbit which would be elliptical, so that the natural action of lunar gravity would pull the command module into an almost circular orbit when it came time for the lunar module to leave the moon and rendezvous with the mother spacecraft. A circular orbit was preferable for rendezvous, and nobody wanted Mike Collins to waste fuel trying to get himself circular. Therefore the preplanned orbit of Apollo 11 — after two LOI burns — was also 54 by 66 nautical miles. The first LOI burn was to put the spacecraft into a skinny ellipse — 61.6 by 169.5 nautical miles. The second LOI burn was supposed to do the rest. At first glance it might have been assumed that this technique would make matters worse, since the elliptical shape had become more pronounced — and more undesirable — during the orbits of two preceding lunar missions. But this effect could be reversed. Lunar gravity is a variable, depending upon latitude, longitude and radius from the lunar center. On Apollo 8 and Apollo 10 the longitude reading at low point ("perilune") was about 100 degrees east, and decreasing. By changing the longitude on Apollo 11 to 85 degrees west, the perilune could be made to increase, and after twenty-five revolutions in the command module Collins would be in an almost circular orbit — without having squandered fuel.

Other navigational changes were made because of some things that happened on Apollo 10 which had not been anticipated. Apollo 10's lunar orbit insertion burn was targeted so that on the first revolution around the moon the spacecraft would pass south of the lunar landing site. As the moon rotated, the landing site would also move, and by the thirteenth revolution it should be directly below the orbital path of the spacecraft. This method had proved effective on all previous flights, manned and unmanned; but it did *not* work on Apollo 10, which in fact missed its hypothetical landing site by five miles. Quite unexpectedly, Apollo 10's orbit actually rotated *with the moon,* so that the landing site did not rotate and move into the position where it was supposed to show up on later revolutions. In addition, the inclination of the orbit decreased slowly so that the orbit became more nearly equatorial. The inclination of Apollo 10's orbit changed two-tenths of one degree in thirteen orbits.

These two related effects had never been noted before. NASA scientists finally concluded that they had been caused by the low inclination of the Apollo 10 orbit — only one degree. Orbital inclinations in previous lunar flights had been no less than twelve degrees. John Mayer said, "If we had had experimental data in an equatorial orbit or nearly equatorial, we would have known this earlier. Similar effects have been observed in earth orbit, but for orbits of high inclination." By now the theory that those "mascons" on the surface of the moon were causing navigational trouble had been discarded; evidently mascon effects, if any, were highly local. But there was a way out of the problem. In addition to inserting Apollo 11 into an elliptical orbit and letting it "degrade" to circular, the flight planners built into their target a bias which would take the spacecraft a little *north* (instead of south) of the landing site on the first revolution. Then the two-tenths of one degree change in orbital inclination would work for rather than against Apollo 11 and, after thirteen revolutions, bring the spacecraft where it was supposed to be — over the site itself.

Two-twenty in the afternoon Houston time, Saturday, July 19. Seventy-seven hours and 48 minutes into the mission, the spacecraft behind the moon and out of communication. Indicated apolune down to 168.5 nautical miles, indicated perilune about the same as before, 61.2 nautical miles. . . . Television pass due soon after reacquisition. . . . Seventy-eight hours, 19 minutes. . . . MARK two minutes from time of predicted reacquisition. . . . One minute. . . . MARK ten seconds away now. . . . We've had AOS by Goldstone. Television is now on. . . .

ALDRIN: Apollo 11. Are you picking up our signals okay?

HOUSTON (McCandless): Apollo 11, this is Houston. Affirmative. We are reading you loud and clear on voice and we have a good clear TV picture — a little bright crater on the bottom of the picture.

ALDRIN: No, that's a . . .

HOUSTON (McCandless): I guess that is a spot on the tube.

ALDRIN: Sorry about that one.

ALDRIN: Houston, Apollo 11. On one of the larger craters on the back side — I noticed a small dark speck on the outer wall and I put the monocular [a kind of spyglass] on it. I was able to see — oh, an area maybe a quarter of a mile in diameter. It was a really fresh-looking, dark-colored pit, and that seems to be in contrast with all the other fresh little craters or holes that you can perceive on the walls of any of these craters. Around this particular one there seemed to be two or three of these — especially the one that caught my attention. Quite remarkable. Over.

HOUSTON (McCandless): Roger. Do you have a location on that one?

ALDRIN: No. . . . I've got several pictures of it, though.

COLLINS: I'd say . . . we're about 95 degrees east, coming up on Smyth's Sea.

HOUSTON (McCandless): Roger. And for your information, we show you at an altitude of about 92 miles above the surface right now.

COLLINS: I'm flying in an SCS minimum impulse, Houston, and it's rather difficult to keep it on a constant rate. The LM wants to wander up and down. I'm not sure if it's in response to mascons or what, but I can get it completely stabilized in rate and let it alone and in another couple of minutes it will have developed its own rate.

HOUSTON (McCandless): This is Houston, Roger.

COLLINS: We're over Smyth's Sea right now . . . sort of a hilly-looking area. It's not like the *maria* at all.

HOUSTON (McCandless): Roger. We copy that about the sea, and it looks like you were just giving us a view of the crater Neper, the large crater on the left, and Jansen on the right.

COLLINS: We think you're close, but no cigar.

HOUSTON (McCandless): Would you care to comment on some of these craters as we go by? . . . We show you at an altitude now of about 110 miles, and of course you'll be considerably lower at the initiation of powered descent.

ARMSTRONG: Okay, there's — on the right side of the screen at the present time there's a triple crater with a small crater between the first and second, and the one at the bottom of the screen is Schubert Y.

HOUSTON (McCandless): Roger. We're seeing the central peak quite clearly now.

ARMSTRONG: Okay, we're zooming in now on a crater called Schubert N — Schubert N. Very conical inside walls, and the bottom of it appears to be flat.

COLLINS: And notice register three has reversed itself and it's heading back the other way now without any pitch thruster firing.

HOUSTON (McCandless): Roger, Mike.

COLLINS: Generally speaking, the tendency seems to be to pull the LM down toward the — the center of the moon as in a gravity gradient experiment. [Collins was trying to figure out what was causing the joined spacecraft to oscillate]. It may have something to do with mascons or it may just be the peculiarity of the DSKY display.

HOUSTON (McCandless): Okay, we observed the behavior of your DSKY and we got the data here to work on it. Let us grind around a little while on it.

ARMSTRONG: Three craters, three horizontal craters, that we now have in the field of view, are immediately underneath the ground track.

HOUSTON (McCandless): And we show you coming up on landmark Alpha One shortly.

COLLINS: Yeah, it is a great bright crater. It is not a large one but an extremely bright one. It looks like a very recent and, I would guess, impact crater with rays streaming out in all directions. . . . The Foaming Sea is easy to see, coming up on it now.

HOUSTON (McCandless): Here we can show you over the Sea of Fertility now and you ought to have Langrenus down south of track a few degrees, about nine degrees south of track.

ALDRIN: The crater that is in the center of the screen now is Webb. . . . It has a relatively flat bottom . . . and you can see maybe two or three craters in the bottom of it on the western wall, the wall that is now nearest the — the camera. Near the bottom of the screen we can see a dimple crater just on the outside, and then coming back toward the bottom of the screen into the left, you can see a series of depressions.

HOUSTON (McCandless): Roger, we are observing the dimple crater now. The central peak we can see on the Orbiter photos doesn't seem to stand out very well here.

ALDRIN: Well, they are not central peaks. They are depressions in the center. . . . We are moving the camera over to the right window now to give you Langrenus.

HOUSTON (McCandless): Roger. . . . We are getting a beautiful picture of Langrenus now with its rather conspicuous central peak.

COLLINS: The Sea of Fertility doesn't look very fertile to me. I don't know who named it.

ARMSTRONG: Well, it may have been named by — a gentleman whom this crater was named after. Langrenus was a cartographer to the king of Spain and made one of the early reasonably accurate maps of the moon.

HOUSTON (McCandless): Roger, that is very interesting.

ARMSTRONG: I'll have to admit it sounds better for our purposes than the Sea of Crises.

HOUSTON (McCandless): Amen to that.

In the living room of her El Lago home Jan Armstrong stared incredulously as Neil came up with his historical allusion and said, "So *that's* what he was doing with the *World Book* in his study!" [1] *Twenty-five minutes into the television pass. . . .*

ARMSTRONG: This is pretty close to ignition point for powered descent. Just passing Mount Marilyn, a triangular-shaped mountain that you see in the center of the screen. . . . And now we're looking at what we call Boot Hill. . . . On the right edge of the screen, Censorinus T.

HOUSTON (McCandless): Roger, and for your information your current altitude is 148 nautical miles above the surface.

ARMSTRONG: I sure thank you.

COLLINS: I'm unable to determine altitude at all looking out the window. I couldn't tell whether we were down to 60 or up at 170.

HOUSTON (McCandless): I bet you could tell if you were down to fifty thousand feet.

ARMSTRONG: We're passing some steep ridges here. . . . The crew of Apollo 10 was very impressed by the steepness of these ridges when they came over them at about fifty thousand feet.

HOUSTON (McCandless): Roger, we can observe they're also steep even from this altitude.

ALDRIN: How's the brightness of the picture you're receiving? Do you think we ought to open up the F stop some as we approach the terminator?

HOUSTON (McCandless): You can go ahead and open it up a stop or two. The automatic light level compensation seems to be working beautifully. . . . We're seeing Boot Hill now.

ARMSTRONG: The next crater coming into the bottom, that's Duke Island right there, and to the left, the largest of the craters, near the center of the picture right now is Maskelyne W. This is a position check during descent at about 3 minutes and 39 seconds, and it's our downrange position check and crossrange position check prior to yawing over face up to acquire the landing radar. Past this point, we would be unable to see the surface below us until getting very near the landing area.

HOUSTON (McCandless): Roger. I imagine you'll get a real good look at that tomorrow afternoon.

COLLINS: The small, well-defined crater is Moltke.

HOUSTON (McCandless): Roger. We can just see, it looks like, a little less than half of its rim right now. . . . Are you wide open on the F stop?

ARMSTRONG: There we are.

ALDRIN: Yes, and it looks like we're just about to get the sun coming into the lens as we'll have to move the camera. . . .

HOUSTON (McCandless): Roger.

COLLINS: We can't see any earthshine or any surface features at all in earthshine now due to the fact that the LM is very bright and is causing our pupils to contract.

ALDRIN: It's a very fantastic view to see the terminator as you look along the edge.

COLLINS: And I think you've got some interesting data on thruster firing versus pitch angle. It looks like that LM just wants to head down towards the surface.

HOUSTON (McCandless): Roger. I have a comment here that says that's what the LM was built for.

COLLINS: I believe.

ALDRIN: And as the moon sinks slowly in the west, Apollo 11 bids good day to you.

HOUSTON (McCandless): Roger. We sort of thought it was the sun setting in the east. [*Seventy-eight hours and fifty-eight minutes into the flight, touchdown only twenty-four hours away. . . .*]

Nassau Bay, Texas

Joan Aldrin had returned from the hairdresser, looking neat and elegant, in time to watch the television transmission. Jeannie Bassett was with her, and they talked about how news photographers had followed them back. "I tried to shake them," Jeannie said. "I went through an amber light once, and three cars followed through on the red." There was some jealousy in the room — between the Aldrin Siamese cat, Makita, and the Archer Siamese, Harvey, aged eighteen years, which Joan's father and stepmother had brought with them. Makita appeared to resent the intrusion but fell asleep on Joan Aldrin's knee. When the television screen showed clear pictures of craters on the moon, Joan looked up and said, "Looks like my pancakes are ready to turn over." There was light laughter when the commentator spoke of Buzz Aldrin as a geologist halfway on to his doctorate. As the camera continued to play over the moon's surface, Joan seemed to lose interest. She continued to read her letters, light another cigarette. Once, when there was a strange noise on the sound track, she looked up to ask, "Is that *wind?*" Her father, Michael Archer, said, "Just think, Joan. You may have a crater named after you." When the screen showed the Sea of Tranquility, Joan sat forward, shoulders hunched, and asked, "Was *that* Buzz?" It was not Buzz. As the camera turned back to the light, just as Apollo 11 was going into the dark of the moon, she said, "Isn't that something?" She switched off the television set and said, "Well, I've had my thrill for the day. Now I just have to get through the next twenty-four hours." She disappeared into the kitchen and came back with a digitalis pill for Missy the dog, pushed it down her throat and stroked her gently: "Come on, now; you didn't swallow. Come along, there's a good girl."

NASA's Bill Der Bing took Andy Aldrin to Mission Control. Uncle Bob Moon went along. Andy and Mr. Moon collected bundles of press handouts, and Andy was shown the signed portrait of his father which hung on the wall of a corridor along with signed portraits of the other astronauts. Andy had a baffled expression on his face. He could not quite relate all the activity and hubbub in Mission Control to his father and to himself. As he walked around, he was not recognized.

In the Collins home Clare Schweickart joined Pat Collins in the bedroom to look over the day's newspaper clippings and watch the television transmission. Dave Scott's wife, Lurton, took the Collins children to a movie — *The Swiss Family Robinson.* Bill Elkins, a lawyer friend from Houston, stopped to leave a pail of boiled shrimp. Tom and Faye Stafford dropped in. The television was showing the lunar surface. "It looks like old home week," Stafford said.

HOUSTON (McCandless): We've been looking at your systems data on playback and everything is looking good. . . . I would like to remind you, though, of a request to perform this burn [the second LOI burn] on the A bank ball valves only, and you are go for LOI two. Also, we have currently in the flight plan you scheduled tomorrow to start entering the LM at about 96 hours GET and we'd like to know if you have any plans to initiate this ingress into the LM earlier. If so, we can call the people in ahead of time. Over.

ARMSTRONG: Well, we didn't have any plans to. No. We just wanted to be ready at that time.

[*Platform alignment for LOI two was now completed. . . . Thirty-three minutes from ignition time, 9 minutes 40 seconds away from loss of signal. . . .*]

HOUSTON (McCandless): Apollo 11, this is Houston. Two minutes to LOS. Your AOS on the other side is 80:33:21 [GET]. . . . Your friendly white team CAPCOM [Charles Duke] will see you when you come out from behind the moon.

Thirty seconds to LOS. . . . Ten seconds . . . out of range now, behind the moon for the third time. . . .

El Lago, Texas

The Staffords had also dropped in on Jan Armstrong, who was preparing for the events of Sunday by taking a cram course in lunar geography. Waving his arms, Stafford said: "When they pitch up in here this Moltke is going to show up so big. Moltke's the one that's really going to stand out. This site has lots better lead-in than site No. 1, although No. 1 is perfectly good." Jan got out a smaller map of Apollo 11's actual landing site and put it on the floor beside the big map. Pointing to the small site map, Stafford said: "Did they tell you they put us five miles south of the site? So they're targeting them to a false point five miles north of the site. It's really pretty good coming here. In the middle you can see things within fifty feet resolution. See that ridge? That can be as much as five hundred feet high. Right in here it looks real good. Here's what may cost them some fuel — when they get down to five hundred feet. Come to a hover and maybe there's a boulder field — he'll have to move over."

"The legs of the LM, I know, are like shock absorbers," Jan said. "Do they adjust or will they sit on an angle?"

"They're on an angle."

"They can't adjust?"

"Yes, but there's no automatic leveling," Stafford said. "Anyway they could take off at an angle."

Jan had heard that on the first LOI the crew had burned two seconds less than the flight plan called for but had used more fuel; what did that mean?

"It means they had a pretty hot engine," Stafford said. "That is good. They've got plenty of fuel — nothing to worry about."

MARK one minute until planned time of ignition, Mike Collins at the controls of the spacecraft. Seventeen seconds of burn time anticipated to change the orbit to nearly circular. . . . Ten seconds. . . . MARK, planned time for ignition. . . .

HOUSTON (PAO John McLeaish): MARK two minutes from time of predicted acquisition of the Apollo 11 spacecraft. During this upcoming pass we will have our second excursion on the part of the Apollo 11 crew into the lunar module. The LM is to be pressurized by a valve in the tunnel hatch and, as a point of interest, will remain pressurized following this period of activation and after the members of the Apollo 11 crew return to the command module. For this period of activation, it's definitely planned that Buzz will go into the LM, and there is a distinct possibility that Commander Neil Armstrong could exercise his option and go into the LM. Our station to acquire as we come around the far side of the moon will be Goldstone. MARK one minute. . . . Thirty seconds. . . . Ten seconds. . . . Goldstone has acquired Apollo 11. . . .

ARMSTRONG: VGY minus .0, VGZ minus .1 . . . The burn time was seventeen seconds. . . . [*Orbital parameters estimated at 66.1 nautical miles by 54.4 nautical miles, right on the button. . . .*]

It was five minutes past five in the Armstrong home; Jan Armstrong banged her fist on the coffee table, upset because the squawk box appeared to have been a little late reporting AOS. Upstairs, where Ricky Armstrong was watching television with his cousin Robb Thiessen, somebody yelled: "Ooh hoo, bases loaded!"

In the living room Tom Stafford got up and stood with his back to the map, Jan beside him. "You're coming in backwards when you land," he said. "After three minutes of burn time you roll over right at Maskelyne crater — it's a huge big rascal. Then left is left and right is right. When he first lights off the engine he's here in the boondocks. And then here's Cape Venus — that was named after Cernan's dog — and here's Bear Mountain. It drops off like the Grand Canyon. High Gate is right here at eight minutes. Low Gate is 10 minutes 8 seconds. From High Gate on down they'll be designating a little bit left, a little bit right."

Jan stooped over the map again. "High Gate is right in here?" she asked.

Eighty-one hours and 16 minutes, a quiet pass on the near side of the moon, very little conversation with the Apollo 11 crew getting ready to enter the LM. . . .

HOUSTON (Duke): Hello, Apollo 11, Houston. We're wondering if you have started into the LM yet. Over.

ARMSTRONG: We have the CSM hatch out, the drogue and probe removed and stowed, and we're just about ready to open the LM hatch now. [*Eighty-one hours and 25 minutes. . . .*]

ARMSTRONG: Okay, Charlie. We're in the LM. The docking index mark is the same.

ALDRIN: There just doesn't seem to be any slow way to get that repress [valve] from closed to AUTO without making a big bang.

HOUSTON (Duke): We copy, Buzz. Thank you much. Out.

COLLINS: Getting water inside the command module for the first time. There's a little puddle on the aft bulkhead sort of like 101 had.

HOUSTON (Duke): We'll be with you in a moment, Mike. Stand by.

Eighty-one hours and 30 minutes now into the flight. The water puddle problem ("like 101 had") was similar to one encountered on Apollo 7, commanded by Walter Schirra. Altitude 54 nautical miles, apolune 65.4, perilune 53.8 — still right on the button. . . . Eighty-one hours and 32 minutes, five minutes away from loss of signal. . . . MARK three minutes, MARK two minutes, MARK one minute. . . . Loss of signal, over the hill and around the back side of the moon for the fourth time. . . .

Once again it was quiet in Mission Control; it was also quiet in the homes of the three Apollo 11 astronauts, where the wives wanted nothing so much as to be alone. Tomorrow, Sunday, July 20, the day of the lunar landing, loomed as a day almost beyond enduring.

For the first time in several days Pat Collins was more or less alone in her own home. Her sister Ellie Golden, from Boston, had prepared a roast beef dinner, and the Collins children were having dinner with the children of Dave and Lurton Scott. They would come home at eight o'clock, and it would be an early bed for everyone. Nobody wanted to think about the possibility of disaster; indeed all the astronauts were a bit sensitive on that subject. But there was that calm, realistic remark that Mike Collins had made, and it still hung in the air, hung in space: "I guess the question that everyone has in the back of his mind is how do I feel about having to leave them on the lunar surface? I don't think that will happen, and if it did I would do everything I could to help them. But they know and I know, and Mission Control knows, that there are certain categories of malfunction where I just simply light the motor and come home without them."

HOUSTON (PAO John McLeaish): Apollo Control, Houston. We're now within two minutes from time of predicted acquisition of the Apollo 11 spacecraft. As we make this pass — near side pass, on the fourth revolution, it will be the first time that we have transferred during this mission to lunar module pilot power, and a communications check will be performed on the lunar module. . . . Less than a minute away now at this time. . . . Standing by now for acquisition. . . . We have acquisition. We are receiving the telemetry data at this time. Hawaii and Goldstone both have acquired.

HOUSTON (Duke): Hello, Apollo 11, Houston. We're standing by.

COLLINS: Okay, Houston. We'll be doing P22 in just a couple of minutes.

HOUSTON (McLeaish): This is Apollo Control, Houston, 82 hours and 38 minutes now into the flight of Apollo 11. Well, we have single contact with Apollo 11 thus far this pass, when Mike Collins identified he was still involved with Program 22, the auto optics landmark tracing activity. . . . Presently we're reading on our orbit displays an altitude of 65.1 nautical miles, apolune 65.3, perilune of 54 nautical miles [just right]. We currently show a velocity on the Apollo 11 spacecraft of 5,318 feet per second [3,625.8 statute mph]. [*Eighty-two hours and 39 minutes, 82 hours and 55 minutes. . . .*]

ALDRIN: Houston, this is Apollo 11. I have a . . .

HOUSTON (Duke): Apollo 11, Houston. You are breaking up badly.

ALDRIN: Roger, I can see the entire landing area looking out the left window. Over.

[*Eighty-three hours and 2 minutes into the flight, apolune 65.3 nautical miles, perilune 53.9. . . . Transfer to lunar module power due momentarily. . . .*]

ALDRIN: Houston, Apollo 11, Apollo 11. Eagle. Over.

HOUSTON (Duke): Roger, Eagle. This is Houston and we read you. Over.

Eagle! Buzz Aldrin's voice was calm, but the word itself was a tocsin as it came tumbling across nearly a quarter of a million miles of space. Used as a call sign for the first time, it would have sounded like a shout even if Aldrin had spoken in a whisper. . . .

Although the practice of naming spacecraft had been dropped with the first manned flight in the Gemini program (when NASA vetoed Gus Grissom's proposal to name his capsule "Molly Brown"), it had to be reinstituted with the flight of Apollo 9 (McDivitt, Scott, Schweickart). This was necessary because this was the first time the command module and the lunar module would fly separately in space, and the two spacecraft had to have call signs for identification in ground-to-air and air-to-ground conversation. The choice was not entirely a matter of taste and aesthetics. Some words have good carrying power in long-distance radio communication and others do not. Two-syllable and three-syllable words are usually best, and they must have enough hard consonants to blast their way through static. Astronaut Scott Carpenter always regretted that he did not take his wife's advice and name his Mercury capsule "Rampart" — after the Rampart Range in his native Colorado. Instead he chose "Aurora," which did not come through well on the communications system. Words that invite slurred pronunciation are out of the question; for instance, "Manhattan" and "Westchester" might well come over the radio as "M'nat'n" and "Wes'che'r." Most one-syllable words are also unsuitable. Up to the time of Apollo 9 the astronauts had commonly called the lunar module "the bug," but "bug" did not have enough carrying power to make it as a call sign. The strange look of the lunar module, with its queer spindly legs, suggested "Spider," a word that had a sharp break between its two syllables, and that was used for Apollo 9 along with "Gumdrop," which was chosen after someone had remarked, semifacetiously, on the command module's physical resemblance to a candy store gumdrop. Besides, the word also had good "carry." Although it sounded facetious, the choice of "Snoopy" for the lunar module to be flown by Tom Stafford and Eugene Cernan on Apollo 10 had a sound logic to it. For two years the famous dog created by Charles M. Schulz had been used by NASA as a safety symbol, and it seemed appropriate to send him on a flight. If Snoopy were going to fly, so should Charlie Brown — and John Young's command module became "Charlie Brown." Tom Stafford informed Schulz of the choices by telephone; Schulz was delighted, although he did wonder whether it was not tempting fate to use Charlie Brown: "We once thought about putting out a Charlie Brown baseball glove, but what kid would want one? He couldn't catch a thing with it." [2]

As in the matter of patch design, there was a considerable sensitivity in the astronaut community about the choice of call signs for the first lunar landing. As long ago as early 1967 — just after the pad fire, before the Saturn V had even launched once, long before the lunar module was due to be ready — Pete Conrad was thinking about the problem. The suggestion "Lem and Abner" [3] was seriously made — and rejected by Conrad, who at that time was a prime possibility to command the first landing mission. "In this instance I believe we should have some dignity," Conrad said. "I agree with some of the stuffy ones about this." How about "Venus"? Conrad liked that, looked up Venus in an encyclopedia and announced: "She won't do." As patroness of Pompeii Venus had somehow got herself associated with prostitution. Finally Conrad said wistfully, "I'd really like a name like 'Intrepid,' one of those lovely names the British give their battleships. I like that particular name not because of the warship angle but because of its symbolism — undaunted, valiant, courageous." The idea was filed away in the back of his mind, and when Apollo 12 was flown, in November 1969, the call signs were "Intrepid" for the lunar module and "Yankee Clipper" for the command module — appropriate for an all-Navy crew.[4]

For the Apollo 11 mission Armstrong, Collins and Aldrin got literally hundreds of suggestions. Discussions on the subject tended to be unproductive. As late as April 1969 the three Apollo 11 astronauts — and their wives — were still rolling "paired" names off their tongues for sound effects: Romeo and Juliet; Antony and Cleopatra; Daphnis and Chloe, the Greek shepherd and shepherdess (Daphne herself was brought up but rejected because she spent half her time trying to escape from the god Apollo); even Amos and Andy. Castor and Pollux were tempting, but as twins they had already exhausted their symbolic usefulness in the Gemini program. The difference in the sizes of the two vehicles suggested David and Goliath, but nobody wanted Goliath slain in this operation. Patricia Collins waged a good fight for Owl and Pussycat but lost. Someone else suggested Majestic and Moon Dancer, but the astronauts thought that one name smacked too much of empire and royalty and that the other name was too frivolous. In the end the astronauts rejected all romantic name-pairing as being inappropriate to the mission.

By now the astronauts were sure in their minds that they wanted the names to reflect a degree of American pride, but within bounds set by taste and dignity. After the patch design had been approved, the problem started to solve itself. "Eagle" for the lunar module came straight off the patch, and then another national symbol occurred to everyone almost at once: "Columbia" for the command module. Armstrong spoke for the crew on that point: "Columbia was also a national symbol, but more importantly the choice was an attempt to reflect the sense of adventure and exploration and seriousness with which Columbus undertook his assignment in 1492. And, of course, there was a tie-in with the Jules Verne exploration book that turned out to be, in some ways at least, an accurate prediction of the technique and details of the Apollo 11 flight." [5] The astronauts had intended to keep the patch design for their flight secret until their last formal press conference in Houston on July 5, then unveil the eagle and the olive branch with a flourish; but they had not intended to keep the call signs a mystery for so long. Final approval of the call signs came from Washington only a half hour before the press conference was due to begin, and only minutes before Armstrong, Collins and Aldrin walked into a three-sided box (with specially rigged ventilation for health security reasons)[6] where they announced that Eagle and Columbia were the names that would go down in the history books.

HOUSTON (Duke): Eagle, be advised. Sounds like a hot mike. Over.

ALDRIN: Yes. . . . Roger. If you read me now, I am in hot mike because I'm in ICS push to talk, and downvoice backup. Over.

HOUSTON (Duke): Columbia, this is Houston. Are you maneuvering to sleep attitude? Over. . . . Eagle, this is Houston. We have lost all the voice and data with Columbia. Would you see if he is maneuvering to sleep attitude? Over.

ARMSTRONG: Hey, Mike. You maneuvering to sleep attitude?

ALDRIN: I don't believe they can hear you, Mike. Are you maneuvering to sleep attitude?

ALDRIN: Houston, Eagle. Columbia has maneuvered to sleep attitude. He's got the high gain antennas — antenna angles set in, and he should be communicating with you. Over.

HOUSTON (Duke): Roger. We don't have him. Stand by.

ALDRIN: Houston, this is Eagle. Roger. Read you loud and clear. How read? Over.

HOUSTON (Duke): Roger. Reading you five by also, Buzz [five by five, loud and clear], and we got the high bit rate. It's looking beautiful for Goldstone.

. . . Hello, Eagle. This is Houston. How do you read — normal voice? Over.

ALDRIN: Eagle, Houston, this is Eagle. Read you loud and clear on S-band, normal voice OMNI. Over.

HOUSTON (Duke): Roger. You're beautiful in this mode, Buzz . . . we're reading you five by. Come in with a short count, and we'd like one back from you. 1 2 3 4 5. 5 4 3 2 1. Houston out.

ALDRIN: Houston, Eagle, You're gorgeous also. 1 2 3 4 5. 5 4 3 2 1. Eagle over.

HOUSTON (Duke): Eagle, Houston. Everybody's happy as a clam with this mode. We'd like to stay here for a little bit. Telemetry looks great, and the voice is great. Over.

[*Eighty-three hours and 27 minutes into the flight, altitude 54.3 nautical miles off the moon, velocity 5,376 feet per second, 3,665 statute mph. . . .*]

COLLINS: Houston, Columbia. Ready to copy TEI 11. Over. [*Finally Mike Collins was coming through directly to the ground; Columbia was on the air.*] All right. I read back. TEI 11 SPS G&N, 37,200 minus 060 plus 047 plus 098. . . .

HOUSTON (Duke): Roger, eleven, and we'd like you to do a waste water dump at 84 hours down to twenty-five percent. . . . And Mike, we'll have LOS in about eleven minutes at 83:44 [GET]. AOS is 84:30. [*Eighty-three hours and 43 minutes into the flight, less than one minute from loss of signal. . . . Loss of signal. . . . Over the hill for the fifth time. . . .*]

The vehicle called Eagle was officially known as LM No. 5. Visually it had nothing in common with any other spacecraft; in fact it was the most highly specialized vehicle built for any kind of manned travel. It could be used only once, for landing on the moon and then getting off the moon, and part of it would be left on the moon for future explorers to seek out, and possibly recover, for its antique value. The ascent stage of the LM would be left in orbit; its thin aluminum skin had no heat shield, and it would burn up if an attempt were made to bring it back into the earth's atmosphere. But even the ascent stage might not be lost permanently. Dave Scott thought of that after the Apollo 9 LM, the first manned LM to fly, had been discarded: "That thing has an estimated life of nearly twenty years. It's a kind of orbiting time capsule. We don't have the technology to recover it now, but within twenty years we should, and it would be rather fun to find it and try to bring it back." [7] Back in 1964, less than two years after Dr. John Houbolt had won his argument for the lunar orbit rendezvous technique to go to the moon, Pete Conrad thought an early model of the LM looked like "an ugly and unearthly bug, with windows for eyes, a front hatch for a mouth and antennas jutting from its shiny skin. It has the innards of a flying machine — but an aerodynamic shell would be useless on the airless moon and therefore has been dispensed with. Each piece of gear has simply been put in the right place to do the right job — aesthetics be hanged." The model Conrad was talking about was made of wood; Grumman Aircraft, in late 1964, was only then cutting the metal for the first engineering mockup. Conrad was following that development closely in those days: "There were countless man hours spent bouncing ideas back and forth across the conference table — and always, always, there was the worry about weight. The astronauts' role in all this was not to design or manage — or even to worry about the physical hazards of a moon flight; that's the job of the space medics. We were there simply to understand every little detail — why it was that way and how it would affect us as the spacecraft drivers. It wasn't easy, because you had two different major contractors, North American building the command module and Grumman the LM, with one of them in California and the other in Long Island. I sat in the middle and could see both sides, but sometimes an outside head could really help the experts. One day I walked cold turkey into a meeting where the engineers were worrying about two electrical switches. If both were turned off by mistake, a vital piece of gear would burn out. I didn't know a switching circuit from a hole in the wall, but I made a suggestion: if one switch didn't have an 'off' button, there would be no problem. It was that simple. Then, seeing that the rest of the discussion was well over my head, I politely excused myself. The ideas and the drawings and the wood were like pieces in a far-out jigsaw puzzle, and their shapes were constantly changing as new ideas were tested. I remember working on the first cockpit that was laid out. There were four windows and two seats, and the first-cut instrument panel didn't look much like what we finally came out with."

The original intent was to start flying manned LMs in 1966, but production kept falling behind schedule — for understandable reasons, since nothing like the LM had ever been conceived before, let alone built. And changes in design kept having to be made, largely because of the weight problem. The early model Pete Conrad was talking about had two docking ports, five legs, four windows and capsule-shaped fuel tanks. By 1965 the front docking port had been made an exit hatch, the body contours had been shaved away, the fuel tanks reshaped, two windows and a leg removed. The seats had also been thrown out — it dawned on someone that in a weightless environment there was no need to sit down. This was the basic design for the LMs flown on Apollo 9, 10 and 11, but the LM never could have been made ready for manned flight in 1966; there were too many problems to be ironed out. The first LM did not fly until the fall of 1967, and it was unmanned. As had been expected, it burned up in the atmosphere; but the test was considered a success and LM No. 2 never flew at all. The vehicle was not even ready for manned flight by mid-1968, and that led to the decision to change the sequence of manned Apollo missions and send Apollo

8 around the moon — without the LM. To a quality control inspector like Grumman's Herman Clark life seemed to be one constant fight: "Just to give you an idea, there's this stress corrosion problem. There's a lot of liquid shimming [joining] done on this. We have two joints. Say you have a gap between those two joints. The maximum we can permit is under two-thousandths of an inch. And we liquid shim it. This is a headache, because the mechanic is under pressure from his lead man to get that job done. If he could just drill it up and put the rivets in, he could have it done in half an hour. But to liquid shim — it might take him all day, because you have to let the liquid cure, or harden, which takes two hours. If a quality control inspector criticizes the job, some guy is on edge. A lot of times he will say, 'What's wrong with the job?' Well, our responsibility is to write a discrepancy on the crab sheet — a minor discrepancy, that is. If it's a major discrepancy we write it on a tag. And then I usually end up in fights with the foreman, who has to side with his mechanics because he is obligated to a schedule. So we get into fights all the time. It's just a constant thing — especially when you run into opinion-type crabs. When it's a cut-and-dried discrepancy, that's no problem. You just do it over."

The fact sheets on the LM as it finally evolved into what the astronauts came to regard as "a pretty good piece of machinery" were deceptively bald, considering all the trouble encountered in building the vehicle, and considering the fractional changes that had to be made up to the time LM No. 5 was delivered for the Apollo 11 mission. (The LM flown by Stafford and Cernan on Apollo 10 was considered suitable to test out flying in a lunar environment, but not to land on the moon and take off again. Always, as Conrad had said much earlier, there was the weight problem.) With its landing legs extended, LM No. 5 had an overall height of 22 feet 11 inches and a width of 14 feet 1 inch. The diagonal diameter, across the landing gear, was 31 feet. The four legs which were to be explosively extended after the LM separated from the command module had foot pads 37 inches in diameter, and three of the four pads had thin "sensing probes" about five feet long to alert the crew to shut down the descent engine when the probes made contact with the lunar surface. A little platform ("the porch") extended from the forward hatch, leading to a ladder with nine steps; Armstrong and Aldrin would have to cling to the ladder rails and simply drop the last three feet. The ascent stage was 21 inches taller than the descent stage and, at 4,804 pounds, had 321 pounds more dry weight. But the descent stage used considerably more propellant — 18,100 pounds as compared to 5,214 pounds allotted to the ascent engine. Additionally, there were 604 pounds of propellant in the reaction control system for maneuvering purposes. Therefore the LM would have a total weight of just over thirty-three thousand pounds when it separated from the command module, and the ascent stage, fuel and all, less than a third of that when it left the moon. This meant the astronauts had to fly the LM differently, depending on its varying weight; Buzz Aldrin once compared it to the difference between driving an old-model motorcar and a modern sports car. With the descent stage left on the moon, and all its fuel consumed, a "hot hand" could make the lighter remnant of the LM, the ascent stage, behave erratically.

It all looked grotesque: a sixteen-ton craft with eighteen rockets, thirty miles of wiring, eight different radio systems and two kinds of radar — one to track the command module and the other to relate the LM to the surface of the moon. The landing radar was tricky to read, but it was an absolutely essential part of the spacecraft; the computer would not do the job by itself. Extrapolation of data following the flight of Apollo 8 showed computer errors of as much as four thousand feet in calculating where Borman, Anders and Lovell were in relation to the moon. On an actual landing that kind of mistake would result in missing the planned touchdown site by a considerable margin. During the last hundred feet of descent the frail LM would be throttled down to three feet per second, and a "hard" contact might break the LM's legs or tip it over, leaving two astronauts stranded on the moon. As Buzz Aldrin once explained it, the technique was to let the landing radar and the computer have an argument with each other. . . .

« That radar ought to lock on at thirty-five to forty thousand feet. Suppose the computer tells us we're at thirty-two thousand. We give that to the radar and the radar comes back and says the computer is a liar; we're at twenty-eight thousand. The computer goes into a sulk and says it will split the difference; call it thirty thousand. The radar takes that and says no you don't, and besides you're down to twenty-seven now. We can keep narrowing the difference, during the automatic part of the descent, until we get down to about five hundred feet, and at some point around about there we have to take over manually. That computer isn't going to dodge boulders.

There is a kind of manual override from about seven thousand feet on down if we need it — if we see that the computer is going to take us toward an undesirable area, a crater area for instance. We had this system of switching attitude control to an attitude hold mode, switch it away from automatic, and then it would say, 'Okay, you're the boss for changes in attitude, and you're also the boss for changes in descent. I'll hold the rate of descent we've got right now, and if you want it slowed up a little bit you hit this little toggle switch, and each time you hit it, it will change the rate of descent one foot per second.' This disturbed me a little bit, because it seemed to me that what we really wanted was to get the feel of the vehicle but not necessarily take over the descent. The way this program was written we were forced to do both jobs at the same time. So I started talking to some of the MIT people, trying to figure some way to change this. We did change it, so by switching over to attitude hold the vehicle will continue with its programmed descent and bring you on down, controlling the throttle. You still have the option of manual attitude control, but you hope you don't have to go manual too soon. The first time you hit the toggle switch it'll say, 'Okay, you've got it now. You've got it for the rest of the way.' I'm not saying that this is the best way, but it is really heartening to me when occasionally I can look at certain things and come up with something else that nobody has thought about. You can't know how happy this makes a person feel.

The simulators can bring in the effects of radar. A long time ago I made a powered descent in a simulator; I didn't really understand what the displays were; the thing was just bringing me in. Neil was doing a lot of this with Pete Conrad from five hundred feet on down. Well, I had a little available time so I sneaked in and tried my hand. It's not something that you do right the first time. You've got to do it many times before you learn the little tricks. But what I did was bring it on in to about one hundred feet, and you had a few funnies every time. If you moved the vehicle a bit too fast — why, something in the simulator computer would get confused and the displays would become erratic. When this happened I would just hit the abort button and *Wham!* — we started lifting off from the descent trajectory. Then I watched the fuel and when it got down to one percent I abort-staged, dropping away the descent engine, lighting the ascent engine to let it fly me automatically into orbit.

Neil will take it down. He has the controls on his side. I don't have them on my side. There's a throttle on my side, but I don't have this rate of descent control, and I don't have this redesignation capability. Near the end of the descent trajectory there is a point at which the spacecraft is intentionally pitched down, in a position where we can begin to see through the bottom part of the window where the landing site that the computer is looking for actually is. The computer is going to tell us where to look on this grid that's superimposed on the left-hand window, and it will initially say something like 62 degrees — 62 degrees down from straight *out* from the LM. If we look there we should see where the computer is planning on landing us. At about 62 degrees there is going to be horizon, and if we measure the angle between the horizon and where the computer thinks we're going to go, we ought to get a fairly reasonable estimate of the altitude. The alternative is to try to measure the rate of motion, but the LM is not in a very good position to observe, through windows, the motion of landmarks and craters going by underneath. It's going to be in this attitude where the engine is pointing forward, and also, to keep the landing radar pointing down, we have to keep a heads-up position. It's not like flying a glider in, where you have a lot of information available and you are landing in familiar terrain. Certainly we're going to know which craters we're going over as we approach the landing site, but whether we can say yes, I know that one down there, and it must be six thousand feet instead of five thousand — I'm not sure that we'll be able to identify that sort of error. You can never pinpoint all these errors as you study data from previous flights. Something that looks like a new phenomenon could be a fly walking around on the antenna at Goldstone.

But during the actual landing there is a fairly neat divison of labor. Neil will be looking more and more out the window with his hand on that 'stick' — a hand controller on the instrument panel which controls yaw, pitch, and roll. He's not able to look much at the displays on board. This is where we have to have a finely tuned teamwork, so that Neil gets the information he needs to transfer whatever he sees into something meaningful. I'll have to relay this information. If Neil has to take his attention away from looking out the window, look down to the keyboard and then back again, this is wasteful. So I'll be giving him the information, and at the same time I'll be looking at the various spacecraft systems to make sure they're operating the way they should. But here I am looking at five or six different gauges, and I don't think we're going to proceed very far in the first landing without good communications. We've got teams and teams of people looking at *each* gauge, back on earth, and they are getting all this information fed to them. So I'm really confirming what a lot of people are getting. »

For the missions of Apollo 9 and Apollo 10, earthbound simulation of the LM's characteristics sufficed. Apollo 9's task was limited to checking out the LM's flying characteristics in a terrestrial orbit, and Apollo 10's flight plan for the LM stopped fifty thousand feet off the lunar surface — enough to check out some of what Tom Stafford called "those unknowns" — those lunar "perturbations" which had seemed so puzzling after the flight of Apollo 8. Each mission was sensationally successful. But for an actual lunar landing there were other unknowns, particularly: how would the LM "handle" in strange gravity, and in situations which no pilot could know about in advance? After all, nobody had landed on the moon — or even tried to.

Simulation of this problem was not only "an art and a science," as Neil Armstrong had said; it was worse than that. There was no building in the world in which one room, big enough for a training version of the lunar module to fly in, could be converted to one-sixth of earth's gravity, and it was out of the question to fly a LM trainer in a water tank. There could be no question of the LM touching down on the moon exactly as a flying simulator would do the job on earth; yet the rehearsal had to take place on earth. There was that problem of the other five-sixths of gravity: to take off vertically from the earth, a vertical thrust equal to or greater than the total weight was needed. If flight simulation were to serve any purpose, it was necessary to compensate for the remaining five-sixths of weight. Someone had the idea of a gantry, under which a LM-type vehicle could perform maneuvers while suspended on a cable which would permanently relieve it of five-sixths of earth weight. The gantry, four hundred feet long and two hundred sixty feet high, was actually built; so was a primitive model of the LM, consisting of a helicopter cockpit, whatever rockets were lying around and a welded tubular structure. Test flights were carried out, but there was good progress on an even earlier idea, some form of free flight. What was needed was a vehicle not attached to a cable at all but equipped with two different lift systems. The first, a jet engine, would supply a thrust equal to five-sixths of the machine's weight. The second system, a propulsive system, the pilot would control himself. This would consist of a clustered group of rockets, as envisioned for the LM itself.[8] This was the formula which won. In January 1963 NASA ordered two lunar landing research vehicles (the LLRV) from Bell Aerosystems at a cost of a little more than three and one-half million dollars.

The LLRV's experience was not a happy one. The basic problem remained: to simulate LM performance in an earthly environment, and in a literal sense that *was* impossible. But Neil Armstrong flew it. . . .

« I guess we should say first that this machine flies in two methods. One, it flies as if it were flying over the moon, and in the other method it does not. That is, it isn't simulating the lunar environment but rather flying like a vertical landing and takeoff aircraft, on earth. Now you use the latter method in taking off, lifting off the surface, and flying to the point where you want to start the trajectory, where you're simulating the lunar descent from about five hundred feet. The vehicle would fly a good bit higher than that, but when we went very much higher we would not be able to simulate the lunar descent. A higher altitude requires a higher velocity. That is, as you are descending to the lunar surface you are also decelerating, and we pass through a five hundred-foot altitude at about the maximum speed of the LLRV, so we start the program at its maximum speed. This corresponds to an altitude of about five hundred feet. So the initial liftoff and climb — that's somewhat like flying a helicopter. It's a matter of flying to the position where you want to start the trajectory and then — what we call setting up the initial conditions. The initial conditions are those characteristics of the point where you want to start working on your problem, those characteristics of the LM that you can duplicate. We're talking about things like speed, altitude and the attitude of the vehicle. At — let's say — seventy feet per second or fifty mph, and then in an attitude that would be nearly level but pitched up a little bit, you would be slowing down at the start of the problem. You would be in effect, then, at the time you started coping with the problem, in an identical relationship with the landing area — how far away the landing area is, what difference in altitude you have, what speed — with what it would be like when you tried to land the real thing on the moon.

The thing that surprises people on their initial flights in a lunar simulation mode is the tendency of the vehicle to float well beyond where you think it's going to go. It takes a good bit of practice to anticipate the distance necessary for you to slow down. Let's say you're approaching the landing area at thirty mph and you want to stop in a particular spot. Everything you've learned on earth will be wrong. If you try to do something the way you do it on earth, as in a helicopter, you will probably overshoot by a couple of hundred feet. So you have to learn to anticipate and start your braking much earlier so that you will stop where you want to stop. Similarly, if you change your mind — if you come to a hover and then change your mind, and decide you want to fly over fifty feet to the left or fifty feet in front of you, it takes a lot of effort to get yourself moving again. The vehicle appears to be sluggish in its translating ability, so it takes a long time and big angles to pick up a little speed to go over fifty feet. Again, it's a matter of anticipating earlier what your requirements will be. We hope to have a minute and a half or two minutes of fuel essentially in hover when we are landing on the moon, but you can chew that up right fast if you change your mind frequently about where you want to go. Anticipation is the key.

The LLRV rocket engines use hydrogen peroxide as a propellant. Hydrogen peroxide is just water with extra oxygen in it. The propellant doesn't burn; it just decomposes into water in the form of superheated steam and oxygen. It's decomposed by being passed over a silver catalyst in the rocket engine chamber. On damp Houston days, and there are a lot of them because we have high humidity in the area, we get a lot of white steam around the vehicle. It comes from this steam being exhausted in the atmosphere. It makes the training vehicle look even stranger than the real LM — a kind of cross between an old-fashioned Stanley Steamer and a calliope. »

Then there was May 6, 1968, and that was a day Neil Armstrong was in the lunar landing research vehicle, the "flying bedstead," at Ellington Air Force Base, near Houston. . . .

« It was my twenty-first flight in the LLRV. I was flying the terminal portion of a simulated lunar landing profile; I lifted the vehicle off the ground and reached an altitude of about five hundred feet in preparation for making the landing profile. I had been airborne for about five minutes and was down to about two hundred feet when the trouble began. The vehicle began to tilt sharply. Afterward this incident was reported as an explosion, but that was erroneous. It's just that there are all the exhaust products of those rocket engines going off, and since there were a lot of engines firing at once people on the ground thought they were seeing an explosion. They were mistaken.

The first sign of trouble was a decreasing ability to control the vehicle. There was less and less response. The trouble developed rather rapidly, but it was not an abrupt stop. It was a decay in attitude control. You have to have attitude control to point the main engines down, the engines that keep you from falling down to the surface. Without attitude control there is no chance to maintain the orientation and keep upright. The vehicle does have two separate systems for doing this, but in this case both systems failed at their common point, when the high pressure helium was supposed to pressurize the propellant to the rocket engines. So we were losing both systems simultaneously, and that's where you have to give up and get off.

My guess is that I ejected at one hundred feet, plus or minus some. We don't have a way of measuring accurately, even from photographs. Seven months after this happened, Joseph S. Algranti, chief of MSC's aircraft operations office, who had been assigned to the team which investigated my accident, ran into a similar situation and ejected — at an altitude a bit lower than the one at which I had to get off. How far the ejection throws you depends on your attitude at the time you leave and also the upward or downward velocity you have at the time. But if you start from an upright attitude at about a hover, it will take you up about three hundred feet. Both of the ejections I am talking about were close to that, perhaps two hundred fifty feet. The chute ejector is automatic, although there is a manual override. I had always thought that I might be able to match the automatic system, but I found out that when I was reaching for the D ring, the automatic system had already fired.[9]

This particular vehicle was the first LLRV; it had made 197 test flights at Edwards before it was shipped to Texas, and it had made about fifty flights at Ellington before the day of the crash. This day it fell straight down. The ejection system threw me somewhat east of the landing point of the machine, but the wind was from the east. At the time my chute opened I was a bit concerned that I might be drifting down into the fireball, because by now the vehicle had crashed and was burning. I started thinking about slipping the chute, but the wind was strong and I actually missed the flames by several hundred feet.

I got up and walked away after I landed. The only damage to me was that I bit my tongue. »

Eighty-four hours and 28 minutes into the flight, less than two minutes away from AOS at the end of the fifth pass behind the moon. . . . Crew preparing for a nine-hour rest period; tomorrow, Sunday, would be the big day — landing day. . . . MARK thirty seconds. . . . Twenty seconds. Ten seconds. . . .

ALDRIN: Houston, Apollo 11.

HOUSTON (CAPCOM Owen Garriott): Eleven, Houston. Roger. Reading you fine. . . . We have several small items to discuss with you here just before you go to sleep.

COLLINS: Go ahead, over.

HOUSTON (Garriott): Okay, eleven. First of all, on our LM systems checks. Everything went fine. . . . I would like to remind you though, tomorrow you may see an ascent pressure light when you activate the MC and W.

ALDRIN: Roger. Understand that. Thank you.

HOUSTON (Garriott): They've also looked at the results of your landmark tracking. The marks all apparently were very good. [*Eighty-five hours and 17 minutes into the flight, spacecraft altitude 56.1 nautical miles off the lunar surface, velocity 5,367 feet per second, 3,659.2 statute mph. . . .*]

COLLINS: We haven't chlorinated the water yet and we haven't changed the lithium hydroxide. We're just still finishing up dinner. [*Eighty-five hours and 40 minutes. . . . One minute and 20 seconds away from loss of signal, over the hill for the sixth time. . . . silence. . . .*]

HOUSTON (McLeaish): MARK one minute from predicted acquisition of signal. We should be acquiring.

COLLINS: Houston, Apollo 11. Over.

HOUSTON (Garriott): Eleven, Houston. Loud and clear here. . . . We believe we've tracked down the reacquisition problem we had on the previous rev. It looks like it was a receiver power supply here on the ground and no problems in the spacecraft at all.

ARMSTRONG: Okay, glad to hear.

HOUSTON (Garriott): Eleven, that really winds things up as far as we're concerned on the ground for the evening. We're ready to go to bed and get a little sleep. Over.

ALDRIN: Yeah, we're about to join you. [*Eighty-six hours and 53 minutes, no more conversation tonight. . . .*]

There would be no extraneous visitors in the astronauts' homes this night; the next forty-eight hours were too critical. Everyone in the Armstrong home had been invited next door for dinner, but Jan was not going. She wanted to be alone. As in the case of Andy Aldrin, Ricky and Mark were having some trouble relating all this tension to what their father was doing. Mark had been at a friend's house all day, and he asked permission to sleep that night with his friend in a backyard tent. Permission was granted. "You mean I can?" he asked.

We go into space because whatever mankind must undertake, free men must fully share.

JOHN F. KENNEDY

What, indeed, is man? Having witnessed this symbolic event, how shall we interpret our lives and our accomplishments?

THE REVEREND DEAN WOODRUFF

10

"An expression of man's self-determination"

It was a few minutes past midnight in Houston when Apollo 11 went "over the hill" for the seventh time; that was about the time Neil Armstrong and Buzz Aldrin went to sleep. Mike Collins stayed awake a little longer, finally dozing off about the time Apollo 11 went around the back side of the moon for the eighth time. *Should have acquisition again at 90 hours and 25 minutes through Honeysuckle Creek, Australia. . . . Three hours and 57 minutes remaining in the sleep period. Crew heart rates running in the forties. Apollo 11 in a lunar orbit with a pericynthion of 55 nautical miles, apocynthion 64.4 nautical miles. Still just right. Velocity 5,363 feet per second, 3,656.5 statute mph. . . . Ninety-one hours and 36 minutes, less than one minute remaining until loss of signal as Apollo 11 goes onto the lunar far side for the eighth time. Two hours remaining in the crew rest period. Four o'clock in the morning Houston time, Sunday, July 20. . . . Spacecraft systems looking good. Forty-five minutes and 28 seconds to next acquisition of signal. All asleep at 91 hours and 37 minutes GET. . . .*

All asleep — the astronauts and their wives, yes; but not quite all. There was Mission Control, where it always looked so calm on the television screen. At four rows of lighted consoles, casually attired young men sat quietly, their eyes flicking back and forth across instruments and dials, pausing to study the changing numbers on a centrally mounted video screen, jotting down quick notes, reaching out to press a button or two. Most of them were around thirty-five years of age. They were all imperturbable — cool, unruffled and definitely not nervous. Those who might have become nervous in a tight situation — the shaky souls — had been weeded out and had not made it to this room. The ones who had made it had to make important decisions, not just decisions that would mean a few hundred million dollars blown one way or another, but decisions that could mean life or death for three men in the spacecraft of Apollo 11. Christopher Columbus Kraft Jr., director of Flight Operations, had built this team, then moved upstairs to let his key people take command and perspire in the limelight of public attention. He said, "I'm like an orchestra conductor. I don't write the music. I just make sure it comes out right."

Between two and six o'clock in the morning of July 20, Houston time, there was not much to do except monitor the consoles and "make sure it comes out right." Occasionally one of the men, or two or three at once, would leave their positions to gather around the small console in the middle for a quiet conference with the man they called FLIGHT — in this case Flight Director Glynn Lunney, head of the "black team." (Head Flight Director Cliff Charlesworth led the "green team," Gene Kranz the "white team," Milt Windler the "maroon team.") With Lunney were BOOSTER (Booster Systems Engineer Don H. Townsend); RETRO (Retrofire Officer Charles F. Dieterich); FIDO (Flight Dynamics Officers Jay H. Greene and Philip C. Shaffer, monitoring and evaluating flight trajectories, verifying maneuver times and results); GUIDO (Guidance Officers William E. Fenner and Stephen G. Bales); SURGEON (Flight Surgeon Kenneth Beers, counting the white dots that blipped across his screen from left to right; they represented the heart and respiration rates of the astronauts, and he had to be ready at all times to diagnose an illness or prescribe a medication); EECOM (Electrical and Communications Systems Officer Seymour Liebergot, assisted by Thomas L. Hatchett from North American Rockwell and James R. Fucci from Philco); TELCOM (Telemetry and Communications Officer Jack Knight Jr., concentrating on the lunar module); OPS (Operations and Procedures Officer Granvil A. Pennington); CAPCOM (Spacecraft Communicator Ron Evans); NETWORK (Network Controller Ernest L. Randall, in liaison with the worldwide tracking network, passing on instructions and talking to his friends in Australia and Spain); AFD (Assistant Flight Directors Harold M. Draughon and David E. Nicholson); PAO (Public Affairs Officer Robert White, known as "Terry"), and about a dozen others — perhaps twenty-five in all.

But Mission Control, for all its complexity and specialization, was only the tip of the iceberg, the central strand of a web that stretched across the nation. There was a staff support room down the hall, manned around the clock by a dozen or more experts. The major contractors (North American Rockwell for the command and service module, Grumman for the lunar module) also were set up in support rooms, maintaining their own "hot lines" to major subcontractors and keeping updated lists of the whereabouts of some forty thousand key scientists and engineers associated with Apollo. Suppose there was a "glitch" involving a component deep inside the command and service module? The problem would be relayed instantly to North American Rockwell's staff support room, and possibly the hot line to Downey, California, would hum. It might even be necessary to find the man who had designed the balky component, get him on the telephone and get his ideas on what to do. Few Apollo problems had had to be taken that far, but — as at Cape Kennedy on the morning of July 16 — there did seem to be a special pressure-cooker atmosphere in Mission Control as the time for the lunar touchdown drew near. "I've never seen things so tense around here," George Low said.

HOUSTON (PAO Terry White): This is Apollo Control, 93 hours 29 minutes Ground Elapsed Time. Some five minutes away from loss of signal on this revolution. . . . Present orbital velocity around the moon 5,370 feet per second [3,661.3 statute mph]. . . . Standing by for Ron Evans's big moment as he makes his call to the spacecraft. . . . Here we go.

CAPCOM (Evans): Apollo 11. Apollo 11. Good morning from the black team.

COLLINS: Good morning, Houston.

CAPCOM (Evans): Good morning. Got about two minutes to LOS here, Mike.

COLLINS: Oh my, you guys wake up early.

CAPCOM (Evans): Yes, you're about two minutes early on the wake-up. Looks like you were really sawing them away.

COLLINS: You're right.

CAPCOM (Evans): Eleven, Houston. For planning purposes, you can go ahead and take the monocular into the LM with you.

COLLINS: Okay. I'll tell them. How are all the CSM systems working?

CAPCOM (Evans): Eleven, Houston. Looks like the command module's in good shape. Black team has been watching it real closely for you.

COLLINS: We sure appreciate that because I sure haven't.

CAPCOM (Evans): Apollo 11. Thirty seconds. AOS will be 94 plus 21. [*Loss of signal at 93 hours and 36 minutes GET. . . . Apollo 11 over the hill for the ninth time. . . . Ninety-four hours and 21 minutes, standing by for acquisition of signal. . . . AOS is confirmed. . . .*]

HOUSTON (White): The descent orbit insertion burn is now scheduled at 101 hours 36 minutes 13.5 seconds. Powered descent initiation at 102 hours 32 minutes 5.1 seconds. . . . Still in the middle of their breakfast period. . . . Members of the white team of flight controllers headed up by Eugene Kranz are drifting into the control room now to relieve the night watch.

ALDRIN: Houston, Apollo 11. We just had a very good view of the landing site. We can pick out almost all of the features we've identified previously.

CAPCOM (Evans): Eleven, Houston. Roger, sounds real fine. And eleven, I have your maneuver PAD and consumables update whenever you want them.

ALDRIN: Stand by a little, please. . . .

COLLINS: Ready to copy.

CAPCOM (Evans): SPS G&N 36 639 your NOUN 48 minus 072 plus 051. . . .

COLLINS: Okay, thank you.

CAPCOM (Evans): Apollo 11, Houston. I have your baseline altitude update now if Buzz is ready to copy.

COLLINS: Go ahead.

CAPCOM (Evans): Roger. Alpha One is 500, that's 500 feet above the landing site.

COLLINS: Houston, Apollo 11. Our crew status report for sleep: commander 5.5 [hours], command module pilot 6.0, lunar module pilot 5.0. Over.

CAPCOM (Evans): The Black Bugle just arrived with some morning news briefs if you're ready. . . . Church services around the world today are mentioning Apollo 11 in their prayers. President Nixon's worship service at the White House is also dedicated to the mission, and our fellow astronaut Frank Borman is still in there pitching and will read the passage from Genesis which was read on Apollo 8 last Christmas. The Cabinet and members of Congress, with emphasis on the Senate and House space committees, have been invited along with a number of other guests. Buzz, your son Andy got a tour of MSC yesterday. Your Uncle Bob Moon accompanied him on his visit which included the LRL [lunar receiving laboratory, in which the astronauts were to be quarantined following their return from the moon].

ALDRIN: Thank you.

CAPCOM (Evans): Roger. Among the large headlines concerning Apollo, this morning there's one asking that you watch for a lovely girl with a big rabbit. An ancient legend says that a beautiful Chinese girl called Chango has been living there [on the moon] for four thousand years. It seems she was banished to the moon because she stole the pill of immortality from her husband. You might also look for her companion, a large Chinese rabbit, who is easy to spot since he is always standing on his hind feet in the shade of a cinnamon tree. The name of the rabbit is not recorded.

ALDRIN: Okay, we'll keep a close eye for the bunny girl.

CAPCOM (Evans): While you're on your way back Tuesday night the American and National League all-stars will be playing ball in Washington. Mel Stottlemyre of the Yankees is expected to be the American League's first pitcher. . . . Even though research has certainly paid off in the space program, research doesn't always pay off, it seems. The Woodstream Corporation, parent company of the Animal Trap Company of America, which has made more than a billion wooden spring mousetraps, reports that it built a better mousetrap but the world didn't beat a door to its path — didn't beat a path to its door. As a matter of fact, the company had to go back to the old-fashioned kind. They said, "We should have spent more time researching housewives and less time researching mice." And the Black Bugle is all completed for the morning.

ARMSTRONG: Thank you very much. We appreciate the news.

Eight o'clock, Sunday morning, Houston time, lunar landing a little more than seven hours away. . . . Ninety-five hours and 25 minutes into the flight, three minutes to loss of signal, AOS at 96 hours 20 minutes. . . . Loss of signal, behind the moon for the tenth time. . . .

About a month before Apollo 11 was due to fly, Buzz Aldrin asked his minister, the Reverend Dean Woodruff of Webster Presbyterian Church, "to research what he had available, to come up with some symbol which meant a little bit more than what most people might be thinking of. What we were looking for — what I was interested in — was something that transcended modern times. Somehow we weren't quite able to work this into something that would have an appeal." Mr. Woodruff did prepare a paper which he called "The Myth of Apollo 11: The Effects of the Lunar Landing on the Mythic Dimension of Man." He concluded:

> Since myth and symbols are so pervasive in the psyche of man, a fairly dramatic and meaningful event is required to have any effect on this dimension of man. The Apollo event will be the kind of occurrence that will reach down to this level in man because this flight will change our view of the world. After Copernicus man could no longer see himself as the center of the world and had to adjust to the new knowledge that things had not been designed especially for him and that he was no longer the center of the universe. And after Darwin man had to make a similar adjustment in understanding himself as a part of nature rather than specially created. Events with such an impact do have an effect on man's self-understanding.
>
> One of the most ancient motifs found in mythology is called "the magic flight." Professor Eliade comments:[1] ". . . It is found everywhere, and in the most archaic of cultural strata . . . the longing to break the ties that hold him in bondage to the earth is not a result of cosmic pressures or of economic insecurity — it is constitutive of man. . . . Such a desire to free himself from his limitation, which he feels to be a kind of degradation . . . must be ranked among the specific marks of man."
>
> Apollo 11 will have the effect of saying to man that he can stand outside his world and view it as a whole. Science, as the achievement of man, has created a worldwide technical civilization and, as yet, has not given birth to any cultural symbols by which man can live. The Apollo event comes at a time when we need a symbol, and need to tap a myth that will graphically express the unending journey outward. Perhaps when those pioneers step on another planet and view the earth from a physically transcendent stance, we can sense its symbolism and feel a new breath of freedom from our current cultural claustrophobia and be awakened once again to the mythic dimension of man.

Nearly nine o'clock in the morning, Houston time, Sunday morning, July 20. . . . Ninety-six hours and 19 minutes into the mission, less than one minute from reacquisition of signal. . . . Buzz Aldrin in the LM beginning the LM power up and checkout, soon to be joined by Neil Armstrong. . . . Stand by for the call to the crew. . . .

HOUSTON (Duke): Hello, Columbia, this is Houston. Do you read? Over. . . . Eagle, did you call me? Over.

EAGLE (Aldrin): Roger, how do you read? Over.

HOUSTON (Duke): Roger . . . a lot of noise on the loop. We think it's coming in from Columbia, but we can't tell. We're unable to raise voice with him. Would he please go to high gain? Over.

EAGLE (Aldrin): Okay. I'll have him go to high gain. . . . I'm up to the point where I turn on the IMU.

HOUSTON (Duke): Roger, stand by. Did you get your high gain to work?

COLUMBIA (Collins): Houston, Columbia. Reading you loud and clear. How me?

HOUSTON (Duke): Roger, about three by [medium clarity and volume]. Mike, we've got a lot of noise in the background. It's clearing up now. Eagle, Houston. Do you read? Over.

EAGLE (Aldrin): Houston, Eagle. About four by four, go ahead.

COLUMBIA (Collins): Houston, Columbia, we have Poo and ACCEPT, and how are you reading me now?

HOUSTON (Duke): Roger, understand. We have Poo and ACCEPT. You're about three by in — on the voice, Mike, over.

COLUMBIA (Collins): Okay, you're coming in loud and clear, and I'm configured for normal voice. If you've got any switch changes, let me know.

HOUSTON (Duke): Roger, we've got the noise somewhere in the system down here, I think. We're working on it. And I've got a 130 landmark update for you, and also a DAP load whenever you're ready to copy. Over. [*"We copy" had better carrying power than "We understand."* . . .]

COLUMBIA (Collins): Stand by. . . . Go ahead.

HOUSTON (Duke): Roger, Mike. Coming at you with the 130, P one is niner 8 37 35, P2 niner 8 42 44, 4 miles north, over. [*"I get my kicks out of those niners," Joan Aldrin had said; "niner" had better carrying power than "nine."* . . .]

HOUSTON (Duke): Columbia, Houston. Did you hit the command reset around after LOS on the last pass? Over.

COLUMBIA (Collins): That's affirmative. When we were having difficulty getting you, Charlie, I pushed the command reset to make sure I had control of high gain.

HOUSTON (Duke): Roger. Thank you much. We're in good shape now. Over.

EAGLE (Aldrin): Houston, Eagle. Can you tell me if you're picking up biomed [medical data] on the commander now? Over.

HOUSTON (Duke): Eagle, Houston. We're not getting any biomed on the commander now. . . . Eagle, Houston. We got the biomed on the commander now. Over.

EAGLE (Aldrin): Very good. Thank you. . . . Houston, Eagle. We're ready for an E-memory dump if you are. Over. [*Armstrong now in the lunar module. . . . LM activation and checkout going well, somewhat ahead of schedule. . . .*]

HOUSTON (Duke): Eagle, Houston. Could you give us a hack on the time that you switched to LM power and also verify that we're on glycol pump one? Over.

EAGLE (Armstrong): This is Eagle. We're on pump one, stand by for the switchover time. . . . The switch time to LM power is 95:54:00 [GET].

HOUSTON (Duke): Roger, copy, Neil. Is Buzz back in the Columbia now? Over.

EAGLE (Armstrong): Yes, he is. [*Thirty-three minutes before loss of signal, Aldrin back in the command module to don his pressure garment before rejoining Armstrong in Eagle. . . .*]

EAGLE (Armstrong): Did you get the time? We're 97:14:20.

HOUSTON (Duke): Roger, copy. . . . Columbia and Eagle, LOS for both spacecraft 97:32, AOS 98:18. . . . Eagle, Houston. We have your gyro torquing angles if you're ready to copy. . . .

EAGLE (Aldrin): Houston, Eagle. Lunar module pilot. How do you read? Over. [*Aldrin now back in Eagle. . . .*]

HOUSTON (Duke): Roger, five by five, Buzz [perfect]. How me? Over.

EAGLE (Aldrin): Oh, loud and clear. I'm going to be going through an ascent battery check. You want to check my biomed briefly? Over.

HOUSTON (Duke): Eagle, Houston. We got a good biomed on you, Buzz. . . . When we go LOS, we'd like you to go OFF on the biomed. [*Four minutes to loss of signal. . . .*]

COLUMBIA (Collins): The capture latch is in the probe engaged in the drogue. Would you like to check them from your side?

EAGLE (Aldrin): All right. Stand by . . .

EAGLE (Armstrong): Mike, the capture latches look good. [*Thirty seconds to loss of signal, both spacecraft looking good as they go over the hill for the eleventh time. . . . Loss of signal. . . . Another forty-five minutes of silence. . . .*]

HOUSTON (Ward): This is Apollo Control at 98 hours 16 minutes. We are now less than two minutes from reacquiring the spacecraft. At this time Armstrong and Aldrin should be completing pressure checks on their space suits. Coming up in this revolution, they will be running checks on the guidance platform of their LM guidance system. They will also be running checks on the reaction control system thrusters and their descent propulsion system, as well as the rendezvous radar. We will also be giving them the go/no go for undocking in the following revolution. All systems performing well. . . . Now about forty-five seconds from reacquiring. . . . CAPCOM Charlie Duke putting in a call to the crew. . . .

COLUMBIA (Collins): Houston, Columbia. Down-voice backup, do you read?

HOUSTON (Duke): Roger, we read you. Columbia, did you call, over?

COLUMBIA (Collins): Affirmative. Calling you on down-voice backup. How do you read me?

HOUSTON (Duke): Roger, better, Mike. . . . We're satisfied with this COMM configuration. Let's stay with where we are. Over.

Eleven in the morning, Houston time; and it was Sunday. . . .

Jan Armstrong spent the morning in her El Lago home, just waiting for the coming afternoon. In Nassau Bay the Collins daughters, Kate and Ann, served their mother breakfast in bed; their father had always brought Pat her breakfast on Sunday when he was home. Then Pat Collins, the three children and Pat's sister, Ellie Golden, attended the ten-thirty Mass at St. Paul's Roman Catholic Church.

The service had begun an hour earlier at the Webster Presbyterian Church, of which Buzz Aldrin was an elder. Joan Aldrin, along with Michael, Jan, Andy and Bob and Audrey Moon, arrived in a NASA station wagon driven by Bill Der Bing. Microphones were thrust into Joan's face as she entered the church. "Let's do this later," she suggested. The church was attractive, light and airy — rather stark and simple. It was also jammed tight with people; ushers were setting out folding chairs to accommodate the overflow. A girl said, to nobody in particular, "I've never seen it so full. I should have stayed home and watched it on television. I just didn't think." The atmosphere was charged with the same kind of emotion which George Low had noted around the consoles of Mission Control. There was tension in the face and in the manner of the Reverend Dean Woodruff; this was not a day for a long sermon, but he had given considerable thought to the form of the service. The scripture reading, in which the congregation joined, was printed on a folding sheet, white words on a black background which showed color pictures (taken by Apollo 8) of the earth as seen from just above the lunar surface. Then there were the words of the opening hymn, No. 84 in the Presbyterian hymnal:

I sing the mighty power of God,
That made the mountains rise;
That spread the flowing seas abroad,
And built the lofty skies.

I sing the Wisdom that ordained
The sun to rule the day;
The moon shines full at His command,
And all the stars obey.

The text of the sermon took up less than three pages of double-spaced typescript, but Mr. Woodruff had gone over every word many times:

Today we witness the epitome of the creative ability of man — and we, here in this place, are not only witnesses but also unique participants. How is it we come to this place at this time to be a part of this event?

What happens today is an expression of man's ability — of man's self-determination. It is the molding of knowledge and theory; it is the channeling of human resources in solving problems; it is dreaming dreams and having visions; it is the concretizing of man's potential. . . .

Self-determination is our way of surviving in the mundane and through the spectacular; it is our existence — our fight with failure, our impetus toward achievement, our struggles with demonic forces; it is the grit of man.

Self-fulfillment is "oughtness" and "giveness" — what we "ought to be" and "what it is given us to be." Fulfillment is the heart and soul of determination. Fulfillment is the meaning and purpose of self-determination.

When these two are put together perfectly we are what we are meant to be. This is what Nietzsche developed in his idea of the "superman." It is not a new biological species but a new kind of man who realizes his capacity for self-transcendence and self-fulfillment. This is what Nietzsche meant when he spoke through Zarathustra: "Bless the cup which is about to overflow, so that the water, golden flowing out of it, may carry everywhere the reflection of thy rapture. Lo! this cup is about to empty itself again, and Zarathustra will once more become a man."

Apollo 11 is the sharply focused symbol of man's power to accomplish — of self-determination. Today Armstrong, Aldrin and Collins will take us to the threshold of the possibility of self-fulfillment. They, as "representative man," will implicitly ask the question, "What is man . . . thou hast made him little less than God, and dost crown him with glory and honor."

As I have said before to this congregation, since World War II we are in the advent of the modern worldwide civilization that is based upon science — for the language and technique of science is the same in every country. If we are, in fact, in the midst of this new worldwide civilization, then the first symbolic event of that civilization is "the bomb." The second symbolic event of that civilization is Apollo 11, the first and most imaginative nondestructive event of a new civilization.

From the back of the church a young soprano sang the anthem, "Let Us Break Bread Together." The service concluded with Communion. As Mr. Woodruff broke the loaf of bread and held it up for view, he pointed out that the loaf was not whole; he did not say what had happened to the missing piece, but the congregation understood that, symbolically, it had gone with Buzz.

There was no formal benediction; Mr. Woodruff suggested that each member of the congregation say his own benediction later on this day, the day of the lunar landing. He did repeat the prayer he had used at the earlier private Communion: "Even as the door to the universe is being opened by this flight, this crew, this man, deliver us all from pride and arrogance and all unrighteousness which make men the enemies of one another. We dedicate unto Thee, Thy servant and our brother, Edwin Aldrin, to represent the Body of Christ, our nation, and all mankind on the first expedition to another planet."

Joan Aldrin waited in line to shake hands with the minister, then ran into a cluster of photographers and had to be escorted to Bill Der Bing's station wagon. By half-past eleven she was home, and the Archers left to attend Mass. When she came back Rosalind Archer said, "I said my Hail Marys for Buzz. I've said so many for him in the last few months and lit so many candles for him. But they didn't have any candles down there today."

HOUSTON (Duke): Eagle, Houston, could you give us an idea where you are in the activation? Over.

EAGLE (Aldrin): Roger, we're just sitting around waiting for something to do. We need a state vector and a REFSMMAT.

HOUSTON (Duke): Roger, Eagle, we'll have the state vectors and the REFSMMAT for you as soon as we get the high gain. Over. [*Communications noisy. . . . Mike Collins preparing to take marks on a landmark near the prime landing site. . . .*]

EAGLE (Aldrin): Houston, Eagle.

HOUSTON (Duke): Go ahead, Eagle. Over.

EAGLE (Aldrin): Roger. In the first of — on page 47 [of the flight plan], step one, we had the guidance control in PGNCS and mode control PGNCS AUTO and of course the circuit breakers are not in on the thrusters yet. So when we started through the DAP and proceeded on NOUN 46 and we're looking at NOUN 47 now, so we've got an RCS TCA light and we've got four out of the eight other bright colored red flags. I think that this is explained by the fact that we are in PGNCS and AUTO and unable to fire the thrusters. Over.

HOUSTON (Duke): Roger, stand by. . . . You are correct. The lights are there and the flags because we haven't closed the breakers yet. Over.

EAGLE (Armstrong): And Houston, Eagle. Are you going to need the high gain before you can look at our GDA position indicator?

HOUSTON (Duke): Stand by. . . . We can see all the throttle data.

EAGLE (Aldrin): I could give you high bit rate on the OMNI if that would help any.

HOUSTON (Duke): Negative. We have all the throttle data we need. You can stay low bit rate. You can proceed through the throttle test, but do not do the gimbal trim, over. Repeat, do not do the gimbal trim.

COLUMBIA (Collins): Boy, you just can't mess with the check point. . . . Auto optics are pointed just a little bit north of crater 130.

HOUSTON (Duke): And Eagle, Houston. Have you deployed the landing gear yet? Over.

EAGLE (Armstrong): That's affirmative.

COLUMBIA (Collins): Houston, Columbia. I've completed my marks. [*Landmark tracking job completed. . . .*]

EAGLE (Aldrin): Houston, Eagle. I believe I've got you on the high gain antenna now.

HOUSTON (Duke): Columbia, Houston. If you go to REACQ on the high gain we can acquire you now. Over. . . . Eagle, Houston. We got some loads for you. . . . We've got both of you on the high gains now. It sounds great now. Over.

COLUMBIA (Collins): Houston, Columbia. Comment on P22. Worked just fine. The target I marked on is a small crater down inside crater 130 as described by John Young.

COLUMBIA (Collins): Eagle, Columbia. Let me know when you come to your RCS hot fire checks so I can disable my ROLL.

EAGLE (Aldrin): Roger, we're right there now. . . . We'd like wide deadband ATT HOLD [attitude hold].

COLUMBIA (Collins): You got it. Are you going to do your hot fire now?

EAGLE (Aldrin): Roger.

COLUMBIA (Collins): Okay, I'm disabling my ROLL. ROLL is disabled. [*Less than ten minutes before loss of signal. . . . Flight Director Gene Kranz talking to his flight controllers about the go/no go decision for undocking. . . . Lunar landing a little more than three hours away. . . .*]

COLUMBIA (Collins): Would you believe you've got thrusters onboard that vehicle?

EAGLE (Aldrin): Houston, Eagle. The RCS hot fire is complete. How do you observe it? Over.

HOUSTON (Duke): Stand by. Eagle, Houston. The RCS hot fire looks super to us. We're all go. . . . Apollo 11, Houston. We're go for undocking. Over.

COLUMBIA (Collins): Roger. There will be no television of the undocking. . . . I'm busy with other things. [*Loss of signal, around the far side of the moon for the twelfth time, 99 hours and 31 minutes into the mission. . . . One hundred hours and 14 minutes. Less than two minutes from reacquisition of signal. . . .*]

HOUSTON (Ward): When next we hear from them the lunar module should have undocked from the command and service module. We're presently about twenty-five minutes away from the separation burn which will be performed by Mike Collins in the command module. . . . The separation maneuver is scheduled to occur at a Ground Elapsed Time of 100 hours 39 minutes 50 seconds; the descent orbit insertion maneuver, which will be performed on the back side of the moon, set for 101 hours 36 minutes 14 seconds, and the beginning of the powered descent at 102 hours 33 minutes 4 seconds. We're now fifty-five seconds from reacquiring Apollo 11. . . .

HOUSTON (Duke): Hello, Eagle, Houston. We're standing by. Over. . . . Eagle, Houston. We see you on the steerable. Over.

EAGLE (Armstrong): Roger. Eagle's undocked.

HOUSTON (Duke): Roger. How does it look?

EAGLE (Armstrong): The Eagle has wings.

Nassau Bay, Texas

The women of the Webster Presbyterian Church brought a cold lunch for the Aldrin family; the frosting of a cake carried an American flag and the legend "We came in peace for all mankind." Joan Aldrin said, very graciously, "I can't thank you enough — you don't know how much it means to me." Andy Aldrin asked his mother, "Do you think we'll all get to go to the White House this year?" Joan said she hoped so, "unless they ask us to dinner at night." Andy had had the idea that an extra satellite might establish spacecraft communications from behind the moon, and Bob Moon got out a map to explain why the suggestion was not really practical. Waiting for reacquisition of signal, Joan grew increasingly restless. A television commentator reported that another baby had just been named Apollo. "*Another* one of them!" Joan exclaimed, fiddling nervously with the TV set. Still waiting for the signal, Andy asked plaintively, "What can I do?" His mother answered quietly, "There's nothing to do." Then she caught Buzz's voice on the squawk box. "That was a good readback, Buzz," she said, and settled back on the couch once more. The Reverend Dean Woodruff came; suddenly the room was once again full of people and bubbling with conversation. Mike Aldrin burst in from the back porch in great excitement: "Mom, do you have a coffee can?" He had a small water snake clutched in one fist; he and Kurt Henize had just caught it. Mr. Woodruff exclaimed, "That's all we want now!" Joan urged Mike to "get out of here with that," and Mike and Kurt went to look for a coffee can in the kitchen.

EAGLE (Aldrin): Go ahead, Houston. Eagle is ready to copy.

HOUSTON (Duke): Roger, Eagle. Coming at you with a DOI pad. . . . Stand by on your readback. If you are ready to copy the PDI data, I have it for you. Over.

EAGLE (Aldrin): Go ahead.

HOUSTON (Duke): PDI pad. . . . Ready for your readbacks. Over.

COLUMBIA (Collins): Neil, I'm maneuvering in ROLL.

EAGLE (Armstrong): Roger, I see you.

HOUSTON (Duke): Columbia, Houston. How do you read?

COLUMBIA (Collins): I've been reading you loud and clear, Houston. How me?

HOUSTON (Duke): Roger, Mike, five by. On my mark seven minutes to ignition. MARK seven minutes.

EAGLE (Aldrin): I got you. Everything's looking real good.

HOUSTON (Duke): You are looking good for separation. You are go for separation, Columbia. Over.

In Nassau Bay Mary Engle, wife of astronaut Joe Engle, had come into the Collins home to supervise the peanut butter and jelly sandwiches for the children's lunch. Waiting for confirmation that the undocking had been accomplished behind the moon, Pat Collins updated her record of which astronaut's wife had sent or brought which casserole or cake. She also made a note to call her brother David in Boston; David's son, two weeks old, was being christened this afternoon — Michael Collins Finnegan. Then voice contact was made, and the inevitable numbers came pouring out of the squawk box. "What are the numbers for?" Kate Collins asked. "They're updating in case they lose contact," her mother answered. "They'll have the numbers on the pad. This makes me nervous." Checking off times in her copy of the flight plan, Pat was interrupted by young Michael: "Mommy, can we have some candy?" Still concentrating on the flight plan, Pat said, "Yes, but don't forget to brush your teeth, children. You know I'm not checking you very well. . . . That's Charlie Duke speaking. Doesn't he have a nice clear voice?" Pat's sister Ellie replied, "That's because he went to school in Massachusetts" — everyone within earshot had been teasing Ellie Golden about her "vodker and tawnic" Boston accent.[2] They were still reading the numbers, and suddenly Pat Collins shouted to the children: "Hey, listen! Did you hear Daddy's 0073's?" Ellie insisted that Pat have some lunch: "Quiet, child. It's crucial."

COLUMBIA (Collins): We're really stabilized, Neil. I haven't fired a thruster in five minutes. . . . I made a small trim maneuver.

EAGLE (Armstrong): Mike, what's going to be your pitch angle at sep [separation]?

COLUMBIA (Collins): Zero zero seven degrees. . . . Is that close enough for you or do you want it to a couple of decimal places?

EAGLE (Armstrong): No, that's good.

COLUMBIA (Collins): I think you've got a fine looking flying machine there, Eagle, despite the fact you're upside down.

EAGLE (Aldrin): Somebody's upside down. [*One minute to the separation burn. . . .*]

COLUMBIA (Collins): Okay, Eagle. . . . You guys take care.
EAGLE (Armstrong): See you later.

El Lago, Texas

When Columbia and Eagle came around following the twelfth trip "over the hill," Jan Armstrong said it again: "Hot dog!" The next item of business was separation — the burn. Jan tapped the flight plan with her pencil and said, "Come on, LM! Make it a good one!" Ken Danneberg, a Denver oilman who had served with Neil Armstrong in Korea, was there. He asked, "How do they know which numbers go VERB and which go NOUN?"

"They know," Jan said.

When Neil said "See you later," Jan's sister Carolyn Trude said, "It must be burn time."

It was.

COLUMBIA (Collins): Houston, Columbia. My DSKY is reading 4.9, in X, 5.0, make it and EMS 105.4. Over. [*Separation performed as scheduled. . . . Separation about eleven hundred feet at the beginning of descent orbit insertion, coming up on the thirteenth trip over the hill. . . . Lunar landing a little more than two hours away. . . .*]
EAGLE (Armstrong): Going right down U.S. One, Mike.

In El Lago Jan was explaining to Carolyn, "It's a maneuver operation." Bill Anders dropped in, and so did Marilyn Lovell. Anders pointed to a lunar strip map on the bed and said, "This is the famous Mount Marilyn." Jan leaned over the map with Anders and pointed to the center.

"Are they right on this line or not?" she asked. She was pointing at the lunar equatorial line.

"It's my opinion that they are," Anders said. "Maybe a little bit off, but I can't tell you which side they'll be on." *Forty-five minutes from DOI, descent orbit insertion, on the back side of the moon, the two spacecraft about half a mile apart. . . .*

Jan Armstrong had another question: "What are the prime mission rules about touchdown? What can go wrong and they can still go?"

"They wouldn't leap right off immediately," Anders said. "They have a two-minute look around. Then it depends on what went wrong."

"Like things they had backups for?" Jan asked.

"Like things they didn't have backups for," Anders answered. "Like, say, your oxygen system blew up . . ."

Astronaut Ron Evans, one of the CAPCOMs, came in, and Jan asked him, "What can they lose and still land?"

"It gets pretty complicated," Evans said.

HOUSTON (Ward): We're coming up on fifteen minutes now until loss of signal with the lunar module. Flight director Gene Kranz has advised his flight controllers to review all their data, take a good close look at the spacecraft, and in preparation for a go/no go decision on the descent orbit insertion . . .

HOUSTON (Duke): Columbia, Houston. We'll have LOS at 101:28. AOS for you 102:15. Over.

COLUMBIA (Collins): Thank you.

HOUSTON (Duke): Eagle, Houston. You are go for DOI. Over.

EAGLE (Aldrin): Roger, go for DOI. Do you have LOS and AOS times?

HOUSTON (Duke): Roger, for you, LOS at 101:28. AOS 102:15. Over. [*Three minutes to loss of signal, both spacecraft looking good going over the hill for the thirteenth time. . . . Loss of signal. . . . Silence. . . .*]

HOUSTON (Ward): We've had loss of signal, and the spacecraft Eagle has been given a go for descent orbit insertion. That maneuver to occur in 7 minutes 40 seconds. . . . The burn duration 29.8 seconds, and the resulting orbit for the LM will be 57.2 by 8.5 nautical miles. When next we acquire the lunar module, it should be at an altitude of about eighteen nautical miles on its way down to a low point of about fifty thousand feet, from where the powered descent to the lunar surface will begin. [*One hundred one hours and twenty-nine minutes into the flight, six minutes to DOI, lunar landing a little more than an hour away. . . .*]

HOUSTON (Ward): This is Apollo Control at 101 hours 35 minutes. We're now less than one minute from the scheduled time for the descent orbit insertion maneuver performed by the lunar module on the back side of the moon. Of course, we don't have radio contact with the spacecraft. . . . Flight controllers standing around in little groups. Not much that we can do at this point. . . . We're now twenty minutes — or twenty seconds, rather — from ignition. . . . It will be a 29.8 second burn, first at ten percent and then forty percent thrust, of the ninety-eight hundred pound thrust descent propulsion system. We should be burning at this time. . . . We should have a cutoff by this time. We would expect to reacquire the command module first. . . .

Charlie Duke was the Houston CAPCOM waiting for reacquisition of signal at 102 hours GET; it was a little more than half an hour to lunar touchdown. The CAPCOM was really the crew's representative in Mission Control; as Duke said: "We're supposed to know how the crew responds, and whether this is good from an operational standpoint or not. So our inputs into the total decision-making process are of an operational nature. To be a good CAPCOM, you've got to know the procedures used by the crew and also the software — in other words, the operational details of how they're flying the spacecraft and also the flow of information to the crew. That's the software, the program in the computer. Or the *programs* — plural." Duke had known since the flight of Apollo 10 that he would have this shift as CAPCOM on Apollo 11. Tom Stafford had asked him to be the prime CAPCOM of Apollo 10, and had got him; therefore, on that flight, he had helped put together the "timeline" for the checkout of the lunar module in lunar orbit. Because of that experience, and because of the knowledge of lunar rendezvous he had acquired during the Apollo 10 mission, Neil Armstrong asked Duke to be the CAPCOM for the lunar orbit phase in which the Apollo 11 lunar module would be checked out; and the normal cycling of Mission Control crews would then make Duke the CAPCOM at lunar landing time. Neil Armstrong recalled having said, "I want you to do it for me, because you probably have a better knowledge of this than anybody that's not flying." Duke himself recalled ruefully that "I would have liked to say that I was on a crew and wouldn't have time to do it. But I wasn't on a crew." Duke was one of those Annapolis graduated astronauts (like Bill Anders, Tom Stafford, Donn Eisele and Jim Irwin) who had gone into the Air Force and then decided to try for the space program. Duke decided to opt for the Air Force after his second Navy cruise ("I got seasick"). He applied to become an astronaut rather late — in 1965, when it became apparent to him at Edwards Air Force Base that there would be no place for him very soon in the program which was flying the rocket-powered X-15. Somewhat to his surprise, getting into the space program turned out to be easier than getting an assignment to fly the X-15, and he was accepted in the spring of 1966. At the time of Apollo 11 Duke was thirty-three years old and had been an astronaut for three years. He was still waiting for a crew assignment, and he sometimes wondered if he would be the first rookie astronaut to fly at the age of fifty. But on the afternoon of Sunday, July 20, he was — from the standpoint of the Apollo 11 astronauts — just about the most important man on the ground at Mission Control. *Acquisition of signal. . . . There they were, dead on time. . . .*

HOUSTON (Duke): Eagle, Houston. If you read, you're a go for powered descent. Over.

COLLINS: Eagle, this is Columbia. They just gave you a go for powered descent.

HOUSTON (Duke): Columbia, Houston. We've lost them again on the high gain. Would you please — we recommend they yaw right ten degrees and then try for high gain again.

COLLINS: Eagle, you read Columbia?

HOUSTON (Duke): Eagle, Houston. We read you now. You're go for PDI. Over . . . MARK, 3:30 until ignition.

Charlie Duke was beginning to get a little upset. It was an easy afternoon to get upset: "It always happens that when we have the critical revolution or the critical pass, we have lousy communications. It just seems like that's our luck. The data kept dropping out. I said to myself, 'Oh, no, here we go again,' because we had a mission rule that said we needed adequate communications and data from the spacecraft before we would commit to powered descent. We had telemetry data in spurts. It was dropping out along with the voice. The high gain for some reason was dithering around up there, and as the signal strength would fall down we'd lose the telemetry.

"Pete Conrad made a suggestion. He was sitting down in the front row, and he came up to the front of the console and said, 'Hey, how about letting them yaw?' So we let them yaw right and it did help. We were getting just enough data to give all the systems people a warm feeling about — well, the spacecraft's ready to go and the guidance looks good. So we committed to descent, although it was an uneasy feeling to know that you were going into the critical phase of the mission with marginal communications.

"We had good communications at ignition when we really needed them, right at the start of the powered descent. We kept communications for a couple of minutes, and then they dropped out again on us. About that time Neil made his big yaw maneuver of a hundred and eighty degrees, yawing around to place the lunar module in an upright, forward-facing position."

Forty thousand feet above the lunar surface now, about ten minutes to go. . . . Time for the lunar module's landing radar to lock on. . . .

San Diego, California

T. Claude Ryan was seventy-one years old and about to retire as chairman of the Ryan Aeronautical Company. As the lunar module hit the forty thousand-foot mark he felt a special thrill, for the landing radar which had to work, starting right now, had been made by Ryan Aeronautical. He had felt the same thrill a little over forty-two years ago when a primitive high-wing monoplane, built by little Ryan Airlines and piloted by a quietly confident young man named Lindbergh, flew nonstop from New York to Paris. The relative simplicity of that flight by *The Spirit of St. Louis,* risky as it appeared at the time, stood in stark contrast to the complexity of this mission. Mr. Ryan had seen it all and remembered it all. He remembered getting that telegram in 1927: "It asked if we could build a plane capable of flying from New York to Paris, nonstop, with a Wright Whirlwind engine. It was very short, that's about all there was to it. And signed, however, by Robertson Aircraft Company. I know why Lindbergh signed it that way. He was thinking that we would know the name Robertson, which we did, and that we would pay a little more attention to it and not think that it was just some crackpot. And we did.

"I happened to be alone in the office at the time with our sole secretary

— we had thirty-five or forty employees. My partner, Franklin Mahoney, was in town, and I got hold of Don Hall, who was our one and only full-time engineer. Don and I figured out that maybe this could be done." (*In 1961 John F. Kennedy would figure out, more or less by instinct, that something could be done — time for a great new American enterprise. . . .*)

At the time it seemed to T. Claude Ryan "a rather ambitious thing" to build a little two-hundred-h.p. airplane which would carry enough fuel to go farther than any airplane had gone before: "However, it did look possible. Franklin Mahoney came back and we talked about it. He was not quite so enthusiastic about giving a favorable answer as I was. But I thought, what harm could it do to tell him yes, we could do it? So we answered in the affirmative."

In his own book, General Charles A. Lindbergh dated the return telegram from Ryan Airlines February 4, 1927: "CAN BUILD PLANE SIMILAR M ONE BUT LARGER WINGS CAPABLE OF MAKING FLIGHT COST ABOUT SIX THOUSAND WITHOUT MOTOR AND INSTRUMENTS DELIVERY ABOUT THREE MONTHS RYAN AIRLINES." Forty-two years later Mr. Ryan remembered the quotation price as being something like ten thousand dollars: "We gave them a price without the engine, without instruments, just the bare airplane." [3]

Then Lindbergh came out — by train. Mr. Ryan did not meet him: "I had flown up to Los Angeles to try to run down an airplane we had sold and had not collected the money for. Mahoney met him at the Santa Fe station and brought him out here. We had some M-1's under construction. The airplanes were made one at a time. There was no production line. And with this one fuselage, it was to be the B-1, the Brougham, we expected to tap a new market and get into the passenger-carrying field for charter work. This first airplane — we had an order for it from Frank Hawks. He was quite a famous pilot and he wanted to fly the first one. He had a deposit on it. We persuaded Frank to forego delivery and let us use the fuselage to convert over and use for Lindbergh's plane."

Lindbergh wanted the airplane delivered in sixty days, not ninety; he had competition. Other people, mostly better financed than he, were trying to fly the Atlantic. Could Ryan Airlines deliver it? Forty-two years later Mr. Ryan answered that question: "You betcha we did! We talked it over with our people, a pretty compact group of people; almost every workman in the plant was taken into the decision. . . . Of course, Lindbergh added a lot to morale because it got to be a challenge to everybody. He went around the plant all the time. He had such a nice personality, was just such a young boyish fellow. They all fell for him. They liked him very much." Mr. Ryan liked Lindbergh too: "He had a very pleasing personality. Not very forward, a little bit bashful in a way, but he was just another pilot as far as I was concerned. I knew he was qualified and knew what he was talking about. I never once thought that he was going about the thing in a foolhardy way. I thought he had an awful lot of nerve, and that he was a brave man, because many times I was talking to him and we discussed and compared notes as to what the percentage of risk had been in making the flight. I had asked him at the time, 'What do you think the percentage of your chance is of success in making it?' I knew that he had a single engine that could stop. He had the weather, which could be impossible, and all these things that could be beyond his control. He turned the question to me and said, 'What do you think the percentage is?' Well, I made it a little better than I really believed, not to be discouraging. I said, 'I think you've got a seventy-five–twenty-five chance of making it.' He said, 'That's just about what I figured.' So he knowingly took his life in his hands on that basis." Many years later Neil Armstrong would say, with a parallel professionalism: "I'm not going to say there's no risk concerned, personal risk. Certainly there are risks, and they occur periodically, but they're not overwhelming risks, and they are far disproportionate to the gain. They are of no significant consideration with respect to the gainful objectives that we're working for. I suspect that on a risk-gain ratio this project would look very favorable compared to those projects that I've been used to in the past twenty years. It's a fact that most people feel the urge to do what they think they're able to do. They would rather work in the area that they feel comfortable in. We feel that we can do this job. We feel comfortable doing it. It's the kind of thing we like to do and want to do."

Mr. Ryan thought the young Charles Lindbergh an outstandingly good pilot: "He could fly an airplane. He had a perfect feel in an airplane, and he had good judgment. He was a master pilot, really." Then, in 1967, he met the astronaut Neil Armstrong. Until then he had never thought about Lindbergh and the astronauts having anything in common; but Lindbergh and Armstrong: "They had quite a bit in common. They both were just completely running over with enthusiasm about what they were doing. They had great self-confidence. And Neil Armstrong had a magnetic personality. Just marvelous. When he went through our electronics plant and saw the working people making the actual landing radars that were going to be so important in landing on the moon, he shook hands with and smiled at them and those people — they would go the limit to do a good job, personally knowing him. Of course, they didn't know he was going to be *the* man, the first man to put his foot on the moon. But they were full of pride that they actually knew him, and I was too. Forty-two years — and we had the privilege of being aboard in this wonderful scientific industry. We had a lot of dreams then, and we had a lot of confidence, but none of us could have visualized that a man would be flying to the moon. That was science fiction, with emphasis on the word fiction."

But not anymore. . . . That landing radar on the lunar module was Ryan's, and this was for real. . . .

In all types of wave motion there is something called the Doppler effect. Discovered by an Austrian physicist named Christian J. Doppler, it is created by the phenomenon of light waves and the color spectrum. The electromagnetic waves which make up a beam of light are all of a specific wave length; each color relates to a different wave length, and all the wave lengths together make up the spectrum. The same effect exists in terrestrial sound motion. As a sound recedes, it has a longer wave length and a lower pitch; as a sound approaches its wave length is shorter and its pitch higher. What was important in all this to the astronauts was the fact that a color could be changed by the movement of the source of light; in fact the slight change in color was proportional to the speed of an object. Ryan Aeronautical had done some pioneering in Doppler type radar. . . .

"We developed it mainly for navigation equipment," Mr. Ryan recalled. "We developed our Ryan Radar Doppler Navigation Systems which were used on a large number of Navy airplanes and some other planes. The development of this technique, and the acquisition of that rather broad and extensive experience, gave us the capability of developing the Doppler radar control for the lunar landing. There was a little parallel here with that M-1 monoplane which was designed for public transport and which was modified to become *The Spirit of St. Louis.* When we were developing our Doppler radar we had no idea that it was the precursor of the system which would help make possible a landing on the moon. But so often that's the way scientific progress is made. You can't see where it's going to lead and you're often surprised about where it does lead.

"But the romance of space travel still has something in common with the romance of flight in the barnstorming era of the 1920's and some of the revolutionary flights of the 1930's. Despite automation and computerization, the individualism has not gone out of the experience. What the astronauts have to know and what they have to do show that it hasn't. The astronauts are still important. The pilot has to be a trained engineer and scientist himself in order to know all this. Lots of these things are automated, but the pilot is there. He understands what is going on and can take over if necessary; in some operations the automated part is just a standby. There's still an awful lot of romance in it, in the combined human brains of all these people. All these computers are just their servants. As in Lindbergh's case in 1927, it seems to me that among the astronauts there is that feeling of wanting to do what's never been done before — an identical spirit."

"12 02, 12 02 . . ." The voice was Neil Armstrong's. Charlie Duke felt as if he had been kicked in the stomach: "This alarm, the 12 02, is called an executive overflow. The computer is informing the crew that it is too busy and saying 'I'm going to restart the program.' And MIT coined a phrase called 'bail-out.' The computer just bails out and starts over at the top but drops off

the things at the bottom which are not important. But we had had this powered descent simulation in June with a 12 02 alarm due to a hardware failure. The people who had written the script for the simulation had programmed the landing radar to fail. This caused the computer to look at the landing radar for information, but the information wasn't available. So the computer kept saying, 'I want it, I'm just going to keep looking for it.' The crew had no control over eliminating this problem between the computer and the landing radar. After it issued the alarm, the computer stopped doing guidance equations; it just kept looking at the landing radar. So the crew kept getting 12 02, 12 02, 12 02, and they could not get the computer out of this lockup. So we had to abort during the simulation.

"After that failure Gene Kranz did a brilliant thing. He called a meeting of all the flight controllers who were involved in this area, the FIDOs and the GUIDOs, and all the MIT and support room people. We spent all day going over every possible program alarm that we could get during the descent. Some of them were mundane little things that you knew could not possibly happen, but it wasn't possible to get a 12 02 alarm either — or so we had thought. A guy named Jack Garman, who is *the* program expert, worked up a presentation on all the program alarms. As he came up with each alarm he'd say, 'Okay, we *can* get this one, and what are you going to do?' We finally decided that if we had those alarms and they did not recur constantly, we were go. If the computer caught up right away every time it got too busy and then started over, we were go for landing. But if the alarms kept recurring, then the computer couldn't perform its other routines and we would have to abort. And Steve Bales, the guidance officer [GUIDO] on this shift at Mission Control, was the one who had to make the go/no go decision.[4] He was in the trench — the first row of Mission Control. He had been following the descent part of the mission for a couple of years and had developed the displays they were looking at.

"As I recall we got five 12 02 alarms and one 12 01 alarm during the powered descent. [As it turned out, the rendezvous radar, not the landing radar, was overloading the computer.] But the computer always fixed itself. However we were concerned, very concerned, at the time. Suppose we had to lift off a couple of hours after touchdown? The computer is busier during ascent than it is during descent! So here we are with a computer that seems to be saturated during descent and my gosh, we might be asking it to perform a more complicated task during ascent. But the alarms were not recurring often enough, *not quite often enough,* to make Steve Bales order an abort. Thank God we had had that meeting after the simulation abort in June. When I heard Neil say '12 02' for the first time, I tell you my heart hit the floor. But there was Steve Bales coming back with the answer instantaneously — we were a go flight. [Bales was "believing" the landing radar.]

And Neil was doing a beautiful job of manually directing the landing. I think he's the greatest pilot in the whole world."

Neil and Buzz Aldrin were both rather busy while this was going on. Armstrong said later . . .

« Supposedly, in this time period, the crew member is supposed to look out and see the landing area, and if any small changes where the automatic system is taking him are required, he initiates those himself manually by putting control inputs into the stick. This was the area where we failed; we got tied up with computer alarms and were obliged to keep our heads inside the cockpit to assure ourselves that we could continue flying safely. So all those good pictures Tom Stafford took for me on Apollo 10, in order to pick out where I was going and to know precisely where I was, were to no avail. I just didn't get a chance to look out the window. In fact, when the problems were less important and I did get a chance to look out about three thousand feet, we had already passed most of the landmarks I had memorized.

When Mount Marilyn went by, before the alarms began, and our ignition occurred at the proper time, we knew that we were going to be approximately right; that is, we weren't going to land on the wrong side of the moon or something like that. The ignition was smooth; the engine's start-up power is only ten percent of thrust, somewhat like an idling motor engine. You can neither feel nor hear that, but you can observe on the instruments that the engine appears to be operating properly. After twenty-six seconds the engine begins to operate at full throttle, and there's no question at that point that you do have a good engine. When we made our final downrange position check, over Maskelyne W, that crater Tom Stafford called 'a big rascal,' we were quite certain that we would land a little long. The old rule is when in doubt, land long. Had we been able to look out the window at five or ten thousand feet I think we would have been able to identify our location more specifically. But then there were those alarms, and there was some difficulty in getting them cleared. But we were able to continue because we had memorized the descent reasonably well, and we could go on the instrument readings even though the computer information was not being displayed to us on the lunar module's DSKY.

When I was finally able to look out we were so low that we couldn't see far enough to identify any significant landmarks. There was a large, impressive crater which turned out to be West Crater, but we couldn't be sure of that at the time. As we neared the surface we considered landing short of that crater, and that seemed to be where the automatic system was taking us. But as we dropped below a thousand feet, it was quite obvious that the system was attempting to land in an undesirable area in a boulder field surrounding the crater. I was surprised by the size of those boulders; some of them were as big as small motorcars. And it seemed at the time that we were coming up on them pretty fast; of course the clock runs at about triple speed in such a situation.

I was tempted to land, but my better judgment took over. We pitched over to a level attitude which would allow us to maintain our horizontal velocity and just skim along over the top of the boulder field. That is, we pitched over to standing straight up. Then the automatic throttle was still giving us a descent rate that was too high, and it was going to get us down to an altitude where we would be unable to look out ahead far enough. That was when I took command of the throttle to fly the LM manually the rest of the way. I was being absolutely adamant about my right to be wishy-washy about where I was going to land, and the only way I could buy time was to slow down the descent rate.

There are three kinds of throttle control on the LM; I chose the semiautomatic version in which I controlled the attitude and the horizontal velocity and let my commands, in conjunction with computer commands, operate the throttle. I changed my mind a couple of times again, looking for a parking place. Something would look good, and then as we got closer it really wasn't good. Finally we found an area ringed on one side by fairly good sized craters and on the other side by a boulder field. It was not a particularly big area, only a couple of hundred square feet, about the size of a big house lot. But it looked satisfactory. And I was quite concerned about the fuel level, although we apparently had a little bit more than our gauge had indicated. It's always nice to have a gallon left when you read empty. But we had to get on the surface very soon or fire the ascent engine and abort. »

Neil Armstrong had noted, as Eagle went by Maskelyne W, that the "rascal" crater was passing underneath two seconds early: "If you applied that same two seconds to the point where ignition took place, when the velocity had been on the order of a mile per second — why, then this would tend to put us about two miles long. Anyway we had had good indications from visual observations, up to three and one-half minutes from ignition, that we were going to be a little bit long. Then we lengthened our trajectory to miss that West Crater, the one we were headed for, and that made us further long. We actually landed about four miles long."

But they landed. . . . Buzz Aldrin's voice came through: *Contact light. . . . Engine stop. . . . Engine arm, off. . . .* Then Neil Armstrong: *Tranquility Base here. . . .*

Tranquility Base! That was not in the flight plan; nobody in Mission Control had known that Neil Armstrong would call it that, although the expression was logical enough. What was more important was the fact that Eagle had landed. Inside Eagle Buzz Aldrin, the man Neil Armstrong once called "my most competent critic," stuck out his gloved hand and shook — hard — with his commander. *Yes,* Neil Armstrong conceded, *there certainly was a sense of relief there. . . .*

. . . And on the ground as well. Among those who had gone unnoticed by Douglas Ward at Mission Control was Dr. John Houbolt, who had argued through the lunar orbit concept which had made it necessary to create the LM. The quiet at Mission Control reminded him of a well-ordered public library. He thought, "By golly, the world ought to stop right at this moment." Then he wondered how many people were not watching at all; it was Sunday, and how many people were out playing golf? The method of getting to the moon had been Dr. Houbolt's idea, but he still felt a sense of incredulity; it seemed unbelievable that this was actually taking place, "within the time we had our sights set on it." He had complete faith that the LM's legs would not sink too deeply into the lunar surface, as some had feared; he was convinced that the subsurface would be hard. (It was.)

Charlie Duke, who had been agonizing over those 12 02 alarms, also found it hard to comprehend: "When I heard Buzz say 'Engine stop' it was hard for me to believe that they were there. Later Chris Kraft told me, 'Boy, Charlie, I thought we were gone when they had those things' [the 12 02 alarms]. On my way home I was saying, 'Well, they've actually done it.' But it sort of boggled the mind to think that the thing had been accomplished. There had been no time to react."

In North Amityville, Long Island, Herman Clark, the quality control inspector who had checked out the ascent engine of Apollo 11's lunar module, found that his own emotional reaction had just begun. Armstrong and Aldrin were on the moon, but the engine he had checked out still had to get them off. And if it failed. . . .

But man had landed on the moon. It had not been easy. All those problems during the descent; all those 12 02 alarms that had nearly driven Charlie Duke and Steve Bales out of their minds; could they have been managed had not men like Neil Armstrong and Buzz Aldrin, with minds of their own, been aboard the lunar module — two trained, intelligent, thinking human beings? The question brought up the whole scientist-engineer argument, the manned *vs.* unmanned argument. Nearly four months later Robert Jastrow, director of the Institute for Space Studies of NASA's Goddard Space Flight Center, addressed himself to this question and came out with a tough answer. With the capability now achieved, he argued, it was cheaper, not dearer, to carry out scientific research on the moon by man rather than with robot instruments. And, "The short history of space flight is filled with examples of manned missions that would have failed without men on board

— and of unmanned missions that did fail for want of a fix that could easily have been supplied by man."[5]

In short, lunar module No. 5, the lunar module of Apollo 11, had soft-landed successfully on the moon because men like Neil Armstrong and Buzz Aldrin were on board.

This blowing dust became increasingly thicker. It was very much like landing in a fast-moving ground fog.

NEIL ARMSTRONG

Touchdown was far smoother than I could have hoped for. I had just a slight feeling of impact.

EDWIN E. ALDRIN JR.

11

"Don't forget one in the command module"

THE FIRST message from Tranquility Base was perhaps the understatement of the flight. The voice was Neil Armstrong's: "I tell you, we're going to be busy for a minute." When Charlie Duke responded, "You are STAY," he was not yet in a position to say for how long; Eagle had to be checked out, and the men in Mission Control had to be satisfied with the data on their consoles. If a serious malfunction should be detected, either by the men in Eagle or by the men on the ground, an immediate liftoff could be ordered and certainly would be ordered. But the checking and the evaluation had to be done at breakneck speed, because the most suitable abort points were three minutes and twelve minutes after touchdown. In that brief time, "STAYS" would be measured in minutes. After passing the twelve-minute abort point, Eagle would have to remain where it was until Columbia had gone behind the moon and come around again. If an emergency warranted an order to abort and leave the lunar surface prematurely, skipping the longed-for moon walk on this mission, Mike Collins had to know about the abort and be prepared to pick up Eagle, possibly as low as fifty thousand feet off the moon.

EAGLE (Armstrong): Houston, that may have seemed like a very long final phase. The AUTO targeting was taking us right into a football field-sized crater, with a large number of big boulders and rocks for about one or two crater diameters around us.

HOUSTON (Duke): Roger, we copy. It was beautiful from here, Tranquility. Over.

EAGLE (Aldrin): We'll get to details of what's around here, but it looks like a collection of just every variety of shapes, angularities, granularities, every variety of rock you could find. The colors vary pretty much depending on how you're looking relative to the zero phase point. There doesn't appear to be too much of a general color at all. However, it looks as though some of the rocks and boulders, of which there are quite a few in the near area — it looks as though they are going to have some interesting colors to them. Over.

HOUSTON (Duke): Roger, copy. Sounds good to us, Tranquility. . . . Be advised there are lots of smiling faces in this room, and all over the world.

EAGLE (Aldrin): There are two of them up here.

COLUMBIA (Collins): And don't forget one in the command module.

HOUSTON (Duke): Rog. That was a beautiful job, you guys. Columbia, this is Houston. Say something. They ought to be able to hear you. Over.

COLUMBIA (Collins): Roger, Tranquility Base. It sure sounded great from up here. You guys did a fantastic job.

EAGLE (Armstrong): Thank you. Just keep that orbiting base ready for us up there now. [*Seventeen minutes on the moon. . . . TELCOM in Mission Control reports that the lunar module systems look good. . . . About twenty-six minutes from loss of signal from the command module. . . .*]

EAGLE (Armstrong): The guys who said we wouldn't be able to tell precisely where we are, are the winners today. We were a little busy, worrying about program alarms and things like that. . . . I haven't been able to pick out the things on the horizon as a reference as yet.

HOUSTON (Duke): Roger, Tranquility. No sweat. We'll figure it out. Over.

But there was sweat. Where were they? Armstrong and Aldrin did not know; Collins did not know; Mission Control did not know. In the Armstrong home in El Lago, Texas, Bill Anders and Neil Armstrong's friend Ken Danneberg started a dollar pool on where Eagle had actually landed. Pointing to the lunar map, Anders said, "I'll give you right here, and I'll take over here." Then Jan Armstrong left the room and Neil's brother Dean said, "When we ask him about it later he'll say, 'a piece of cake.'" Ricky Armstrong said, "Usually when you ask him something he just doesn't answer."

What Armstrong and Aldrin could see out the window was not of much help. The lunar horizon was only about four miles away, as seen from inside Eagle, and to a six-foot man standing on the lunar surface it would appear to be less than one and one-half miles away. One thing was certain. The whole area was much more acceptable than the one to which the computer had been taking them.

EAGLE (Armstrong): You might be interested to know that I don't think we noticed any difficulty at all in adapting to one-sixth G.

HOUSTON (Duke): Roger, Tranquility. We copy. Over.

EAGLE (Armstrong): Out of the window is a relatively level plain created with a fairly large number of craters of the five to fifty-foot variety and some ridges, small, twenty or thirty feet high, I would guess, and literally thousands of little one and two-foot craters around the area. We see some angular blocks out several hundred feet in front of us that are probably two feet in size. . . . There is a hill in view, just about on the ground track ahead of us, difficult to estimate but might be half a mile or a mile.

COLUMBIA (Collins): Sounds like it looks a lot better now than it did yesterday at that very low sun angle. It looked rough as a cob then.

HOUSTON (Duke): Columbia, Houston. We have a P22 update for you, if you're ready to copy. Over.

COLUMBIA (Collins): At your service, sir.

HOUSTON (Duke): T1 104:32:18, T2 104:37:28, and that is four miles south.

COLUMBIA (Collins): Do you have any idea where they landed, left or right of center line? Just a little bit long. Is that all we know?

HOUSTON (Duke): Apparently that's about all we can tell.

Not that people were not trying to find out more. Within ten minutes after touchdown, the eager men of the Mapping Sciences Laboratory at the Manned Spacecraft Center were studying enormous enlargements of lunar maps. One of Mapping Sciences' experts, Lew Wade, was at a console in Mission Control. "All indications were that they were long," Wade recalled. "But the various guidance systems didn't agree. PNGCS [the primary navigation, guidance and control system] put them down a bit north of the planned site. AGS [the abort guidance system] put them in the middle, and the first MSFN [manned space flight network] report had them to the south." As more and more information came in from telemetry and analysis, one big map in the laboratory accumulated more and more round dots which were "estimated" landing sites. Finally there were fourteen of them, all within a four to five-mile radius. A lot of them were clustered just west of West Crater, but no one knew which was the real site. It was not critical to the success of the mission, or even to later rendezvous, to know precisely where Eagle was. But the mapping scientists were getting stubborn because their quarry was proving to be so elusive. Besides, Mike Collins in particular, and the terrestrial world in general, were demanding an answer to the simple question: "Where?"

Off in a little science support room at Mission Control, an eight-man geological team headed by Dr. Eugene Shoemaker, who at that time was attached to the United States Geological Survey, went over its own moon maps. Dr. Shoemaker's team had a direct telephone open to another room in another building, where three more men and a woman from the USGS's Center for Astrogeology in Flagstaff, Arizona, were huddled over *their* moon maps. But there was a difference in how the mapping scientists and the geologists went about solving the riddle. While mappers strained their ears for telemetry, the geologists listened for verbal descriptions from the crew. In the first few minutes there was not much to go on. But when Neil Armstrong mentioned "a relatively level plain with a fairly large number of craters," the geologists had a clue. They leaned over their maps. Dr. Shoemaker said later, "Tracking information indicated that Neil had flown past the middle of the landing ellipse and was several kilometers downrange. We knew that he had flown over a blocky-rim crater, that he had seen rays of ejecta as he passed over, and that the landing pattern had been rather like a fish hook. There were maybe six craters which could fit his description, but once we knew he was downrange we narrowed it to two. I believe that it was Marita West, over in building No. 2, who first suggested that the crater Neil had described was West Crater. All of us came to the same conclusion pretty rapidly."

EAGLE (Armstrong): I'd say the color of the local surface is very comparable to what we observed from orbit at this sun angle, about ten degrees sun angle . . . it's pretty much without color. It's gray and it's a very white chalky gray, as you look into the zero phase line, and it's considerably darker gray, more like ashen gray as you look up ninety degrees to the sun. The — some of the surface rocks in close here, that have been fractured or disturbed by the rocket engine plume are coated with this light gray on the outside, but when they've been broken they display a dark, very dark gray interior and it looks like it could be country basalt. *This from Tranquility Base — and the earth's own ocean beds were made of dark, heavy rock, basalt. If our earth were lifeless and waterless its ocean beds would still show dark, like the moon's* maria. . . .

Earlier in the week Coach Vince Lombardi of football's Washington Redskins, a Roman Catholic who attended Mass every morning, stopped practice to show his players a side of himself few of them had ever seen. He asked them to pray for Armstrong, Collins and Aldrin. "This is something transcending what we all do," said Lombardi, who had a reputation for eating football players alive. "I'm conscious that there are a hell of a lot of things more important than football — you know how much I like football — and this is one." At Walter Reed Army Hospital PFC Jim Hess of Scranton, Pennsylvania, aged twenty years, muttered: "Brave men, brave men." He had to mutter because he had been wounded in Vietnam and his jaw was wired shut. The prayers said for the astronauts on the Sunday itself, both publicly and privately, were beyond counting. So was the television audience, although there were some estimates. The Nielsen rating experts figured that 29,410,000 American households had their TV sets on during the lunar landing, and that at least another ten million households were staying up to watch at least part of the moon walk. In Prague the streets were deserted as crowds jammed the beer gardens and the hotel lobbies to watch television. Hasty calculations in Japan indicated that as many as seventy million of Japan's one hundred million population watched the landing, stayed up for the moon walk, or did both. Many Tokyo cabdrivers made it a point to say "Roger" when signing off on their two-way radio communication; the normal signoff was *Ryokai* — literally, "Okay." A Japanese stockbroker complained: "When I called my customers to pass on hot tips, they would invariably say they were busy watching. I ended up watching myself." In Chicago the radio talk host Jack Altman thought the young people calling in seemed "quite jaded by the whole thing. They can't tell the difference between Apollo 11, 10, 9 and 8." Altman's observation had something in common with that of Canon Michael Hamilton of Washington Cathedral. Canon Hamilton thought, "The older people are getting a bigger bang out of all this than the younger ones. The youngsters have grown up with astronauts and space. Older people remember when it was all just a dream." One member of the older generation, Emperor Hirohito of Japan, was called at his villa in the mountains of Nasu by a chamberlain, to be told that the lunar module had landed. Hirohito said, "Yes, I know." He had been watching television in his room. Then he said, in a matter-of-fact way, that he would cancel an outing he had scheduled that morning to collect plant specimens in the woods. He wanted to watch the moon walk that would take place Monday afternoon, Japan time.

But where were they? . . .

HOUSTON (Duke): Columbia, Houston. Two minutes to LOS. You're looking great going over the hill. Over.

COLUMBIA (Collins): Okay, thank you. I'm glad to hear the systems are looking good. Do you have a suggested attitude for me? This one here seems all right.

HOUSTON (Duke): Stand by.

COLUMBIA (Collins): And let me know when it's lunchtime, will you?

HOUSTON (Duke): Say again.

COLUMBIA (Collins): Oh, disregard.

HOUSTON (Duke): Columbia, Houston. You got a good attitude right there. *Loss of signal from the command module. . . . Heart rates for Neil Armstrong during the powered descent to the lunar surface: at the time the burn was initiated, Armstrong's heart rate was 110. At touchdown, he had a heart rate of 156 beats per minute. . . . Heart rate now in the nineties. . . . One hundred three hours and thirty-two minutes into the flight. . . .*

In El Lago, Texas, Jan Armstrong had crushed out her cigarette when Neil came on the squawk box to explain why the landing was long. She sat on a bed as Neil described the colors he could see outside the window of the lunar module; her son Ricky came to sit beside her. He had a little American flag in his hand, and he wanted to go outside and wave it. "Okay," Jan said. "Wave it high." In the Aldrin home in Nassau Bay Joan Aldrin remembered the box of cigars Jan Armstrong had left with her the day of the swimming party and started passing them around. Her father said, "You should keep those." She replied, "No, you're supposed to take one." Joan watched Jan Armstrong's televised press conference and then realized she had to get ready for her own. Young Michael, who disliked being photographed, asked how long it would take; when his mother said about ten minutes, he said, "That's outrageous!"

EAGLE (Aldrin): Houston, Tranquility Base. Does somebody down there have a mike button keyed? It sounds like somebody banging some chairs around in the back room.

HOUSTON (Duke): Roger. . . . Stand by. . . . Tranquility, Houston. We got the MSFN relay in. You're hearing the noise suppression device. We'll try to take it out. . . . It ought to be a little quieter up there now. We disabled the MSFN relay.

EAGLE (Aldrin): Okay, I think the noise has stopped now. Thank you, Charlie. [*Five o'clock in the afternoon, Houston time; Eagle had been on the moon for more than an hour and a half, and the word was still STAY. . . . Columbia on the near side of the moon now, Mike Collins back in communication. . . .*]

COLUMBIA (Collins): You said four miles long. Is that correct, Houston?

HOUSTON (Duke): That's affirmative, Columbia.

EAGLE (Aldrin): Houston, Tranquility Base is ready to go through the power down and terminate the simulated countdown.

HOUSTON (Duke): Roger, stand by. . . . You can start your power down now. Over.

EAGLE (Armstrong): Our recommendation at this point is planning an EVA with your concurrence starting at about eight o'clock this evening, Houston time. That is about three hours from now.

HOUSTON (Duke): Stand by. . . . We will support it. We're go at that time. . . . You guys are getting prime time TV there.

EAGLE (Armstrong): Hope that little TV set works, but we'll see.

HOUSTON (Duke): Columbia. Did — how did Tranquility look down there to you? Over.

COLUMBIA (Collins): Well, the area looked smooth, but I was unable to see him. I just picked out a distinguishable crater nearby and marked on it . . . it looks like a nice area, though.

HOUSTON (Duke): Columbia, Houston. We have your LOS at three minutes — AOS will be 106 [hours] plus 11. Over. [*Over the hill again for Mike Collins, still all alone, traveling at 5,367 feet per second, about thirty-six hundred fifty statute mph. On the moon, the crew preparing to eat lunch, about thirty minutes allotted for that. . . .*]

EAGLE (Aldrin): Houston, Tranquility. Over.

HOUSTON (Garriott): Tranquility, Houston. Go ahead.

EAGLE (Aldrin): Roger. This is the LM pilot. I'd like to take this opportunity to ask every person listening in, whoever and wherever they may be, to pause for a moment and contemplate the events of the past few hours, and to give thanks in his or her own way.

Buzz Aldrin had a special reason for wanting a little quiet; he had something very personal to do now, inside Eagle's cabin. After Mike Collins in Columbia had passed overhead, on the first lunar revolution following Eagle's touchdown, it was apparent that Eagle was going to be on the moon for some time. Aldrin had a small table in front of the abort guidance computer, and he had that little wine chalice, the one Dean Woodruff had given him, in his personal preference kit. He did not have a Bible with him, but he had written down some passages he liked, including one his minister used during the Communion service at Webster Presbyterian Church: *I am the wine and you are the branches. . . . Without Me you can do nothing at all.* . . . Aldrin requested air-to-ground silence and celebrated one of the strangest Communions in the history of the Christian religion. He said later, "I would like to have observed just how the wine poured in that environment, but it wasn't pertinent at that particular time. It wasn't important how it got into the cup; it was important only to get it there. I offered some private prayers, but I found later that thoughts and feelings came into my memory rather than words. I was not so selfish as to include my family, or so spacious as to include the fate of the world. I was thinking more about our particular task, and the challenge, and the opportunity that had been given me. It was my hope that people would keep this whole event in their minds and see, beyond minor details and technical achievements, a deeper meaning behind it all — a challenge, a quest, the human need to do these things."

One hundred five hours and forty-three minutes into the mission, about ten minutes from reacquiring signal with Columbia. . . . Mike Collins, meanwhile, alone with all those switches and buttons. . . . He was alone, but he did not expect to be bored. . . .

« I didn't expect to be bored for two reasons. One, there are always things to look at in there, to make sure that everything is working properly. Two, the only space flight I had ever been on kept me so ruddy busy for three days that I really looked forward to a chance to relax for a little while and look out the window — get some assessment of what it's all about.

And there are those great handfuls of switches. Around four hundred of them, if you include the plungers, the ratchets, the handles and the knobs. The most common is a thing very similar to a light switch, except that a light switch only goes on and off. Most of these switches have a third position — a center position. Center will be "off." Up will turn it on main bus A, which gets all its voltage from one source. If you turn it down it disconnects the switch from A and puts it on main bus B, where it gets all its current. These electrical busses are built-in redundancies — if one doesn't work, you try another. So most of these switches are three-position switches. Some are two-position; either you want the switch on or off, and if you want it on you don't care whether it's main A or main B. I should have counted those switches one day, put it on my list of things to do. . . .

Most of them, you set and leave alone. Some of them you set before you are launched down at the Cape, and you never touch the dratted things unless something goes wrong. Others have special purposes. For example, there are three switches to inflate three float bags that you use to change the vehicle from upside down to right side up in the water. Obviously you can go all the way to the moon and back and never even look at these switches or have to think about them until after splashdown. They don't matter. On the other hand there are some that you use continuously.

For conversation with the DSKY you need a special language. I have a little black book wherein live all my secrets: the VERB key words, the NOUN key words and the numbers. The DSKY also has a 'key release,' a 'proceed' and an 'enter.' During the course of a rendezvous you'd probably push those buttons, if you counted each individual push as a unit, seven or eight hundred times. So this thing gives you a hell of a workout.

And you do have to be careful not to hit one of them by accident. You have to be careful when you move around if for some reason you're inside an inflated suit. You can bash half a dozen of them and you would know it immediately if you were in your underwear. But in that pressure suit you would never know it. You'd never feel it. You're so clumsy, and there's so much force required to move inside the suit that — everything is WHAM! I could bump right into you and maybe I wouldn't know it. I must have my suit on during the rendezvous, because if we should have to make an EVA transfer it would be a hell of a thing to have to go put the suit on in a big hurry. It's easy to go without the helmet and the gloves, because you can put them in a bag nearby and get them on in a hurry if you need them. But the suit takes fifteen minutes or so, depending on how lucky you are, whether your zippers get stuck or anything like that. And there's nobody there in the command module to help. As a matter of fact it's a problem to get the suit on by yourself, because the zipper starts right in the small of your back. You can reach the small of your back when you're sitting in normal clothing, but with that suit on your arm only goes so far and you can't reach the zipper — not very easily, anyway. It's a little like a girl trying to dress for a formal dance in a telephone booth. »

On earlier manned Apollo flights the command module pilot had occupied the center couch, between the flight commander and the lunar module pilot. But on this flight it seemed to Mike Collins that he should *not* stick to the center seat; in fact he was hardly ever going to be there, and this word got to the flight planners. Ted Guillory teased him about this one day: "I was telling him he wanted to sit in the left seat because he didn't want to sit between those guys with all those moon germs. He said, 'Yeah, I want a big plastic bag to put over my head so I won't have to go to the lunar receiving lab with them.' But your training is affected by the seat you are assigned to. The logic was this: if the LM couldn't make it back and Mike had to fly the command module home by himself, he had to know how to make the transearth injection, the transearth midcourses, and reentry. Okay: he had to train for those things, so he might as well be in the left seat most of the time. That left Neil free to concentrate on other things. Bill Anders came back from C prime [Apollo 8] and said that the command module pilot could do his job faster and better from the right couch at launch time. So we evolved a plan for Neil to launch from the left seat, with Mike in the right. That put Buzz in the middle seat where he had done most of his training anyway, as backup command module pilot for Apollo 8, and they could swap out after transposition and docking."

Mike Collins had a further explanation. . . .

« Usually if you want to look out the window and fly the vehicle, you fly from the left seat. But if you want to do any navigating you have to go down to the lower equipment bay, to the navigator's station. If you want to fool around with the internal health of the machine, ask how it's feeling today and take its pulse and temperature and all that stuff, you do most of that over in the right seat. The plumber's delight is over on the right-hand side. That's where the fuel cells are and the communications, cooling and radiator systems. The center seat is sort of like an aisle, when you're in there by yourself. There's not much reason to tarry in the center. There's a DSKY that's located just above the center seat, but you can reach it from the left seat or you can use the other DSKY, which is down in the lower equipment bay.

Of course I spent a lot of time in the command module simulator by myself when Neil and Buzz were off in the lunar module simulator or doing something else. It's a lot more difficult to slide around in the simulator, because you have to get handholds everywhere to hold yourself up, hold your back off the seats as you change positions. In zero G you never have to worry about that. Even in one G it's no problem to slither from one side to another. Most of the trips are from the couch down into the lower equipment bay. The only thing that gets in the way is the hand controller, the control stick that you use to fly the vehicle. It's mounted on a pole about two feet long and it has an elbow joint. You can release the elbow and the thing drops down out of the way. Whenever you want to go from the left seat down to the lower equipment bay you have to go around or over it or through it, so you just drop it out of the way and sort of go around it. The controller is on the right-hand side of the left seat, where a gearshift would be on an automobile. There's another one on the left that's more like a throttle. The equivalent in a car is to find the throttle on the left of the gearshift. In a spacecraft it's the thing you push to go forward or backward, or up or down, or left or right. The one on the right, where you would find the gearshift in a car, is more like a steering wheel in function. If you want to nose down you push it down. If you want to move up you pull it up. If you want to roll around this way you move it this way, and if you want to yaw around you do it left or right that way.

But it's hard to make any firm general rules. Take those fuel cells. They are funny things. It's not that they either work or don't work. They are like human beings; they have their little ups and downs. Some of them have bad days and then they sort of cure themselves. Others are hypochondriacs. They put out lots of electricity, but they do it only bitterly with much complaining and groaning, and you have to worry about them and sort of pat them and talk to them sweetly. The Apollo fuel cells are stronger and have fewer frailties than the Gemini fuel cells had. Every time you started up those Gemini fuel cells they behaved a bit differently. After some time people got to know how many start-ups the fuel cells liked, and on the ground prior to a flight they would start them up a precise number of times, so that when you went to fly them they were all tweaked up to a peak of perfection. We don't have those problems with the Apollo fuel cells. But we do have all sorts of little sticky valves and things, and no two fuel cells are exactly alike.

In general there's a lot more redundancy built into this spacecraft than Gemini had. Not only that, but there's a lot of manual control over redundancy. And this makes for a lot of switches and valves and levers. You can sit down with a piece of paper and play designer. You can say here's an oxygen tank, and I want to have a pipe going out of that so you can sniff oxygen. So you have one tank and you have one line. Then somebody says well, maybe we need two tanks. So now you have two tanks and two lines. Then somebody says suppose you had a leak over here and you lost all this oxygen. Wouldn't it be nice if you could get it from that tank over to the alternative line? So now you've got two tanks, two lines and a crossover tube with its own valve. Then somebody else says suppose you got a leak over there? You wouldn't want to leak oxygen from both tanks out through one place, so we'll put a check valve in each one of those lines. But suppose the check valve gets clogged up? Well, we'll put in two check valves. But what if you want to bypass the check valve? You have to put a line around it. So now you've got four check valves and a crossover valve, and you've got a bypass line here and a bypass line there, and each one of those needs a valve. Now just to get oxygen out of a tube we've got seven valves; it grows like Topsy. And each one of those valves has a little switch and a lever, and you've got to remember which is which.

In my mind I've tried many times to compare this to flying a jet aircraft. I'd say the command module, although it's a lot smaller, probably is about as complex as a B-52 bomber, which has a crew of six. A lot of these things could be done completely automatically, without us ever being aware of the things we are in there fiddling with. You could argue this point either way. On Mercury, you know, they were very worried about how man would take to space flight. Maybe his heart would explode, and his veins wouldn't know what to do, and he would lose consciousness, and he would get tearaway or breakaway phenomena. So for Mercury, they said: we'll make the systems on board as automatic as possible, and the man will just sit there and go along for the ride. But man functioned all right during the Mercury flights, and the design pendulum swung back sharply. The end of Mercury was about the time that they were sitting down to write the design specifications for Apollo, even though manned Apollo flights were a long way off. The Apollo vehicles were built not to do all these things automatically, but to have the guy inside monitoring and checking that pressure before he put that diverter valve over to a certain position. Consequently we've got our fingers in all kinds of pies around the interior of that spacecraft — dials we have to read, switches we have to switch, heaters we have to turn on and off, crossovers, a lot of things that *could* be done automatically. One school of thought says that's wonderful because the human being can apply intelligence to this, and everything is more reliable. Then another camp says no, the man spends all his time potting around with useless housekeeping chores and he doesn't really have enough time left over to tend to the important things. Where the truth lies I don't know. Hopefully we've hit a fairly good balance. »

HOUSTON (Garriott): Tranquility Base, Houston. We'd like some estimate of how far along you are with your eating and when you may be ready to start your EVA prep. Over.

EAGLE (Armstrong): I think that we'll be ready to start EVA prep in about a half hour or so. [*One hundred five hours and fifty-one minutes into the flight, well past six in the evening, Houston time, normal timeline for an*

EVA prep two hours more or less. . . . MARK one minute from reacquiring Columbia. . . .]

COLUMBIA (Collins): Houston, Columbia. How do you read?

HOUSTON (McCandless): Columbia, Columbia. This is Houston.

EAGLE (Armstrong): Houston, this is Tranquility base. We are beginning our EVA prep.

Neil Armstrong was aware that his heartbeat had gone up considerably at the beginning of the powered descent to the lunar surface, and he said, much later, "I'd really be disturbed with myself if it hadn't." When he found time to look out the window of the lunar module, he found the view comparable to looking at a nighttime scene well lighted for photography. . . .

« The sky is black, you know. It's a very dark sky. But it still seemed more like daylight than darkness as we looked out the window. It's a peculiar thing, but the surface looked very warm and inviting. It looked as if it would be a nice place to take a sunbath. It was the sort of situation in which you felt like going out there in nothing but a swimming suit to get a little sun. From the cockpit, the surface seemed to be tan. It's hard to account for that, because later when I held this material in my hand, it wasn't tan at all. It was black, gray and so on. It's some kind of lighting effect, but out the window the surface looks much more like light desert sand than black sand.

We wanted to do the EVA as soon as possible. We had thought even before launch that *if* everything went perfectly and we were able to touch down precisely on time; *if* we didn't have any systems problems to concern us; *if* we found that we could adapt to the one-sixth G lunar environment readily — then it would make more sense to go ahead and complete the EVA while we were still wide awake and not try to put that activity in the middle of a sleep period. In all candor, we didn't think that an early EVA was a very high probability. But as it turned out we did land on time; there were no environmental or systems complications, so we chose to request permission to go ahead. »

More than three and one-half hours after landing. . . . Mike Collins still trying to locate the lunar module with his sextant. . . .

COLUMBIA (Collins): Do you have any topographical clues that might help me out here? AUTO optics is tracking between two craters. One would be long at eleven o'clock. The other would be short and behind him at five o'clock. And they're great big old craters. Depressions.

HOUSTON (McCandless): Stand by. . . . Columbia, this is Houston. The best we can do on topo features is to advise you to look to the west of the irregularly shaped crater and then work on down to the southwest of it. . . .

Another possibility is the southern rim of the southern of the two old-looking craters. Over.

COLUMBIA (Collins): Roger, Houston, Columbia. No joy. I kept my eyes glued to the sextant that time, hoping I'd get a flash of reflected light off the LM, but I wasn't able to see any of him in my scan areas that you suggested.

HOUSTON (McCandless): Roger. On that — southern of the old craters, there's a small bright crater in the southern rim of — one slot would put him slightly to the west of that small bright crater, about five hundred to one thousand feet. Do you see anything down there? Over.

COLUMBIA (Collins): It's gotten bad now, Bruce, but I scanned that area that you talked about very closely and no, I did not see anything.

HOUSTON (McCandless): Columbia, this is Houston. Are you aware that Eagle plans the EVA about four hours early? Over.

COLUMBIA (Collins): Affirmative. When's hatch open time in GET estimated?

HOUSTON (McCandless): Roger. Somewhere around 108 hours. We'll have an update for you on that a little later.

COLUMBIA (Collins): Okay, I haven't heard a word from those guys, and I thought I'd be hearing them through your S-band relay.

HOUSTON (McCandless): Roger. They're about page Surface 27 in the checklist, proceeding in good time.

COLUMBIA (Collins): Glad to hear it. If you'll excuse me in a minute I'm going to have a cup of coffee.

HOUSTON (McCandless): I think we have some more coordinates for you on the LM location. Over.

COLUMBIA (Collins): Ready to copy.

HOUSTON (McCandless): Roger, Mike. POPPA .2 and 6.3. . . . On your next pass, Columbia, rather than performing a P22 as such, we would like you to look in the vicinity of the coordinates that we gave you which is our best analysis based on MAP physics trajectory. And we also have another set of coordinates that we would like you to search in the vicinity of. This last one being based on an interpretation of the geological features that were seen by the crew on their way down. The coordinates of this second site are MIKE .7 and 8.0. . . . And if you can find the LM then by all means track it or make a note of where it was and we can track it on the next rev. [*One hundred seven hours and twenty-three minutes into the flight, loss of signal as Mike Collins went over the hill again, still not sure where Eagle had landed.*]

Nassau Bay, Texas

By now, a little more than four hours after touchdown, some people on the ground were fairly sure that they knew where Eagle was. They were the geologists. Arnold Brokaw had said that it didn't matter whether the men knew or did not know where they had landed on the moon: "We can figure that out later." And they did. Having agreed that the crater Neil Armstrong had mentioned earlier really was West Crater, the geologists settled on a certain point on their map. It would turn out that their guesswork was correct within two hundred meters. The geologists translated their guess into lunar longitude and latitude and asked Mission Control to relay the information up to Mike Collins. For some reason the message was not sent to Collins. Dr. Eugene Shoemaker said later, "No one wanted to believe us. I won some bets on it, though, and I collected!"

By eight o'clock in the evening, Houston time, the families of the astronauts were also "collected." They, at least, were ready to watch Neil Armstrong and Buzz Aldrin do what no man had ever done — walk on the moon. . . . But it would not happen right now, after all. . . .

HOUSTON (PAO Jack Riley): This is Apollo Control at 107 hours 52 minutes. We're sixteen minutes away from acquisition of Columbia on its seventeenth revolution of the moon. We do not at this time have a good estimate for the start of the EVA. We'll have to wait until Eagle's crew, Neil Armstrong and Buzz Aldrin, give us some more information about how they're coming along. . . . Indications are now that they are running on the order of thirty minutes behind the nominal time preparations line. Maybe a bit longer. . . .

Jan Armstrong sat on the floor facing the coffee table, staring at the television set, and laughed. She said, "It's taking them so long because Neil's trying to decide about the first words he's going to say when he steps out on the moon. Decisions, decisions, decisions!" Nothing seemed to be happening, so she left the room.

In the Collins home, in Nassau Bay, there was an atmosphere of jocularity. The tension of touchdown had evaporated. There was a prevailing note of impatience: why were they taking so long? *One hundred eight hours and two minutes. Latest report that the crew is getting the electrical checkout indicating they are about forty minutes behind the timeline. . . . Due to reacquire Columbia in six minutes. . . .*

And then there was more conversation, between Eagle and Columbia:

TRANQUILITY: "Help you on that? Feed it or disconnect or what? Say again? Switch PLSS electrical umbilical to — on the F type — have to get up straight — up. Radio COMM. . . . All set? . . . I'll put my antenna up. Okay, how do you read now?" "Okay." "Okay, I think that's going to be better." "You read me all right now?" "Yes." "Okay. That sounds pretty good. . . . 1 2 3 4 5. 5 4 3 2 1. That sounds pretty good. . . . Better keep it pretty close to your mouth though."

HOUSTON (Riley): We have acquisition of Columbia.

Pat Collins thought it was all a bit like labor pains: "God, they come and they go. All this time and they are just getting ready to do it. When are they going to start the countdown?" She sat in front of the television and nibbled at a plate of lasagna. Nothing much was coming out of the squawk box. Barbara Gordon said, "Surely they're saying something? Their voltage must be sick or the antenna is down. We request that you open your mouths!" Clare Schweickart was there; it was getting on toward nine o'clock in the evening, and she asked, "Do you think they're getting tired with the prep? Can they stop if they want to and rest?" Rusty Schweickart consulted the flight plan and started to give a commentary. Pat Collins questioned one of his explanations. "Is that normal?" she asked. "As long as it's normal, don't bother to tell us. If it's abnormal, leave." Astronaut Joe Engle's wife Mary volunteered: "Today my English pen pal called me from London. I've written to her for twenty-five years and I've never met her. She said excitement was running so high in London that she just had to call and wish us Godspeed. She couldn't understand why I sounded so calm and I couldn't explain it to her." Someone else said that Mike Collins would be on the back side of the moon when Armstrong and Aldrin came out of the lunar module. "Well, he'll be back around again," Rusty Schweickart answered. "I hope so. It's going to be a long night for those guys if he isn't."

In the Aldrin home, when it was learned that Neil and Buzz hoped to step out of the LM around eight o'clock, or as soon thereafter as possible, there was some banter about upstaging the late show and preempting prime television program time. Joan Aldrin was utterly relaxed; she listened to some old Duke Ellington records. She seemed to be in love with the world. "You know," she said, "this is so corny — but I love every minute of it." At one point she turned to Audrey Moon and said, "I have thought sometimes, privately, and may even have said so to Buzz, that he was so caught up in the mechanics of all this, that he really hadn't realized the significance of what he was doing — but now I really think he did! This man I married is full of surprises." Joan had not eaten all day; then, about eight o'clock in the evening, she helped herself to a plate from the buffet table. At a quarter past eight Mike Aldrin burst in: "Have we missed anything?" Mike and his friend Kurt Henize had let their water snake go and had been chasing bullfrogs.

They had not missed anything; indeed half a billion people watching television would have to wait a bit.

Until fairly recently it had not occurred to Neil Armstrong that his first words, when he stepped on the moon, would have a critical importance for history. But it seemed to him that every person he had met in the past three months had asked him what he was going to say or had made a suggestion. In fact Armstrong had had hundreds of suggestions, including passages from Shakespeare and whole chapters of the Bible. As Eagle's astronauts fell behind schedule on their electrical checkout, with the job of depressurizing the lunar module's cabin yet to begin, there was some teasing banter in their homes about Neil taking a lot of time to make up his mind on what first to say. The truth was that Armstrong and Aldrin had a lot of work to do, and it was taking them longer to do it than they and Mission Control had estimated in advance. Aside from the systems checkouts, there was the matter of donning the backpacks which would keep them alive when they stepped on the moon. The backpack looked — and was — a piece of equipment straight out of Buck Rogers; and although its moon weight was only a little over twenty pounds (as compared to about a hundred and twenty-five pounds earth weight), it was awkward to put on, and each astronaut had to help the other.

COLUMBIA (Collins): Houston, Columbia on high gain. How do you read?

HOUSTON (McCandless): Roger, Columbia. Reading you loud and clear on the high gain. We have enabled the one-way MSFN relay that you requested. The crew of Tranquility Base is currently donning PLSSs. The LMP has his PLSS on, COMM checks out and the CDR is checking his COMM out now. Over.

COLUMBIA (Collins): Sounds good. Thank you kindly.

The PLSS — portable life support system — was the central element in the backpack. The whole pack was actually a three-piece model of ingenuity which had been in the making since October 15, 1962, and which had consumed two and one-half million man-hours in design, testing and production. It was first used in space by Rusty Schweickart during his 38-minute space walk on the flight of Apollo 9 — on March 6, 1969. The first model of the pack, made by the Hamilton Standard division of United Aircraft Corporation in Windsor Locks, Connecticut, used oxygen gas for cooling the astronaut in his space suit; then in September 1964 there was a design change to water cooling. The water-cooled pack used by Schweickart passed its flight qualification tests less than two months before Schweickart used it in space.

Built into the comparatively small PLSS (26 inches high, 17.8 inches wide and 10.5 inches deep) were several systems: a primary oxygen supply, pumped into the space suit automatically to maintain a pressure of 3.9 pounds per square inch and provide oxygen to breathe; an oxygen ventilating system to keep the oxygen circulating, cool and dehumidify it and remove carbon dioxide and contaminants; a water support loop circulating cool water at four pounds per minute through a network of tubing built into the space suit's undergarment; and a communications system to provide primary and backup dual voice transmission and reception.

On top of the PLSS was the forty-pound OPS — the oxygen purge system. This backup system carried a thirty-minute supply of oxygen for emergency use either on the moon or during an unscheduled space walk. Unlike the PLSS, it could not be left on the lunar surface. Suppose for some reason the lunar module and the command module were unable to dock successfully on the return trip? Then both Armstrong and Aldrin would have to make an extravehicular transfer to Columbia, using the OPS to breathe. And they would have to hurry to do it in thirty minutes. The backpack had a third element: a small remote control unit which was to be worn on the chest area of the space suit. It contained water pump and oxygen fan switches, a four-position communications selector, dual radio volume control and a mount for the lunar surface camera. Each backpack carried enough water and oxygen to permit four hours of extravehicular activity, and each could be recharged, if necessary, by the lunar module's own environmental control system. Each was built to catch and disperse metabolic heat generated by an astronaut at an average rate of sixteen hundred British Thermal Units an hour — roughly the equivalent of heat generated by a man shoveling sand. It would sustain, for a time, a peak rate of two thousand BTUs an hour — equivalent to a man sawing wood or walking rapidly at more than four miles an hour.

The hard part was getting the thing on, and Armstrong and Aldrin had to accomplish this before they could depressurize the cabin and step on the moon. First the astronaut, in his pressure suit but for the moment without helmet and gloves, mounted an OPS on top of his PLSS and checked all functions — oxygen supply pressure, regulators, the heater circuit. Then he hoisted the OPS and the PLSS together to his back and secured them by two shoulder harnesses and two waist harnesses, clipped to rings on his space suit. Then he got the helmet and gloves on, connected the pack's oxygen, water and electrical umbilicals — hoses — in the general heart area of his suit. Now he could open the pack's oxygen valve and turn on the fan; then he was independent of Eagle's environmental control system, so he could unhook the hoses which connected him with Eagle's system and use the same connectors to hook up the OPS. Then, and only then, was he ready to walk on the moon. The danger remained of a fall, an unforeseen accident of some kind, that could rip the space suit and tear it open; but edges on Eagle's metal appurtenances — e.g., the descent ladder — had been rounded off with that in mind. Additionally, each unit of the backpack had been covered with a thermal insulator of fire-resistant Beta cloth, and aluminized Kapton covered the pack to restrict heat leakage in and out. Over the whole pack there was a fiber glass cover to protect the pack from damage by micrometeoroids, or an accidental bang during the ladder descent to the lunar surface. But the theoretical danger of a suit tear remained, although Buzz Aldrin preferred to minimize it. . . .

« That is something you don't want to see happen. The thermal protective covering is, I think, more than adequate. The pressure garment is more than adequate for any type of fall you are likely to encounter, and on top of this you have this fairly thick and bulky protective layer.

If there were such an accident, I'm not sure that we really know what would happen. We used to think that there would be a foaming up of the blood, and it would boil. We don't think so now; after all, the fluids in the body are not exposed to the outside. All we know is that with a big loss of pressure you would lose consciousness in a short period of time, a matter of seconds. But it isn't necessary to look at this problem in disaster terms. There is a more rational way to look at it. You try to understand the needs of the body for certain amounts of pressure, the need to be able to breathe in a partial pressure to sustain consciousness. Then you build a space suit to do that job. Throughout the Apollo program we were getting smarter about how to build elbows and shoulders into the space suit to provide more mobility and still keep high reliability for decompression. As for the element of risk, that is present at every stage in a mission, and you can't eliminate it entirely. You just don't dwell on it unnecessarily. »

HOUSTON (McCandless): Neil, this is Houston. We're reading you loud and clear. Break, break. Buzz, this is Houston through Tranquility. Over.

EAGLE (Aldrin): Roger, Houston. This is Buzz through Tranquility. How do you read? Over.

HOUSTON (McCandless): We're reading you loud and clear, Buzz. Out.

EAGLE (Aldrin): And are you getting a signal on the TV? Over.

HOUSTON (McCandless): That's affirmative, Neil. . . . You trailed off down into the noise level, Neil. Over.

HOUSTON (McCandless): Columbia, this is Houston. Are you reading Tranquility all right on the relay? Over.

COLUMBIA (Collins): I believe so. I haven't heard anything from them lately. They've been breaking up, but up until about three minutes ago I was reading them loud and clear. [*Nearly nine in the evening, Houston time. . . .*]

EAGLE (Armstrong): My antenna's scratching the roof. . . . Do we have a go for cabin depress?

EAGLE (Aldrin): I hear everything but that. Houston, this is Tranquility. We're standing by for a go for cabin depress. Over.

HOUSTON (McCandless): Tranquility Base, this is Houston. You are go for cabin depressurization. Go for cabin depressurization.

In El Lago and Nassau Bay the families were getting restless. In the Collins home Clare Schweickart asked her husband, "Are they getting the hatch open, Rusty?" He said, "No, they're still depressurizing the cabin. Getting the hatch open takes forever. They'll get it open, but it's a lot of work." Kate Collins asked her mother if Neil and Buzz were out yet. "You'll know when they're out," Pat Collins said. "I'll tell you." Pat Collins's friend, the Reverend Eugene Cargill, remarked, "I keep saying my prayers and thanking God for the static." There was a lot of static, and with Mike Collins in Columbia about to go over the hill again, the conversation at Tranquility was not very clarifying: "Takes a while for the water separator." "I don't understand. Suit fan No. One circuit breaker opened." "Thank you." "Got it." "Let me do that for you." "Lock one." "All of the . . . [*static*]." "[*Static*] . . . locked and lock-lock." "Did you put it . . ." "Oh, wait a minute." "I'll try it on the middle." "All right, check my [*static*] valves vertical." "Both vertical." "That's two vertical." "Okay."

EAGLE (Aldrin): Sure wish I had shaved last night.

COLUMBIA (Collins): I don't know if you guys can read me on VHF, but you sure sound good down there.

EAGLE (Aldrin): How's the count now, Houston?

HOUSTON (McCandless): Buzz, this is Houston. The count is very good. You are coming through loud and clear, and Mike passes on the word that he is receiving you and following your progress with interest.

. . . . And still more from TRANQUILITY: "Do we move it?" "Okay, we can stow this." "All right, prep for EVA. First you connect the water hose. Okay, let me get your — okay, now we should be able to stow these up."

EAGLE (Aldrin): Houston, Buzz here. Over.

HOUSTON (McCandless): Go ahead, Buzz. This is Houston.

EAGLE (Aldrin): Roger. Our COMM just seemed to clear up a good bit. Did CSM just go over the hill?

HOUSTON (McCandless): Negative. He's been over the hill, I figure, for a minute or so. . . . Correction. He should be losing contact with you in about a minute.

. . . . TRANQUILITY again: "Okay." "All right, verify your diverter valve open." "What position?" "Diverter valve up." "Diverter valve up." "Just a minute here." "Switch on." "Mine's running also, and it's cooling all right." "Get them both on, verify cooling. Why don't you bend down and let me stow that?"

HOUSTON (McCandless): Columbia, this is Houston. . . . Were you successful in spotting the LM on that pass? Over.

COLUMBIA (Collins): Negative. I checked both locations and no joy.

HOUSTON (McCandless): Okay, if you'd like to look again next pass we have a different set of coordinates based on the onboard P57 solution of the

LM. These are Echo .3 and 4.8. . . . [*Mike Collins was still trying to find them, but through no fault of his own he was looking in the wrong places. . . .*]

Honeysuckle Creek, Australia

The homes of the Apollo 11 astronauts had no monopoly on tension. In a sleepy valley called Honeysuckle Creek, named for the plant which abounds in that rugged area twenty-five miles from the Australian national capital, Canberra, some one hundred technicians were gathered around two giant screens. They were waiting to receive — and relay to the rest of the world — the first television pictures of man on the moon. Because of the relative positions of the earth and the moon at the time Armstrong and Aldrin were to step out of the lunar module and onto the moon, the television relay was Honeysuckle's job. Honeysuckle, built by NASA in 1966 at a cost of some two million dollars [Australian], was actually part of a chain of three Australian tracking stations. Backing up Honeysuckle's 85-foot diameter antenna "dish," rising 114 feet in the air, were facilities at nearby Tidbinbilla and at Parkes, more than one hundred miles away in New South Wales, where an even bigger antenna — 210 feet, largest in the southern hemisphere — had been borrowed from the Commonwealth Scientific and Industrial Research Organization (CSIRO) to receive signals from Apollo 11 during its journey to the moon. Both the Honeysuckle and Parkes antennas were ready to start receiving — and relaying — pictures when the moon walk began. There was a discernible tension among the technicians, most of them — like their Houston counterparts — surprisingly young, many of them wearing beards. They had tracked other missions, but, said station administrative officer Bernie Scrivener, "Somehow this seemed to be so much more important. Since the Apollo 7 mission, this was the day for which everyone on the station had worked and trained." Supervising technician Laurie Turner, who had to process data related to the heartbeats of the astronauts during the moon walk and on the condition of their backpacks, confessed to a feeling of genuine anxiety: "Not for the safety of the astronauts. I think we all had faith in the technology which put them there, but I was anxious for myself. Would the tests and simulations we had performed *really* be like the thing itself? Or would critical adjustments or repatching of circuits be necessary to get the vital data back to Houston?" And the week had not been without its crises. On the third day of the mission Tidbinbilla messaged: "We have lost both our transmitters — suspected power failure." It turned out that only one transmitter had been lost, but a prodigious effort had to be made to repair the other — the nearest available spare parts were at the Woomera rocket range, nearly a thousand miles away in the desert. Hard work and the cooperation of a domestic airline got the job done less than twenty-four hours before the lunar landing. As the staffs of all three stations waited for the moon walk, Dr. E. G. ("Taffy") Bowen, a world figure in radio astronomy who directed CSIRO's radiophysics division at Parkes, remarked, "Our only worry now is the weather or some event we can't foresee." Shortly afterward there was bad news: a serious squall was approaching from the northwest. A few minutes later it hit Parkes's giant dish with a 35-mph gust, threatening to put the whole station out of commission. Dr. Bowen made no effort to conceal his worry as the dish shook under the wind's impact. But it held, and the tracking continued as the world waited.

EAGLE (Aldrin): It took us a long time to get [the cabin pressure] all the way down, didn't it?

EAGLE (Armstrong): Yeah.

The long time required was one of the things that surprised Armstrong most. . . .

« That's a part of the exercise we never get to duplicate in any tests or simulations. I think it took so long partly because the backpacks were operating in the cockpit and adding some exhaust gas to the atmosphere which was stabilizing out just above a vacuum. So it took us much more time than we had expected to get our sublimators in the backpacks operating. Again, we had never really done that in simulations. You have to be in a vacuum to do that. We had trained for this aspect of the mission in a chamber and under somewhat different conditions. »

HOUSTON (McCandless): Neil, this is Houston. What's your status on hatch open? Over.

EAGLE (Armstrong): Everything is go here. We're just waiting for the cabin pressure to bleed so — to blow enough pressure to open the hatch. It's about .1 on our gauge now.

EAGLE (Aldrin): Sure hate to tug on that thing [the hatch handle]. Alternative would be to open the overhead hatch too.

Past nine-thirty in the evening, Houston time. . . .

In El Lago, hearing Neil quote the pressure at .1, somebody asked if the hatch could now be opened. Jan Armstrong nodded sagely and said, "There's not that much pressure," then went through the flight plan again and added, "Make up your mind what you're going to say." Her son Ricky said, "Knowing Dad, you can't tell. It'll be something good, though."

HOUSTON (McCandless): We're showing a relatively static pressure on your cabin. Do you think you can open the hatch at this pressure?

EAGLE (Armstrong): We're going to try it.

EAGLE (Aldrin): The hatch is opening.
EAGLE (Armstrong): The hatch is open!

ARMSTRONG: "Is my antenna out? Okay, now we're ready to hook up the LEC here." ALDRIN: "Now that should go down . . . [*static*] . . . Put the bag up this way. That's even. Neil, are you hooked up to it?" ARMSTRONG: "Yes. Okay. Now we need to hook this." ALDRIN: "Leave that up there." ARMSTRONG: "Yes." ALDRIN: "Okay, your visor down? Your back is up against the . . . [*static*]. . . . All right, now it's on top of the DSKY. Forward and up, now you're there, over toward me, straight down, relax a little bit. . . . Neil, you're lined up nicely. Toward me a little bit. Okay, down. Okay, made it clear . . . [*static*] . . ."

Ten minutes before ten o'clock, Houston time. . . . In the Collins home, in Nassau Bay, Rusty Schweickart shouted: "It's open! Now they're like two bulls in a china shop. Don't bump into anything! Just find the ladder, Neil baby!" CBS commentator Walter Cronkite wondered why it was taking so long. "Because he doesn't have eyes in his rear end, that's why," Schweickart answered. Clare Schweickart said, "I know what Neil's first word from the moon will be: 'Arwk, hut, up, blurp.' " Rusty again answered, "If he doesn't find that ladder soon the first words on the moon will be something like what Gene Cernan said." "Not Neil," Clare said. "He'd say it in fewer words." Listening to the television commentary, Clare said, "He's on his knees." Father Cargill responded, "I think that's very appropriate." Schweickart encouraged, "Okay, Neil baby, pull the string!" Jan Armstrong grumbled, "He's still trying to think of what to say."

ALDRIN: "Move. Here, roll to the left. Okay, now you're clear. You're lined up on the platform. Put your left foot to the right a little bit. Okay, that's good. Roll left." ARMSTRONG: "Okay, now I'm going to check the ingress here." ALDRIN: "Okay, not quite squared away. Roll to the — roll right a little. Now you're even." ARMSTRONG: "That's okay?" ALDRIN: "That's good. You've got plenty of room." ARMSTRONG: "How am I doing?" ALDRIN: "You're doing fine. . . . Okay, do you want those bags?" ARMSTRONG: "Yes. Got it. . . . Okay, Houston, I'm on the porch." . . . [*Nine ladder steps to the moon. . . .*]

HOUSTON (McCandless): Roger, Neil. . . . Columbia, Columbia, this is Houston. One minute and thirty seconds to LOS. All systems go. Over. . . . [*Goodbye to Mike Collins for another forty-five minutes. . . . Neil Armstrong on the porch at 109 hours 19 minutes and 16 seconds GET. . . .*]

And from TRANQUILITY:

ARMSTRONG: "You need more slack, Buzz?" ALDRIN: "No, hold it just a minute." ARMSTRONG: "Okay." [*Twenty-five minutes of PLSS time expended. . . .*] ALDRIN: "Okay, everything's nice and straight in here."

ARMSTRONG: "Okay, can you pull the door open a little more?" ALDRIN: "Did you get the MESA out?" ARMSTRONG: "I'm going to pull it now." ARMSTRONG: "Houston, the MESA came down all right."

HOUSTON (McCandless): Roger, we copy, and we're standing by for your TV. [*So were millions of other people . . . people looking at giant screens in Trafalgar Square, London, and in Seoul, Korea. . . . Not in Harry's Bar in Paris, where there was no television set — but only one topic of conversation. . . .*]

HOUSTON (McCandless): Neil, this is Houston. You're loud and clear. Break, break. Buzz, this is Houston. Radio check and verify TV circuit breaker in.

EAGLE (Aldrin): Roger. TV circuit breaker's in. LMP reads loud and clear.

HOUSTON (McCandless): And we're getting a picture on the TV.

EAGLE (Aldrin): Oh, you got a good picture. Huh?

In the briefing room at Honeysuckle Creek, Australia, where the picture was seen first — it was Neil Armstrong's leg — a voice in the back broke the silence: "Who said we'd miss it?" There was an anonymous answer from somewhere else in the room: "We've got it, all right." Ed Renouard, the scan converter technician at Honeysuckle Creek, breathed a little easier: "When I was sitting there in front of the converter, waiting for a pattern on the input monitor, I was hardly aware of the rest of the world. I remembered how smoothly Apollo 10 had gone. But that was Apollo 10 — straightforward commercial TV which needed only proven techniques. Today we really fired 'The Beast' in earnest, and suddenly all those long hours of training seemed worth it. I heard Buzz Aldrin say 'TV circuit breaker in,' and suddenly on the screen I saw the sloping strut of the lunar module's leg against the moon's surface." Bernie Scrivener remembered, "I just paused and wondered. It was hard to tear yourself away from that screen and get back to your job."

HOUSTON (McCandless): Okay, Neil, we can see you coming down the ladder now.

EAGLE (Armstrong): Okay, I just checked — getting back up to that first step, Buzz, it's not even collapsed too far, but it's adequate to get back up. . . . It takes a pretty good little jump. . . . I'm at the foot of the ladder. The LM footpads are only depressed in the surface about one or two inches. Although the surface appears to be very, very fine-grained, as you get close to it. It's almost like a powder. Now and then, it's very fine. . . . I'm going to step off the LM now. . . . [*I had thought about what I was going to say, largely because so many people had asked me to think about it. I thought about that a little bit on the way to the moon, and it wasn't really decided until after we got onto the lunar surface. I guess I hadn't actually decided what I wanted to say until just before we went out. . . .*]

In El Lago Jan Armstrong bent down to explain to her son Mark why his father had to move around slowly: "The suit is very heavy and the sun is very hot, like Texas. He has to be careful because if he tears his suit it could be very, very bad." Mark rubbed his eyes. At 9:55:10 P.M., Houston time, Mark pointed to the television screen. His father was coming down the ladder of the lunar module. "I don't see it," Mark complained. When Neil described the fine-grained soil, Mark asked, "How come I can't see him?" Jan Armstrong smiled and encouraged her husband from a quarter of a million miles away: "Be descriptive." In Nassau Bay Joan Aldrin clapped her hands and said, "Look, look! Gee . . . I can't believe this." She turned to her daughter Jan and asked, "Do you suppose your brothers *really* know what is going on right now?" Michael and Andrew were upstairs watching television — alone. Joan Aldrin said, "You know, you really don't miss nearly as much as I think you do." Jan replied, a bit coldly, "Oh, I pay attention." In the Collins home there was a babble of conversation: "I see something moving. I can't stand it." "This is science fiction." "It looks like the North Pole." "Let's listen for his first words. . . ."

At 9:56 P.M., Houston time, Neil Armstrong stepped out of the dish-shaped landing pad and onto the surface of the moon: "THAT'S ONE SMALL STEP FOR A MAN, ONE GIANT LEAP FOR MANKIND." [1]

The quality of the television transmission delighted everyone in the world — everyone, perhaps, except Jon Eisele, age four, the son of the astronaut Donn Eisele. He had been put to bed and was sound asleep long before Armstrong and Aldrin stepped onto the moon. But his mother Harriet,[2] who was working for a television company during the moon flight, left word with her neighbor and best friend, Faye Stafford, the wife of Tom Stafford, to wake the children up when it came time to watch TV from the moon. Faye Stafford awakened Jon and plunked him on the floor in front of the television set. Jon sat there for about five minutes and said, "I can't see, the picture is all fuzzy. I'm going *outside* to see." He went out, looked up at the moon, came back and said, "I still can't see. Where's the flashlight?"

In Houston there was a new class of temporarily unemployed: the long-distance telephone operators. When the moon walk was on, they had little to do. The number of long-distance calls handled by Houston operators through the Southwestern Bell exchange between 9 P.M. and midnight on July 20 was 6,408, compared with 8,968 during the same three hours on the preceding Sunday.[3]

At Honeysuckle Creek in Australia, where it was 12:56 P.M. Monday, July 21, there was a sign on the notice board in the operations room which bore the title of the pop song, "Fly Me to the Moon." Someone had written in two additional words: ". . . and back."

Moving carefully with his bulky suit and backpack, Aldrin backs out of Eagle and starts down the ladder to join Armstrong on the moon's surface. Armstrong had already been out almost twenty minutes and took this picture while coaching Aldrin on just how to place his feet.

Aldrin pauses on the ladder's last step as Armstrong, talking to him by radio, points out "That's a good step. About a three footer." Aldrin reported it was "a very simple matter to hop from one step to another."

At ladder's end but still inside the saucer-shaped footpad of the LM, Aldrin practices the jump required to get back up the three feet to the first step. One of his assigned tasks was to calculate the difficulty of this task, before turning to the tiring work of deploying experiments on the surface.

ABOVE LEFT: *detail of Aldrin setting up the passive seismic package, designed to detect lunar tremors caused by meteorite impact or internal activity. The extended panels are to catch the sun's light and convert it into power to radio instrument readings to earth.* OPPOSITE: *Aldrin sets up a banner of aluminum foil facing the sun, to catch nuclei of atoms blowing from the solar wind. After a couple of hours on the surface the foil was folded up, stowed in one of the rock sample boxes and brought home to be analyzed by a laboratory in Switzerland.* ABOVE: *half-silhouetted against the startlingly near lunar horizon, Aldrin sets up the passive seismic experiment. Behind him are the LM, the laser reflector experiment, and the American flag especially fitted with rigid wires to make it appear to wave in the airless lunar atmosphere.*

Apollo 11 commander Neil Armstrong grins through whiskers and tired eyes after his walk on the moon. He is back in the LM, wearing the "Snoopy helmet," which contains radio headset for voice communication with earth.

TOP LEFT: *On lunar surface, Buzz Aldrin salutes the American flag. Though surface seemed soft underfoot, it turned unexpectedly resistant just a few inches down, and astronauts had difficulty shoving the flagstaff down far enough to make it remain erect.*

OPPOSITE: *Special gold-plated visor of lunar helmet was designed to protect astronauts from harsh, unfiltered light on the surface. It also proved to be an excellent mirror; here Aldrin's visor reflects image of Armstrong at work, of the LM itself, even of some of the deployed scientific equipment.*

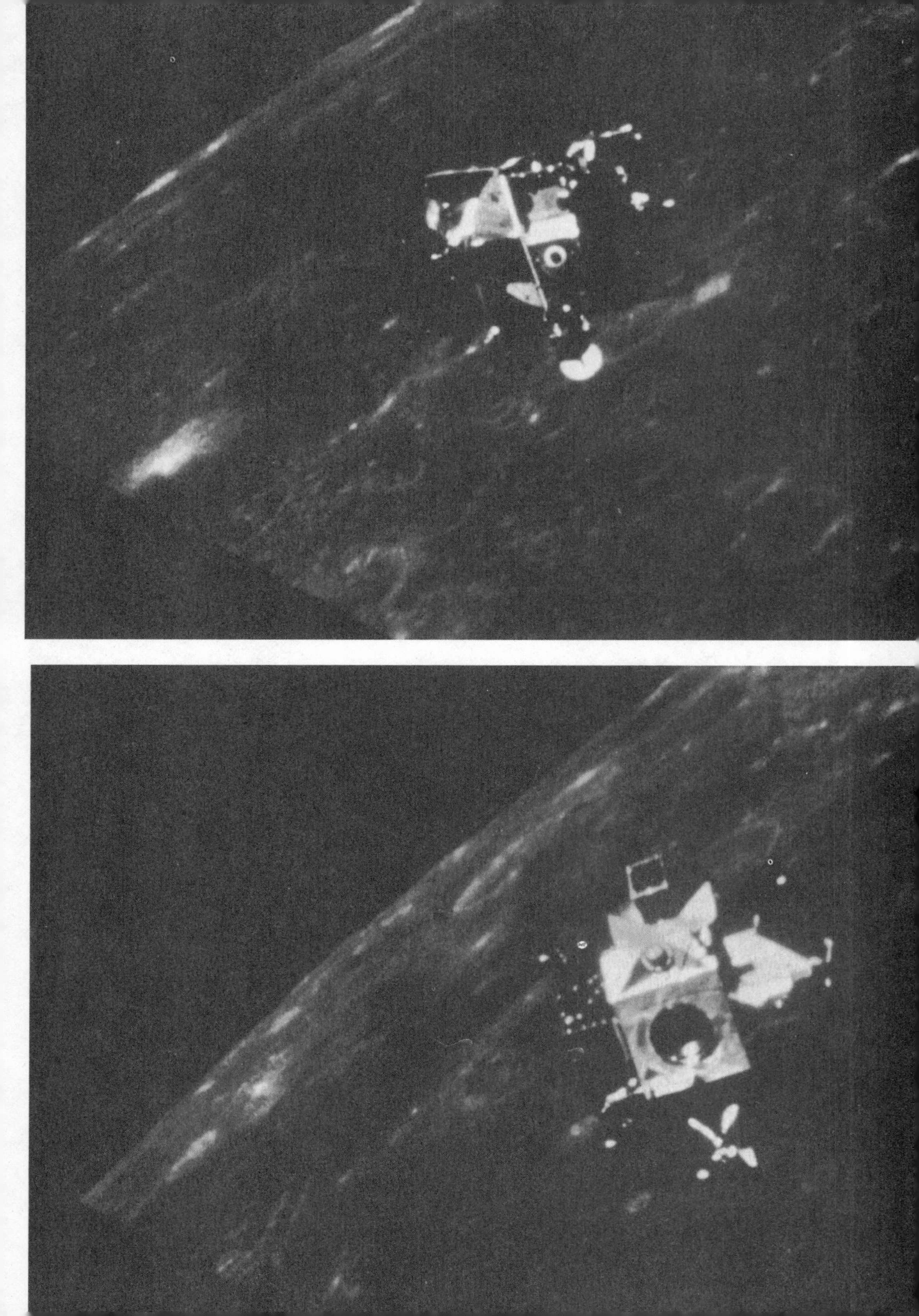

OPPOSITE: *Eagle rises from the lunar surface toward a rendezvous with Columbia. Picture was taken by Mike Collins from one of the spacecraft's windows. Lunar horizon is at top left.* BELOW LEFT: *moving closer, the lunar module has now spun slowly so that the hollow drogue, which fits onto the command module's conelike probe to dock the two craft, is pointed toward Columbia.* BELOW: *seconds before docking and the reunion of the Apollo 11 crew, sixty-nine statute miles above the lunar surface, Eagle maneuvers in close. Propeller-like device, bottom, is its antenna. Note globe of earth hanging in the black sky upper left.*

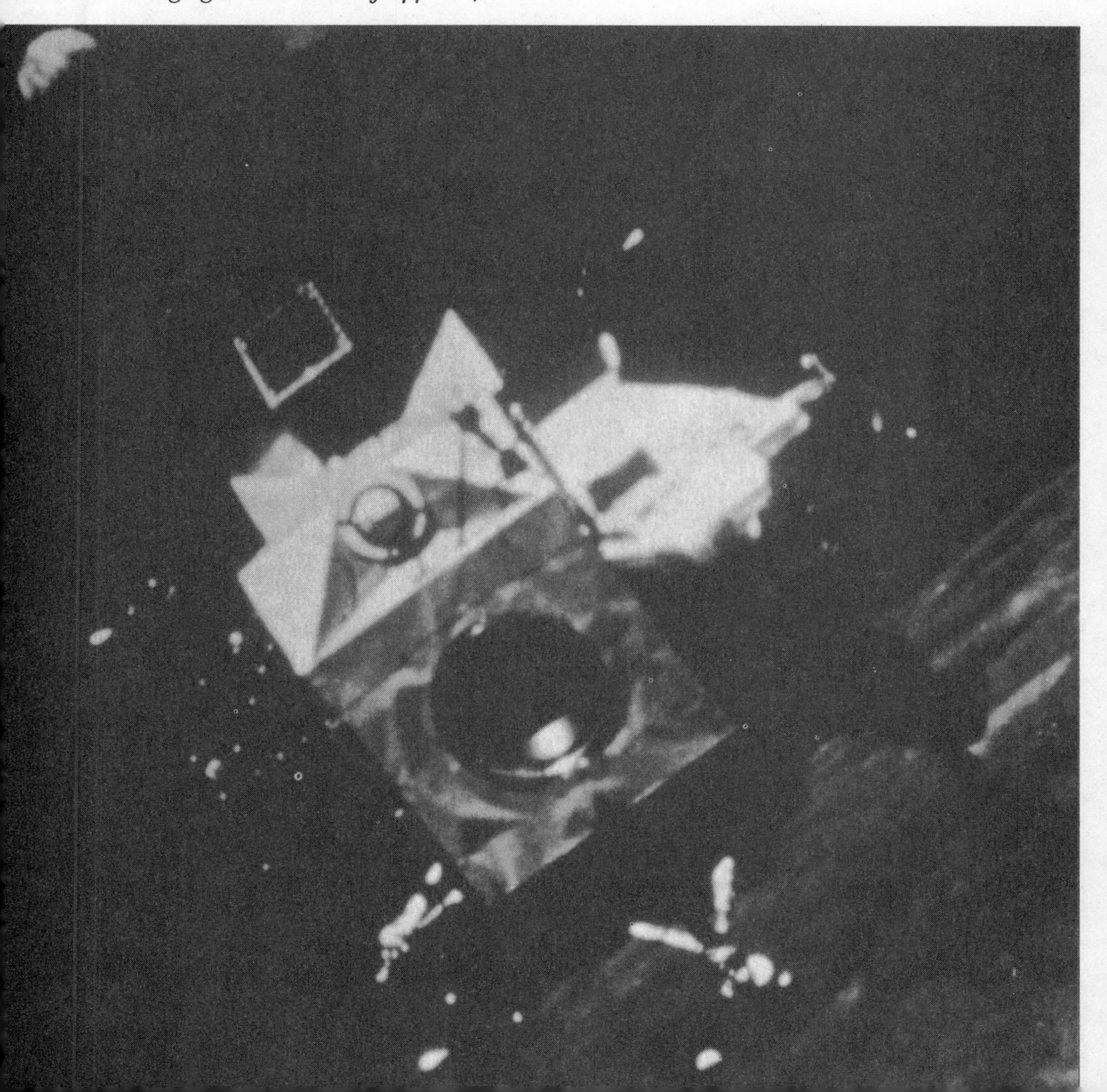

NAVY
66

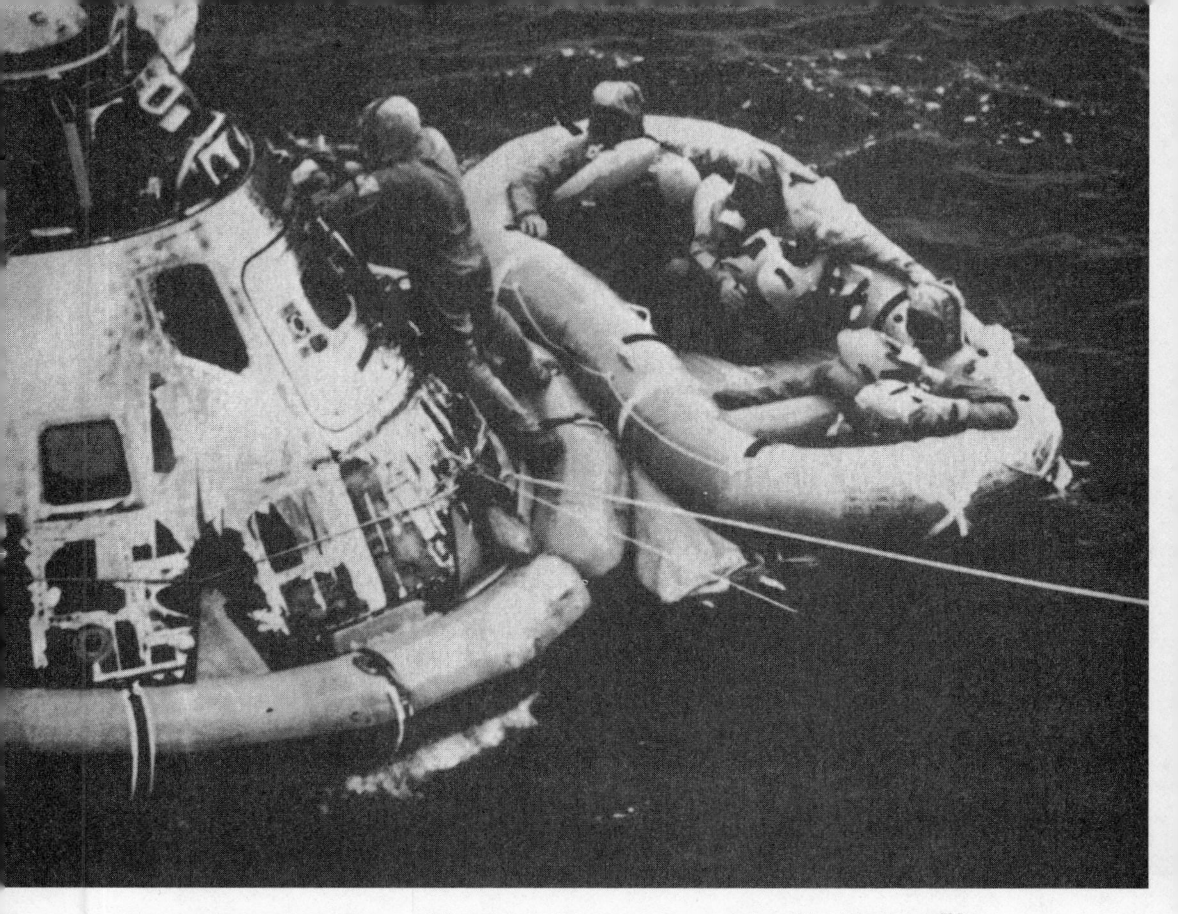

ABOVE: *crewmen bounce in rubber life raft, wearing special BIGs (biological isolation garments), gas mask-like breath filters and life jackets after splashdown July 24. Frogman is swabbing down hatch area of spacecraft to kill any moon germs which might have been released from inside when hatch was open.*

LEFT: *Navy rescue helicopter swings a crewman aloft in cage made of nylon cording. Cage, called a "Billy Pugh" for its inventor, replaced the old and dangerous arm sling with which early astronauts were plucked from the sea. Raft at right contains frogmen who helped put flotation collar on the spacecraft and helped crewmen out of hatch. Crew raft is behind spacecraft, barely visible in this picture.*

Spacecraft Columbia being hoisted onto deck of recovery vessel Hornet *after splashdown, with flotation collar visible and the big balloons which were inflated after landing to lift it from "Stable 2" position (upside down) to more comfortable "Stable 1" (right side up). Spacecraft went into quarantine — along with its crewmen and all its contents — for three weeks at the Lunar Receiving Laboratory near Houston.*

Showered, shaved — all except Mike Collins's new moustache — and freshly dressed in earth coveralls, Armstrong, Collins, and Aldrin beam at President Nixon from inside the Mobile Quarantine Facility, a trailer-like isolation unit, on deck of the recovery ship Hornet.

It has a stark beauty all its own. It's like much of the high desert of the United States.

NEIL ARMSTRONG

I don't believe any pair of people had been more removed physically from the rest of the world than we were.

EDWIN E. ALDRIN JR.

12

"Magnificent desolation"

IT WAS TIME for Neil Armstrong to walk on the moon, and as Armstrong waited for Aldrin to follow him out in nineteen minutes his first reaction to the environment was a favorable one. He was able immediately to discard the theory, once widely held, that the windless surface of the moon was overlaid with a dangerously deep coating of dust in which men and man-made machines would founder. The lunar module's footpads had made only a shallow penetration, and Neil's boots sank only a fraction of an inch: "Maybe an eighth of an inch, but I can see the footprints of my boots and the treads in the fine sandy particles." And he could move around: "There seems to be no difficulty. . . . It's even perhaps easier than the simulations at one-sixth G. . . . It's actually no trouble to walk around. The descent engine did not leave a crater of any size. There's about one foot clearance on the ground. We're essentially on a very level place here. I can see some evidence of rays emanating from the descent engine, but very insignificant amounts. . . . Okay, Buzz, we're ready to bring down the camera."

"I'm all ready," Aldrin answered. . . . "Okay, you'll have to pay out all the LEC [lunar equipment conveyor]. It looks like it's coming out nice and evenly."

"Okay, it's quite dark here in the shadow and a little hard for me to see if I have good footing. I'll work my way over into the sunlight here without looking directly into the sun."

"Okay, it's taut now," Aldrin said. "Don't hold it quite so tight."

Armstrong looked up at the LM and messaged: "I'm standing directly in the shadow now, looking up at Buzz in the window. And I can see everything clearly. The light is sufficiently bright, backlighted into the front of the LM, that everything is very clearly visible."

"Okay, I'm going to be changing this film magazine," Aldrin said.

"Okay," Armstrong said. "Camera installed on the RCU bracket [remote control unit]. I'm storing the LEC on the secondary strut. . . . I'll step out and take some of my first pictures here."

HOUSTON (McCandless): Roger, Neil, we're reading you loud and clear. We see you getting some pictures and the contingency sample.

"This is very interesting," Armstrong said. "It's a very soft surface, but here and there where I plug with the contingency sample collector, I run into a very hard surface. . . . I'll try to get a rock in here. Here's a couple."

"That looks beautiful from here, Neil," Aldrin said from Eagle's cabin.

"It has a stark beauty all its own. It's like much of the high desert of the United States. It's different, but it's very pretty out here."

In El Lago Jan Armstrong was ticking off the minutes, but not out of any particular safety concern; her concern was still the one she had expressed much earlier: would they be able to do all they had been assigned to do on this first lunar landing mission? But there was tension in the small talk. Jan volunteered, "Buzz will not come down until Neil gets the contingency sample — which will fall out of his pocket." There was some discussion about how to tell who was saying what; Armstrong and Aldrin had voices of similar timbre, and even people who knew them well had had difficulty with voice identifications throughout the flight. Ricky Armstrong suggested, "You can recognize Daddy — he always says 'Uhhh.' " In Nassau Bay Pat Collins commented: "Look at Neil move. He looks like he's dancing — that's the kangaroo hop." Waiting for Buzz to come out, Joan Aldrin reacted to Armstrong's physical descriptions of the lunar surface with amazement: "He *likes* it!" Later Neil Armstrong said. . . .

« The most dramatic recollections I had were the sights themselves. Of all the spectacular views we had, the most impressive to me was on the way to the moon, when we flew through its shadow. We were still thousands of miles away, but close enough so that the moon almost filled our circular window. It was eclipsing the sun, from our position, and the corona of the sun was visible around the limb of the moon as a gigantic lens-shaped or saucer-shaped light, stretching out to several lunar diameters. It was magnificent, but the moon was even more so. We were in its shadow, so there was no part of it illuminated by the sun. It was illuminated only by earthshine. It made the moon appear blue-gray, and the entire scene looked decidedly three-dimensional.

I was really aware, visually aware, that the moon was in fact a sphere, not a disc. It seemed almost as if it were showing us its roundness, its similarity in shape to our earth, in a sort of welcome. I was sure that it would be a hospitable host. It had been awaiting its first visitors for a long time. »

"You can really throw things a long way up here," Armstrong said. "That pocket open, Buzz?"

"Yes, it is, but it's not up against your suit though. Hit it back once more. More toward the inside. Okay, that's good."

"That in the pocket?"

"Yes, push down," Aldrin said. "Got it? No, it's not all the way in. Push it. There you go."

"Contingency sample is in the pocket," Armstrong confirmed. " . . . Are you getting a TV picture now, Houston?"

HOUSTON (McCandless): Neil, yes we are getting a TV picture. . . . You're not in it at the present time. We can see the bag on the LEC being moved by Buzz, though. Here you come into our field of view.

ALDRIN: "Okay. Are you ready for me to come out?" ARMSTRONG: "Yes. Just stand by a second. I'll move this over the handrail. Okay?" ALDRIN: "All right. That's got it. Are you ready?" ARMSTRONG: "All set. Okay, you saw what difficulties I was having. I'll try to watch your PLSS from underneath here. . . . Okay. Your PLSS is — looks like it is clearing okay. The shoes are about to come over the sill. Okay, now drop your PLSS down. There you go. . . . About an inch clearance on top of your PLSS. . . . Okay, you're right at the edge of the porch. . . . Looks good." ALDRIN: "Now I want to back up and partially close the hatch, making sure not to lock it on my way out." ARMSTRONG: "A particularly good thought." ALDRIN: "That's our home for the next couple of hours and I want to take good care of it."

Watching television in Nassau Bay, Rusty Schweickart encouraged: "Don't close it all the way, Buzz. What do you mean? You forgot the key?" As Buzz started down the ladder Jan Armstrong wondered aloud, "Wouldn't that be something if they locked themselves out?" In the Aldrin home the astronaut Fred Haise, watching Armstrong move across the screen, laughed and said, "That's the fastest I've ever seen Neil move!" He warned Joan Aldrin, "Buzz is about to come out now, Joan." She said, "It's like making an entrance on stage." But as seconds passed with no sight of Buzz, she said, "You see, he's doing it just like on Gemini 12. He's going to explain every single thing he does." Then Buzz's legs did appear. Joan screamed and kicked her own legs up in the air. As he came slowly down the ladder she said, "He's going to analyze every step at a time."

ALDRIN: "It's a very simple matter to hop down from one step to the next." ARMSTRONG: "Yes, I found it to be very comfortable, and walking is also very comfortable. You've got three more steps and then a long one."

ALDRIN: "Okay, I'm going to leave that one foot up there and both hands down to about the fourth rung up." ARMSTRONG: "There you go." ALDRIN: "Okay. Now I think I'll do the same." ARMSTRONG: "A little more. About another inch. There you got it. That's a good step, about a three-footer." ALDRIN: "Beautiful view." *Ten-fifteen in the evening, Houston time. . . . They were now both on the moon, and for a few seconds they were awed in spite of themselves. . . .*

"Isn't that something?" Armstrong asked. "Magnificent sight out here."

"Magnificent desolation," Aldrin said. And later . . .

« Neil and I are both fairly reticent people, and we don't go in for free exchanges of sentiment. Even during our long training we didn't have many free exchanges. But there was that moment on the moon, a brief moment, in which we sort of looked at each other and slapped each other on the shoulder — that was about the space available — and said, 'We made it. Good show.' Or something like that. . . . »

Now there was the work to do. But for all the high drama that had attached to Eagle's touchdown, and to Neil Armstrong's first step onto the moon (no matter how many lunar landings there might be to come, that emotional tension would never be felt in quite the same way again), the lunar exploration part of Apollo 11's flight plan had been pared down to some bare essentials. It was, to pick up Neil Armstrong's own words, "one small step"; and it reflected General Phillips's observation, before launch, that we were going to go to the moon more than once. The original flight plan had called for the astronauts to leave the lunar module, collect the soil samples which might provide clues concerning the moon's origin, return to the LM for a sleep period, then go out again to deploy something called ALSEP (Apollo Lunar Surface Experiments Package), which was actually a sophisticated geophysical station created by the Bendix Corporation. The concept of ALSEP had evolved from a meeting of the National Academy of Sciences' Space Science Board at Woods Hole, Massachusetts, in the summer of 1964, which settled on fifteen basic questions about the nature of the moon — its internal structure, its geometric shape, its surface composition and internal energy, if any; its dominant processes of erosion, its age and its history of dynamic interaction with the earth; its cosmic and solar radiation flux and the nature of its magnetic fields, if any — etc., etc. But by late 1968, with the lunar module itself still unproved in space, the cooler heads in NASA had concluded that this sounded too ambitious for the first lunar landing. Also, the ALSEP added weight to the descent stage, which reduced hover time and there was uncertainty about the terrain and the total requirements for exact hover time. Armstrong and Aldrin were assigned to go to the moon, land, come back and report on the nature of the environment; other explorers would follow them, just as other explorers had followed Columbus and Dr. Livingstone. There should be no question of them wandering off into the lunar hills or climbing down into a lunar crater, out of communication with the earth; until we knew more about the environment, it seemed the better part of wisdom for them to stay reasonably close to Eagle — within fifty to one hundred feet. And, until we knew more about the difficulties of performing physical tasks on the surface of the moon, it seemed wise not to give the astronauts too many tasks to perform. Aware of the unknowns in what could only be called "space sickness," but which had been characteristic, up to now, of Apollo missions, NASA's medical community argued for conservatism and expressed the hope that Armstrong and Aldrin would not be on the lunar surface more than two hours; there would be time enough later for longer and more complicated tasks. Therefore ALSEP, which would require at least half an hour for deployment on the surface, became EASEP (Early Apollo Scientific Experiments Payload). It used ALSEP-proved components, but its two packages could be removed and deployed within ten minutes. One was the so-called Passive Seismic Experiment (PSE), which had as its basic unit a suspended weight that tended to remain immobile as the experiment moved with the motions of the moon. A subsystem would measure seismic activity on or in the moon and transmit data to the earth. The other part of EASEP was the Laser Ranging Retro-Reflector Experiment (LRRR), which would be set up as a target for earth-based laser beams. It had no electronics; it looked rather like a honeycomb, because of its one hundred cylindrical cavities, each containing a fused silica prism shaped like the inside of a hollow cube sliced in half. Light hitting one of these "corner reflectors" (a flat reflecting surface would be useless, since the angle of incidence would equal the angle of reflection) would bounce off each of the three faces of the corner and be reflected straight back to its earthly source. Thus earth-based laser beams could get, finally, an accurate measurement of the distance from the earth to the moon, hopefully within inches.[1] Both EASEP packages were about knee-high with their carrying handles, although the PSE's antenna, when deployed to send data back to the earth, reached just above a man's waist. Both had carrying handles, and in lunar terms, were fairly light (that weight problem again!) — the PSE had a lunar weight of a little more than sixteen pounds, and the LRRR a lunar weight of about eleven pounds. Both would be left on the moon to intrigue the scientists, but a third instrument would not. That was a sheet of aluminum foil, about the size of a big window shade, to be hung from a pole, facing the sun, for the purpose of trapping gas particles. At the end of their EVA Armstrong and Aldrin would roll it up, return it to the lunar module and bring it back to the earth for laboratory analysis.

Yet the highest scientific priority was moon rocks. The "contingency sample" was already in Neil Armstrong's suit pocket, but there were also two sample containers, shaped like a fisherman's tackle box, to be filled with about fifty pounds (earth weight) of lunar specimens which were geologically priceless. One container was designed to carry about twenty pounds of a bulk sample scooped into one large bag; the second container had about fifteen bags to be filled with documented lunar samples and a subsurface "core sample." If Apollo 11 brought back to the earth nothing more than fifty pounds of rocks, if even the slimmed-down EASEP packages had to be abandoned without deployment, some scientists would feel cheated, but the scientific community as a whole would have considered the voyage more than worthwhile. The question of rocks, as a matter of fact, was involved in the whole manned flight argument. Could not, one school argued, an unmanned vehicle land on the moon, scoop up rocks and take off again? (Indeed, was that what the Soviets' Luna 15, now in orbit around the moon, intended to attempt?) Rocks were serious business, and Armstrong and Aldrin would get at the business right away. . . . Along with two other pieces of serious business: the American flag and that plaque, both of which would be left on the moon. . . .

ALDRIN: "Reaching down fairly easy. Can't get my suit dirty at this stage." [2] ALDRIN: "The mass of the backpack does have some effect on inertia." ARMSTRONG: "You're standing on a rock, a big rock there now." ALDRIN: "This pad sure didn't sink much." ARMSTRONG: "No, it didn't . . . I wonder if that right under the engine is where the probe might have hit . . . Yes, I think that's a good representation of our sideward velocity at touchdown there." ALDRIN: "Can't say too much for the — for the visibility here in the LM shadow without the visor up . . . [*static*] . . ."

HOUSTON (McCandless): Try again, please, Buzz — you're cutting out.

ALDRIN: "I say, the rocks are rather slippery. . . . Very powdery surface when the sun hits. The powder fills up all the very little fine porouses. My boot tends to slide over it rather easily. . . . About to lose my balance in one direction and recovery is quite natural and very easy . . . And moving arms around — Jack [Schmitt, astronaut-scientist], doesn't lift your feet off the surface . . . We're not quite that lightfooted. . . . Got to be careful that you are leaning in the direction you want to go, otherwise you . . . [*static*] . . . In other words, you have to cross your foot over to stay underneath where your center of mass is. Say, Neil, didn't I say we might see some purple rocks?" ARMSTRONG: "Find the purple rocks?" ALDRIN: "Yes, they are small, sparkly . . . [*static*] . . . I would make a first guess, some sort of biotite. We'll leave that to the lunar analysts, but . . . [*static*] . . ." [Biotite is more commonly called mica.]

In Nassau Bay, watching her husband on the moon, Joan Aldrin blew kisses, laughed and was close to tears — all at the same time. As he advanced across the lunar surface in his kangaroo hop, she asked, "How can you be serious about what you're doing when you're doing *that?*" Watching with Pat Collins, Barbara Gordon, wife of the astronaut Dick Gordon [Apollo 12], thought the movements resembled those of a person who was trying to skate on ice for the first time. Sue Bean, wife of Apollo 12's Al Bean, was impressed by the fact that you could see footprints. In El Lago Jan Armstrong heard about the purple rocks, looked pleased and went back to her flight plan. Ricky Armstrong wanted to know, "How can you tell Daddy from Mr. Aldrin?" *And all this time Mike Collins was behind the moon, out of communication. . . .*

HOUSTON (McCandless): Neil, this is Houston. . . . We're getting a new picture. You can tell it's a longer focal length lens, and for your information, all LM systems are go. Over.

ALDRIN: We appreciate that. Thank you. . . . Neil is now unveiling the plaque. . . .

ARMSTRONG: For those who haven't read the plaque, we'll read the plaque that's on the front landing gear of this LM. First there are two hemispheres. . . . Underneath it says, "Here Men from the planet Earth first set foot upon the Moon, July 1969, A.D. We came in peace for all Mankind." It has the crew members' signatures and the signature of the President of the United States. Ready for the camera? I can . . . Do you want to pull out some of my cable for me, Buzz?

ALDRIN: Houston, how close are you able to get things in focus?

HOUSTON: (McCandless): This is Houston. We can see Buzz's right hand. It is somewhat out of focus. I'd say we're approaching down to probably about eight inches to a foot behind the position of his hand, where he is pulling out the cable.

In Nassau Bay, watching the camera pan around the lunar landscape, Buzz Aldrin's uncle Bob Moon remarked that it looked like west Texas. Joan Aldrin said wonderingly, "I've never seen anything like that before." Audrey Moon asked, "Who has?" The room broke up in relaxed laughter.

ARMSTRONG: "Something interesting. In the bottom of this little crater here. It may be . . ." ALDRIN: "Keep going. We've got a lot more . . . Forty, fifty feet. Why don't you turn around and let them get a view from there and see what the field of view looks like?" ARMSTRONG: "Okay." ALDRIN: "You're backing into the cable. . . . Turn around to your right, I think, would be better." ARMSTRONG: "I don't want to go into the sun if I can avoid it." ALDRIN: "That's right — yeah."

ALDRIN: Okay, Neil, I've got the table out and a bag deployed.

ARMSTRONG: Straight south.

HOUSTON (McCandless): Roger, and we see the shadow of the LM.

ARMSTRONG: Roger, the little hill just beyond the shadow of the LM is a pair of elongated craters about — that will be the pair together, is forty feet long and twenty feet across, and they're probably six feet deep. We'll probably get some more work in there later. [*One hour and 7 minutes expended. . . . Aldrin was now erecting the solar wind experiment, that aluminum foil window shade. . . .*]

ALDRIN: Incidentally, you can use the shadow that the staff makes to line this up perpendicular to the sun. . . . Some of these small depressions, of the boot toes, when moved slowly, produce clods of about three inches. It could suggest exactly what the Surveyor pictures showed when the scoop pushed away a little bit. You get a force transmitted through the upper surface of the soil and about five or six inches of . . . [*static*] . . . breaks loose and moves as if it were caked on the surface though in fact it really isn't.

ARMSTRONG: I noticed in the soft spots where we had footprints nearly an inch deep that the soil is very cohesive, and it will retain a slope of probably seventy degrees . . . [*static*] . . . footprints.

COLUMBIA (Collins): Houston, AOS . . . Houston. Columbia in high gain. Over.

HOUSTON (McCandless): Columbia. This is Houston reading you loud and clear. Over. [*Collins was back on the near side of the moon, in communication again. . . .*]

COLUMBIA (Collins): Yes. How's it going?

HOUSTON (McCandless): Roger. The EVA is progressing beautifully. I believe they are setting up the flag now.

COLUMBIA (Collins): Great!

HOUSTON (McCandless): I guess you're about the only person around that doesn't have TV coverage of the scene.

COLUMBIA (Collins): That's all right. I don't mind a bit. How is the quality of the TV?

HOUSTON (McCandless): Oh, it's beautiful, Mike. Really is.

COLUMBIA (Collins): Oh gee, that's great. Is the lighting halfway decent?

HOUSTON (McCandless): Yes, indeed. They've got the flag up, and you can see the Stars and Stripes on the lunar surface.

COLUMBIA (Collins): Beautiful. Just beautiful. . . . [*Neil Armstrong now on the lunar surface for about forty-five minutes, Buzz Aldrin there for about half an hour. . . .*]

It was about ten minutes of eleven, Sunday evening, in El Lago, Texas. It was time for Armstrong and Aldrin to get the American flag in position to be left on the moon.[3] They had brought along a hammer, plus scoops and tongs to collect the lunar samples; and now the hammer came in handy. The windless condition of the moon's surface had been anticipated; the flag would hang limply if it were not supported, so this flag had a metal strip woven into its upper edge that would lock at a right angle to the staff. If it could not "fly" on the moon, it could at least stand there with all fifty stars showing. What had not been anticipated was what nobody could have known: that beneath the powdery surface lay a compact subsurface of some kind. Now the hammer was used to pound the staff of the flag deep enough to make the flag stand. It stood.

There was a different kind of restlessness in the astronauts' homes: everyone wanted to see the men get on with it. In El Lago, watching Buzz try out his kangaroo hop, Jan Armstrong said, "Dad is back there guarding his heart." Ricky Armstrong agreed: "Dad wants to see how long he can stand still." In Nassau Bay, Joan Aldrin sympathized with Mike Collins: "He doesn't know what's going on, poor Mike — isn't that terrible?" Then the flag appeared. There were claps and cheers and shouts of "There it goes!" Joan complained, "I can't tell one from the other." The flag ceremony seemed to go on and on, and at Mike Collins's house Rusty Schweickart mourned, "I can see all the scientists of the world on the edges of their chairs and saying: 'For God's sake get this over and get those rocks!' " Sue Bean's mother, Mrs. Edward B. Ragsdale, thought one of the astronauts looked headless: "Now which one is that?" . . . *Five minutes of eleven, Sunday night in Houston. . . .*

HOUSTON (McCandless): Tranquility Base, this is Houston. Could we get both of you on the camera for a minute, please?

ARMSTRONG: Say again, Houston.

HOUSTON (McCandless): Roger. We'd like to get both of you in the field of the view of the camera for a minute. . . . Neil and Buzz, the President of the United States is in his office and would like to say a few words to you. Over.

ARMSTRONG: That would be an honor.

HOUSTON (McCandless): Go ahead, Mr. President, this is Houston. Out.

THE PRESIDENT: Hello, Neil and Buzz, I am talking to you by telephone from the Oval Room at the White House. And this certainly has to be the most historic telephone call ever made. I just can't tell you how proud we all are of what you — for every American, this has to be the proudest day of our lives. And for people all over the world, I am sure that they too join with Americans in recognizing what an immense feat this is. Because of what you have done, the heavens have become a part of man's world. And as you talk to us from the Sea of Tranquility, it inspires us to redouble our efforts to bring peace and tranquility to earth. For one priceless moment, in the whole history of man, all the people on this earth are truly one — one in their pride in what you have done. And one in our prayers, that you will return safely to earth.[4]

ARMSTRONG: Thank you, Mr. President. It's a great honor and privilege for us to be here representing not only the United States but men of peace of all nations — and with interest and a curiosity and men with a vision for the future. It's an honor for us to be able to participate here today.

THE PRESIDENT: And thank you very much; and I look forward — all of us look forward — to seeing you on the *Hornet* next Thursday.

ARMSTRONG: Thank you.

ALDRIN: I look forward to that very much, sir.

ALDRIN: "I noticed several times in going from the sunlight into the shadow that just as I go in, there's an additional reflection off the LM that — along with the reflection off my face onto the visor — makes visibility very poor just at the transition — sunlight into the shadow. I think we have so much glare coming off my visor that my . . . [*static*] . . . Then it takes a short while for my eyes to adapt to the lighting conditions." ALDRIN: "But inside the shadow area visibility, as we said before, is not too great, but with visors up we can certainly see what sort of footprints we have and the general condition of the soil. Then, after being out in the sunlight a while, it takes — watch it, Neil. Neil, you're on the cable!" ARMSTRONG: "Okay." ALDRIN: "Lift up your right foot. Right foot. Your toe is still hooked in it." ARMSTRONG: "That one?" ALDRIN: "Yes, it's still hooked in it. Wait a minute. Okay, you're clear now." ARMSTRONG: "Thank you." ALDRIN: "Now, let's move that over this way." [*"That" was the scoop for the bulk sample collection. . . . Work to do. . . .*] ALDRIN: "The blue color of my boot has completely disappeared now into this — still don't know exactly what color to describe this other than grayish-cocoa color. It appears to be covering most of the lighter part of the boot . . . very fine particles. . . ." [*One hour and a half expended on the support systems, the backpacks. . . .*]

HOUSTON (McCandless): Columbia, this is Houston. Over.

COLUMBIA (Collins): Houston, Columbia on Delta.

HOUSTON (McCandless): Roger. You should have VHF AOS with the LM right about now, with VHF LOS about forty minutes fifteen seconds. [*Heart rates on Armstrong and Aldrin between ninety and one hundred, right on the predicted number of BTUs expended in their work. . . . But neither Mike Collins nor anyone else knew precisely where they were. . . .*]

ALDRIN: "As I look around the area, the contrast in general is . . . [*static*] . . . looking down sun, zero phase, it's a very light-colored gray, light gray color. I see a halo around my own shadow, around the shadow of my helmet. Then as I look off cross-sun, the contrast becomes strongest . . . [*static*] . . . in that the surrounding color is still fairly light as you look down into the sun. . . ." ALDRIN: " . . . considerably darker in texture . . ." [*Armstrong on the television screen, carrying the lunar scoop. . . .*] ALDRIN: "Right in this area, there are two craters. The one that's right in front of me now — as I look off in about the eleven o'clock position from the spacecraft. About thirty to thirty-five feet across. There are several rocks and boulders about six to eight inches across . . . many sizes . . . [*static*] . . ."

HOUSTON (Riley): Neil is filling the bulk sample bag attached to a scale. You can see him in the picture. Buzz is behind the LM at the minus Z strut. That's the landing gear directly opposite the ladder. . . . Buzz is making his way around the LM photographing it from various angles, looking at its condition on all sides. Neil is still occupied with the bulk sample. One hour and forty minutes' time expended on the PLSSs now.

ALDRIN: "How's the bulk sample coming, Neil?" ARMSTRONG: "Bulk sample is just being sealed."

COLUMBIA (Collins): Houston, Columbia.

HOUSTON (McCandless): Columbia, this is Houston. Go ahead. Over.

COLUMBIA (Collins): Roger. No marks on the LM that time. I did see a suspiciously small white object — whose coordinates are . . . [*static*] . . . right on the southwest rim of a crater. I think they would know it if they were in such a location. It looks like their LM would be pitched up quite a degree. It's on the southwest wall of the smaller crater.

In Nassau Bay, Pat Collins heard her husband say he had seen a bright light and laughed. "It's Venus, Mike," she said. "He always sees Venus. It must be in his horoscope." As the minute hand of the clock moved toward midnight, Pat's sister Ellie yawned. "And high time they went back, and don't slam the door," she said. Pat reflected aloud, "After all the conversations over the past few months — 'We won't be able to do this and we won't be able to do that.' Here they are with no problems. We were all so worried about the fatigue factor, and it's not even there."

HOUSTON (McCandless): Neil and Buzz, this is Houston. To clarify my last, your consumables are in good shape at this time. The thirty minutes reference was with respect to the nominal time. Over.

ALDRIN: Roger, I understand that.

ARMSTRONG: "I don't note any abnormalities in the LM. The quads seem to be in good shape. The primary and secondary struts are in good shape. Antennas are all in place. There's no evidence of problem underneath the LM due to engine exhaust or drainage of any kind." ALDRIN: "It's very surprising, the surprising lack of penetration of all four of the foot pads. I'd say if we were to try to determine just how far below the surface they would have penetrated you could measure two or three inches. Wouldn't you say, Neil?" ARMSTRONG: "At the most, yes. But that Y strut there is probably even less than that." ALDRIN: "I'll get a picture of the plus Y strut taken from near the descent stage, and I think we'll be able to see a little bit better what the effects are. . . . There's one picture taken in the right rear of the spacecraft looking at the skirt of the descent stage. . . . On descent, both of us remarked that we could see a very large amount of very fine dust particles moving out. It was reported beforehand that we would probably see an out-gassing from the surface after actual engine shutdown but, as I recall, I was unable to verify that. . . . This is too big an angle, Neil." [Aldrin was taking a picture of the earth above the LM.] ARMSTRONG: "Yes, I think you are right." ALDRIN: "We're back at the minus Z strut now. The stereo pair we're taking of the pad will show the very little force of impact that we actually had." ALDRIN: "And Neil, if you'll take the camera, I'll get to work in the SEQ bay." ARMSTRONG: "Okay, I'm taking some close-up pictures of that rock." ALDRIN: "I was saying that Houston — when you stop and take a photograph or something and then want to start moving again sideways there's quite a tendency to start doing it with just gradual sideways hops, until you start getting moving. Can you see us underneath the LM over at the SEQ bay, Houston?" . . . [The scientific equipment bay contained scientific instruments — e.g., the laser reflector and the passive seismometer — to be left on the moon.]

HOUSTON (McCandless): Yes, indeed, Buzz. We can see your feet sticking out underneath the structure of the LM descent stage. . . . Now we can see you through the structure of the minus Z secondary strut. [*An hour and a half of lunar surface time for Neil Armstrong. . . .*]

ALDRIN: Houston. The passive seismometer has been deployed manually.

HOUSTON (McCandless): Roger.

ALDRIN: "In the manual deployment of the LR cubed [the laser reflector is sometimes called the LR cubed vehicle because of the three R's in its full name] . . . the spring that is at the end of the string pulled off of the pin's head. However, I was able to reach up and get hold of the pin's head and pull it loose. It will be deployed manually, also . . ." ARMSTRONG: "And the panorama is complete and . . . got the LM at seven-thirty position at about sixty feet. . . ." [*Neil Armstrong at the left of the television screen. . . . Houston wanting to know if they had a "good area" picked out. . . .*] ARMSTRONG: "Well, I think straight out on that rise out there is probably as good as any. It'll probably stay on the high ground there and . . ." ALDRIN: "Watch it, the edge of that crater drops. . . ." ARMSTRONG: "That's kind of a drop-off there, isn't it?" ALDRIN: "Take a couple of close-ups on these quite rounded large boulders. . . ." [*Aldrin coming into view on the right carry-*

ing the two EASEP experiments. . . .] ARMSTRONG: "About forty feet out — I'd say out at the end of that next. . . ." ALDRIN: "It's going to be a little difficult to find a good level spot here." ARMSTRONG: "What about next to the ridge there? Wouldn't that be a pretty good place?" ALDRIN: "All right. Should I put the LR-cubed vehicle here?" ARMSTRONG: "All right." ALDRIN: "I'm going to have to get on the other side of this rock here." ARMSTRONG: "I would go right around that crater to the left there. Isn't that a level spot there?" ALDRIN: "No, I think this right here is just as level." ARMSTRONG: "These boulders look like basalt and they have probably a two percent white mineral to them, the white crystals. And the thing that I reported as the vesicular[5] before, I'm not — I don't believe that anymore — I think that small craters . . . they look like little impact craters where shot . . . BB shot . . . has hit the surface." ALDRIN: "Houston. I have the seismic experiment flipped over now and I'm leveling it, but I'm having a little bit of difficulty. . . ." [*Houston: You're cutting out again, Buzz . . .*] ALDRIN: "I say I'm not having too much success in leveling the PSE experiment." ARMSTRONG: "The laser reflector is installed and the bubble is leveled and the alignment appears to be good." ALDRIN: "Hey [Houston]. You want to take a look at this BB [a leveling device in the PSE] and see what you make of it?" ARMSTRONG: "I found it pretty hard to get perfectly level, too." ALDRIN: "That BB likes the outside. It won't go to the inside." ARMSTRONG: ". . . But that little cup is convex now instead of concave." ALDRIN: "I think you're right. . . . Houston, I don't think there's any hope for using this leveling device to come up with an accurate level. It looks to me as though the cup here — that the BB is in, is now convex instead of concave. Over . . ." [*Houston: Press on. If you can level by eyeball, go ahead. . . . The problem was to level the passive seismic monitor; try to do it now by visual . . .*] ARMSTRONG: There you go. Good work." ALDRIN: "Good show." ARMSTRONG: "Hey, stop, stop! Back up!" ALDRIN: "Houston . . . the right-hand solar array deployed automatically. The left-hand I had to manually — bending the bar at the far end. All parts of the solar array are clear on the ground now."

HOUSTON (McCandless): Buzz, this is Houston. I understand that you did successfully deploy both solar arrays. Over.

ALDRIN: That's affirmative.

HOUSTON (McCandless): We've been looking at your consumables, and you're in good shape. Subject to your concurrence, we'd like to extend the duration of the EVA fifteen minutes from nominal. We will still give Buzz a hack at ten minutes prior to heading in. Your current elapsed time is two plus twelve. Over. [*Two hours and twelve minutes of time expended on the PLSSs. . . . Time to collect the core tube sample, the subsurface sample. . . .*]

HOUSTON (McCandless): This is Houston. If you're still in the vicinity of the PSE, could you get a photograph of the ball level? Over.

ARMSTRONG: "I'll do that, Buzz." ALDRIN: "Right. We'll get a photograph of that. Houston, what time would you estimate we could allow for the documented sample? Over." ARMSTRONG: "Oh, shoot! Would you believe the ball is right in the middle now?" ALDRIN: "Wonderful. Take a picture before it moves."

HOUSTON (McCandless): Neil, this is Houston. We're estimating about ten minutes for the documented sampling. Over . . . Columbia, Columbia, this is Houston. Over.

COLUMBIA (Collins): Go ahead, Houston, Columbia.

HOUSTON (McCandless): Roger. Like you to terminate charging battery Bravo at 111 plus 15 [GET]. Over.

COLUMBIA (Collins): How about right now?

HOUSTON (McCandless): Roger. Buzz, this is Houston. You've got about ten minutes left now prior to commencing your EVA termination activities. Over.

ALDRIN: Roger, I understand. . . . [*Armstrong on the lunar surface now for nearly two hours. . . . Aldrin collecting a core tube sample on the television screen. . . .*]

ALDRIN: I hope you're watching how hard I have to hit this into the ground to the tune of about five inches, Houston.

HOUSTON (McCandless): Roger.

ALDRIN: It almost looks wet. . . .

In Nassau Bay, Joan Aldrin was thinking that there had been an unduly long time of inaction — and then there was movement on the television screen. "There they are," Joan said. "What *are* they doing?" Astronaut Gerald Carr answered: "He has to go to the bathroom — he's looking for a bush!" Fred Haise said, more seriously, "Buzz is getting the second rock box out now. . . . There's the first core tube — Buzz is going to pound it into the ground. . . . Don't hit your hand now, Buzz. . . ."

HOUSTON (McCandless): Neil and Buzz, this is Houston. . . . Neil, this is Houston. We would like you all to get two core tubes and the solar wind experiment. Two core tubes and the solar wind. Over.

ALDRIN: Get the next one. Maybe you can clear away the rocks a little bit.

ARMSTRONG: Yes, I'll take care of it.

HOUSTON (McCandless): Buzz, this is Houston. You have approximately three minutes until you must commence your EVA termination activities. Over.

ALDRIN: Roger. Understand.

HOUSTON (McCandless): Columbia, this is Houston. You have approximately one minute to LOS. Over.

COLUMBIA (Collins): Columbia, Roger. [*Over the hill again. . . .*]

HOUSTON (McCandless): And — do you plan on commencing your sleep on the back side of this pass? If so, we'll disable uplink to you while we're talking to the LM. Over.

COLUMBIA (Collins): Negative that.

HOUSTON (McCandless): Neil, this is Houston. After you've got the core tubes and the solar wind, anything else that you can throw into the box would be acceptable.

ARMSTRONG: "Right-o . . . [*static*] . . ." ALDRIN: "I got the cap." ARMSTRONG: "Got the cap?" ALDRIN: "They both have got caps on them." ARMSTRONG: "Okay." ALDRIN: "And, you want to pick up some stuff, and I'll . . . [*static*] . . ." [*Armstrong picking up rocks, Aldrin retrieving the solar wind experiment. . . . Two hours and twenty-five minutes on the backpacks, time to finish and get back into Eagle with an adequate safety margin. . . .*]

HOUSTON (McCandless): Neil and Buzz, this is Houston. We'd like to remind you of the close-up camera magazine before you start up the ladder.

ALDRIN: Okay. Got that over with you, Neil?

ARMSTRONG: No, the close-up camera's underneath the MESA, so I'll have to pick it up with the tongs. I'm picking up several pieces of really vesicular rock out here now.

ALDRIN: You didn't get anything in those environmental samples, did you?

ARMSTRONG: Not yet.

ALDRIN: I don't think we'll have time.

HOUSTON (McCandless): Roger, Neil and Buzz. Let's press on with getting the close-up camera magazine and closing out the sample return container. We're running a little low on time.

ALDRIN: Roger. . . . Okay, can you quickly stick this in my pocket, Neil, and I'll get on up the ladder. I'll hold it and you open the pocket up.

ARMSTRONG: . . . [*static*] . . . Just hold it right there, Okay. Let the pocket go.

ALDRIN: Adios amigo. . . . Anything more before I head on up, Bruce?

HOUSTON (McCandless): Negative. Head on up the ladder, Buzz. [*Aldrin climbed up the ladder, entered the LM, left the hatch open and was ready to receive the rock boxes. . . .*]

In Nassau Bay, Gerald Carr said the obvious: "There goes the rock box." Joan Aldrin said, "God bless the rock box! I feel as if I'd lived with that rock box for the last six months. . . . It's like a Walt Disney cartoon, or even a television show — it's all too much to believe or understand." Pat Collins had a somewhat different reaction. When someone remarked on how casually the astronauts' wives seemed to accept the fact of their husbands spinning off to the moon, she replied, "In a way we're all conditioned. We've been conditioned and we haven't even realized it, but it's been going on for a long time."

ALDRIN: "How are you coming, Neil?" "Okay." "Get the film off of that." ARMSTRONG: "I will. I got it." ALDRIN: "Okay, I'm heading on in. . . . And I'll get the LEC all ready for the first rock box." ARMSTRONG: "I've got the Hasselblad [camera] magazine hooked to the SRC [sample return container]." ALDRIN: "About ready to send up the LEC?" ARMSTRONG: "Just about." ALDRIN: "Okay. That's got it clear." ARMSTRONG: "Oh, oh, the camera came off. I mean the film pack came off." ALDRIN: "All right. Just ease it down now. Don't pull so hard on it. All right. Let it go." ARMSTRONG: "While you're getting that, I've got to get the camera." [*Consumables still in good shape. . . .*] ALDRIN: "How's it coming, Neil?" ARMSTRONG: "Okay. I've got one side hooked up to the second box and I've got the film back on." ALDRIN: "Okay. Good." ARMSTRONG: "I've got silt from on the LEC. It's kind of falling all over me while I'm doing this." ALDRIN: "Kind of like soot, huh?" ARMSTRONG: "It looks like . . . [*static*] . . . down here." ALDRIN: "I think my watch stopped, Neil. . . . No, it didn't either. Second hand." ARMSTRONG: "Okay. Stand by a minute. Let me move back." ALDRIN: "Okay, easy. All right, easy in the hatch now. . . . Okay. I'll get it the rest of the way."

HOUSTON (McCandless): Neil, this is Houston. Did you get the Hasselblad magazine?

ARMSTRONG: Yes, I did. And we got about — I'd say, twenty pounds of carefully selected, if not documented samples.

HOUSTON (McCandless): Houston. Roger. Well done. Out. [*One hundred eleven hours and thirty-seven minutes into the mission, both Armstrong and Aldrin now off the lunar surface. . . .*]

ALDRIN: "Okay, now start arching your back. That's good. Plenty of room. You're all right, arch your back and move your head up against . . . [*static*] . . . Roll right just a little bit. Head down. And in good shape." ARMSTRONG: "Thank you. I'm bumping now?" ALDRIN: "Now you're clear. You're rubbing up against me a little bit. . . . Okay. Now move your foot, and I'll get the hatch. . . . Okay, the hatch is closed and latched. And we're . . . [*static*] . . . secure." ARMSTRONG: "Okay. Now we turn the feed-water valve. And I got your PLSS antenna stowed." ALDRIN: "Okay. Feed-water valve closed, and your antenna's stowed . . . [*static*] . . ."

Cabin pressure coming up again . . . 2.7, .8, .9 pounds per square inch . . . 3 pounds . . . 4 pounds . . . 4.8 pounds. . . . Another fifteen min-

utes required to switch back to the lunar module's interior communications system. . . .

Back in the LM, Neil Armstrong thought it all over. . . .

« My impression was that we were taking a snapshot of a steady-state process, in which rocks were being worn down on the surface of the moon with time and other rocks were being thrown out on top as a result of new events somewhere near or far away. In other words, no matter when you had been to this spot before, a thousand years ago or a hundred thousand years ago, or if you came back to it a million years from now, you would see some different things each time, but the scene would generally be the same. I could only guess at the nature of this steady-state process, or whether there was more than one kind of process involved in the evolution of the moon to the state in which we found it. I thought that if there were several kinds, most of the processes seemed to be external, but the nature of the materials involved here would seem to indicate that some kind of internal process had gone on at some time on or inside the moon. In any event that first hour on the moon was hardly the time for long thoughts; we had specific jobs to do. Of course the sights were simply magnificent, beyond any visual experience that I had ever been exposed to. Of course I thought about the magnificence of the whole thing, but that's difficult to capture in a simple description. »

HOUSTON (McCandless): Neil, this is Houston. If you read, we suggest you unstow one PLSS antenna so we can have communications. Over.

EAGLE (Armstrong): Houston, do you read? [*Time to reacquire Mike Collins. . . .*]

HOUSTON (McCandless): Columbia, Columbia, this is Houston. Over.

COLUMBIA (Collins): Roger. . . . How do you read?

HOUSTON (McCandless): Roger, Columbia. This is Houston. We're reading you loud and clear on OMNI Charlie. The crew of Tranquility Base is back inside their base, repressurized, and they're in the process of doffing the PLSSs. Everything went beautifully. Over.

COLUMBIA (Collins): Hallelujah!

It was past twelve-thirty in the morning, Houston time, and the astronauts had been up and at work for more than nineteen hours; yet they could not go to sleep quite yet. The cabin of Eagle had to be depressurized once again before jettisoning redundant equipment, including the two PLSSs, onto the lunar surface; while this was going on Armstrong and Aldrin would get their oxygen from the hoses which now reconnected them with the lunar module's life support system; the oxygen supply of the two OPSs had to be held in reserve in case they failed to dock successfully with Columbia and had to walk through space to the command module.

EAGLE (Aldrin): We've probably got another half-hour's worth of picture taking, and I guess we could run through an eat cycle.

HOUSTON (McCandless): Roger. That sounds fine to us.

EAGLE (Aldrin): Well, it will be a little crowded in here for a while.

COLUMBIA (Collins): Houston, Columbia. You got the new coordinates? [*Collins was still looking for Eagle on the lunar surface. . . .*]

HOUSTON (McCandless): Roger. Latitude 00.691, that would be plus 00.691, and longitude over 2 is plus 11.713. . . . Grid coordinates kilo decimal 9, 6 decimal 3, on LAM 2.

COLUMBIA (Collins): Thank you. One of these grid squares is about as much as you can stand on a single pass.

HOUSTON (McCandless): We understand this is intended to be your last P22. We don't want to use up too much fuel in this effort.

COLUMBIA (Collins): I say again, I am maneuvering to the P52 attitude, and do you want a crew status report?

HOUSTON (McCandless): Roger, and go ahead with your crew status report.

COLUMBIA (Collins): Roger, no medication, radiation 100.16 . . .

EAGLE (Armstrong): The weight of the RCU was twelve ounces by itself without the bag, and the weight of the CDR's PLSS water was twelve and one-half ounces. That's reading zero with the bag on.

HOUSTON (McCandless): This is Houston. We copy. And for your information the new LM weight after jettison of equipment, including lithium hydroxide canister, is 10,837 [pounds].

HOUSTON (McCandless): Columbia, this is Houston. . . . Couple of quick flight plan updates here. . . . And we have a little less than two minutes to LOS. . . . And Columbia, if it's agreeable with you, we'd like for you to stay awake until we have one successful acquisition on the high gain antenna, and I guess you can plan on turning in shortly after AOS in this next pass. Over.

COLUMBIA (Collins): Okay.

Before the mission Mike Collins had speculated on whether he would be able to sleep at all when he was alone in the command module. . . .

« There is a problem in that when you go around the back side there is a period of forty-five minutes or fifty minutes when the ground is unable to give you any telemetry, so you're completely in the dark, literally and figuratively, and with nobody awake to watch the candy store. This sort of rubs you the wrong way, and I'm not sure whether I'll come to some mental accommodation with that and sleep like a log or whether I won't. I may be dead tired and just say, 'Well, you've got it, and I'm going to sleep,' and not be concerned about it. On the other hand I can see where I might be pussyfooting around in there and fretting. I might be dozing for half an hour and then start looking at all those crazy fuel cells and everything. Part of it depends on how the flight has been going up to that point. If we've been plagued with a bunch of little problems I expect that I'll be less happy about really conking out. On the other hand, if everything has been going along all right, probably I'll get some sleep. . . . »

One hundred thirteen hours eighteen minutes into the flight. . . . Loss of signal from Columbia. . . .

HOUSTON (McCandless): On your next depressurization, it's acceptable to use the overhead hatch dump valve in addition to, or instead of, the forward hatch dump valve to speed up the depressurization of the cabin.

EAGLE (Aldrin): All right, you've got the DSKY.

HOUSTON (McCandless): Roger. Your T13 time is 124:22:02. Over. . . . And do you have a time estimate for us until you're ready to start cabin depress? Over.

EAGLE (Aldrin): Fifteen minutes, maybe? [*In the Houston MSC, Deke Slayton went on the CAPCOM. . . .*]

HOUSTON (Slayton): I want to let you guys know that since you're an hour and a half over the timeline, and we're all taking a day off tomorrow, we're going to leave you. See you later.

EAGLE (Armstrong): I don't blame you a bit.

HOUSTON (Slayton): I really enjoyed it.

EAGLE (Armstrong): Thank you. You couldn't have enjoyed it as much as we did.

HOUSTON (Slayton): Roger. It sure was great. Sure wish you'd hurry up and get that trash out of there, though.

EAGLE (Armstrong): Well, we're just about to do it. [*Columbia reacquired, Mike Collins back following another trip over the hill. . . .*]

HOUSTON (McCandless): Columbia, this is Houston. Over. . . . We've successfully reacquired high gain antenna. Unless you have some other traffic with us, I guess we'll bid you a good night and let you get some sleep, Mike. Over.

COLUMBIA (Collins): Okay. Sounds fine. [*Good night to Mike Collins and Columbia at 114 hours 6 minutes. . . . Cabin being depressurized at Tranquility Base . . . three and one half pounds . . . one and one half pounds . . . Less than a pound of pressure . . . a tenth of a pound . . . .04. . . . There went something; it looked like a portable life support system. . . .*

Then the other PLSS. . . . All over so quickly; cabin being repressurized now . . . One pound per square inch . . . two pounds . . . three pounds . . . four pounds. . . . Leveling off at 4.8 pounds. . . .]

EAGLE (Armstrong): Houston, Tranquility Base. Repress complete.

HOUSTON (McCandless): Roger, Tranquility. We observed your equipment jettison on the TV and the passive seismic experiment recorded shocks when each PLSS hit the surface. Over.

EAGLE (Armstrong): You can't get away with anything anymore, can you?

HOUSTON (McCandless): No, indeed.

EAGLE (Aldrin): Houston, Tranquility. Have you had enough TV for today?

HOUSTON (McCandless): Tranquility, this is Houston. Yes, indeed; it's been a mighty fine presentation. [*Off the air at 114:25:47 GET, just short of three o'clock in the morning, Houston time. . . . And still no sleep for the very weary. . . .*]

HOUSTON (McCandless): Columbia, Columbia, this is Houston. . . . Sorry to bother you, Columbia. Two things. We request that you select ten-degree dead band in your DAP in accordance with the procedures on Foxtrot 9–7 in your checklist. And secondly, we would like to leave a display on the DSKY that is not one that is cycling, being continuously updated. What you would have when you get through winding the dead band would be a static display, and that will be satisfactory. Over.

COLUMBIA (Collins): Okay.

HOUSTON (McCandless): Roger. Good night again. [*Finally, Mike Collins could get some sleep. . . .*]

HOUSTON (McCandless): Tranquility, this is Houston. We also have a set of about ten questions relating to observations you made, things you may have seen during the EVA, that we can either discuss a little later on this evening or sometime later in the mission. It's your option. How do you feel? Over? [*Past three in the morning in Houston, families in bed, Armstrong and Aldrin still going after nearly twenty-two hours. . . .*]

EAGLE (Armstrong): I guess we can take a couple of them up now.

HOUSTON (McCandless): Okay, and your friendly green team here has pretty well been relieved by your friendly maroon team, and I'll put Owen [Garriott] on with the questions.

EAGLE (Armstrong): Okay. Thank you, Bruce.

HOUSTON (Garriott): Tranquility, Houston. First question here is your best estimate of the yaw on the — of the LM as compared to the nominal of crew flight plan. Over.

EAGLE (Armstrong): We got thirteen degrees left on the ball, and I think that's probably about right. Looking at the shadow and so on, we're probably about thirteen degrees left of the shadow.

HOUSTON (Garriott): Roger, that's thirteen degrees left of the shadow. And

— next question relates to the depth of the bulk sampling that you obtained near the first part of the EVA, and any changes in composition that you might have observed during the bulk sampling interval. Over.

EAGLE (Armstrong): I'm not sure I understand that question, but we got a good bit of the ground mass in the bulk sample, plus a sizable number of selected rock fragments of different types.

HOUSTON (Garriott): Roger, Neil. One of the implications here is the depth from which the bulk sample was collected. Did you manage to get down there several inches or nearer the surface? Over.

EAGLE (Armstrong): We got down some from as much as three inches in the area where I was looking at . . . [*static*] . . . Later on, other types and other areas, where I got just a short distance — an inch or two — and couldn't go any further.

HOUSTON (Garriott): Next question: the second SRC was packed rather hurriedly due to the time limitations, and wonder if you would be able to provide any more detailed description of the samples which were included in the second SRC. Over.

EAGLE (Armstrong): We got two core tubes and a solar wind, and about half of a big sample bag full of assorted rocks, which I picked up hurriedly from around the area. I tried to get as many representative types as I could.

HOUSTON (Garriott): Roger, Neil. Next topic here relates to the rays which emanate from the DPS engine burning area. We were wondering if the rays emanating from the — beneath the engine are any darker or lighter than the surrounding surface. Over.

EAGLE (Aldrin): The ones that I saw back in the aft end of the spacecraft appeared to be a good bit darker. And of course, viewed from the aft end — why, they did have the sun shining directly on them. It seemed as though the material had been baked somewhat and also scattered in a radially outward direction, but in that particular area this feature didn't extend more than about two, maybe three, feet from the skirt of the engine. Over.

HOUSTON (Garriott): I understand, Buzz, that these were — this was the appearance of the material which had been uncovered by the rays that appeared darker for two or three feet extending outward. Is that correct?

EAGLE (Aldrin): No, I wouldn't say it was necessarily material that had been uncovered. I think some of the material might have baked or in some way caused to be more cohesive and perhaps flow together some way — I don't know. Now, in other areas, before we started traveling around out front — why, we could see that small erosion had taken place in a radially outward direction, but it had left no significant mark on the surface other than just having eroded it away. Now it was different back in the — right under the skirt itself. It seemed as though the surface had been baked in a streak fashion. . . . But that didn't extend out very far. Over.

HOUSTON (Garriott): Did either of the solar panels on the PSE touch the surface of the moon during deployment? Over.

EAGLE (Aldrin): I think that corners did touch just when it was deployed, but both of them didn't come out at the same time. It unfolded a little unevenly, and of course the terrain that it was on was a little bit — not quite as level as it was — as I would have liked to have it. I think that two corners did touch to about one inch, three-fourths to half an inch deep and maybe along the bottom it might have been maybe three inches.

HOUSTON (Garriott): Roger. Understand the description there, and the next subject — the two core tubes which you collected, how did the driving force required to collect these tubes compare? Was there any difference? Over.

EAGLE (Aldrin): Not significantly. I could get down to about the first two inches without much of a problem, and then as I would pound it in . . . I think the total depth might have been about eight or nine inches. But even there it didn't for some reason — it didn't seem to want to stand up straight.

HOUSTON (Garriott): We did copy your comments prior to the EVA of your general description of the area. We wonder if either of you would have any more lengthy description, or more detailed description, of the general summary of the geology of the area. Over.

EAGLE (Armstrong): We'll postpone our answer to that one until tomorrow. Okay?

HOUSTON (Garriott): You commented, Neil, that on your approach to the landing spot you had passed over a football field-sized crater containing rather large blocks of solid rock perhaps ten to fifteen feet in size. Can you estimate the distance to [this crater] from your present position? Over.

EAGLE (Armstrong): I thought we'd be close enough so that when we got outside we could see its rim back there, but I couldn't. But I don't think we're more than a half mile beyond it. That is, a half mile west of it.

HOUSTON (Garriott): Okay, well . . . Unless you have something else that will be all from us for the evening. Over. . . . [*But not quite all. . . .*]

HOUSTON (Garriott): Tranquility Base, Houston. Over.

EAGLE (Armstrong): Go ahead, Houston.

HOUSTON (Garriott): Roger. A few more verifications, here. Can you — will you verify that the disc with messages was placed on the surface as planned, and also that the items that are listed in the flight plan, all of those listed there, were jettisoned? Over. [Garriott was referring to the tiny disc which had messages from heads of state, the Apollo 1 patch, the two medallions for the Soviet cosmonauts, the little gold olive branch — all inside a small Beta cloth bag left on the moon.]

EAGLE (Armstrong): All that's verified.

HOUSTON (Garriott): Roger. Thank you, and I hope this will be a final good night.

The flight directors at Mission Control found it difficult to unwind in the small hours of Monday morning, July 21. At 2:30 A.M. Cliff Charlesworth was asked to pass along a message to Bruce McCandless, who was still on duty as CAPCOM. The wife of one of McCandless's neighbors was due to have a baby, and she had gone into labor. But with other things on his mind, Charlesworth got the message garbled in such a way as to puzzle McCandless. At 3 A.M. McCandless called his wife Bernice to ask with alarm, "Hey, I just got word that you're in labor. What's going on?"

One hundred fifteen hours and fifty minutes into the mission, nearly four-thirty in the morning, Houston time. . . . Mike Collins sound asleep in Columbia. . . . Neil Armstrong resting in Eagle but apparently not asleep. . . . No biomedical instrumentation for Buzz Aldrin. . . .

And back inside Eagle's cabin, there was time, finally, for some stock-taking. There was that film pack that had had to be recovered; when it slipped off the lunar module's conveyor belt Neil Armstrong had been more surprised than annoyed. . . .

« I thought I had latched that hook, and I didn't think there was any way for it to come off, but when it did I wasn't worried. I was quite certain that I could pick it up again and send it up with the next rock box, which I did. Fortunately the pack fell right by the side of the ladder, and I had no trouble bending down. You can actually just fall over on your face like a dead man, right down to the surface, and push yourself back up if you have something to grab, like the ladder. I didn't have to do that, of course; I just held onto the ladder and leaned over at an angle of about forty-five degrees, maybe a little more than that, and picked it up with my glove. I could just push myself right back up into a standing position at that point.

I don't know just what the temperatures were outside. I've heard it guessed that they were only zero to one hundred degrees Centigrade. I really wasn't aware of any temperatures inside the suit. And at no time could I detect any temperature penetrating the insulated gloves as I touched things — the LM itself, things in the shadow, things in the sunlight, the tools, the flagpole, the TV camera, the rocks that I held.

Back inside Eagle, once we had repressurized the cabin and removed our helmets, there was one more little surprise. There was a decided odor in the cockpit. It smelled, to me, like wet ashes in a fireplace. It seemed presumptive to jump to the conclusion that the lunar material we had taken into the cabin was causing this odor, but we had to guess that to be the case. Of course, there was the possibility a vacuum condition had had some kind of effect on the wiring insulation inside the spacecraft; that could possibly have caused odors to originate from within the spacecraft itself. But it seemed the less likely possibility. »

As he and Neil Armstrong sought to get some sleep for the first time in twenty-four hours (Aldrin on Eagle's floor, Armstrong leaning back against the aft part of the cabin, essentially lying on the ascent engine), Buzz Aldrin found himself wondering how long those footprints would linger on the surface of the moon. . . .

« The moon was a very natural and very pleasant environment in which to work. It had many of the advantages of zero-gravity, but it was in a sense less *lonesome* than zero G, where you always have to pay attention to securing attachment points to give you some means of leverage. In one-sixth gravity, on the moon, you had a distinct feeling of being *somewhere,* and you had a constant, though at many times ill defined, sense of direction and force.

One interesting thing was that the horizontal reference on the moon is not at all well defined. That is, it's difficult to know when you are leaning forward or backward and to what degree. This fact, coupled with the rather limited field of vision from our helmets, made local features of the moon appear to change slope, depending on which way you were looking and how you were standing. The weight of the backpack tends to pull you backward, and you must consciously lean forward just a little to compensate. I believe someone has described the posture as 'tired ape' — almost erect but slumped forward a little. It was difficult sometimes to know when you were standing erect. It felt as if you could lean farther in any direction, without losing your balance, than on earth. By far the easiest and most natural way to move on the surface of the moon is to put one foot in front of the other. The kangaroo hop did work, but it led to some instability; there was not so much control when you were moving around.

As we deployed our experiments on the surface we had to jettison things like lanyards, retaining fasteners, etc., and some of these we tossed away. The objects would go away with a slow, lazy motion. If anyone tried to throw a baseball back and forth in that atmosphere he would have difficulty, at first, acclimatizing himself to that slow, lazy trajectory; but I believe he could adapt to it quite readily.

Technically the most difficult task I performed on the surface was driving those core samplers into the ground to get little tubes of lunar material for study. There was a significant and surprising resistance just a few inches down. But this resistance was not accompanied by a strong supporting force on the sides. What this meant, quite simply, was that I had to hold on to the top of the core tube extension while I was hitting it with the hammer to drive it down into the ground. I actually missed once or twice. It wasn't a question of visibility. In bringing the hammer down, I tended to disturb my own body position and my balance. One explanation for the strange degree of resistance may be that, having already been compressed by the lack of atmosphere, it has been continually pounded by meteorites. This pounding probably has compacted that lower material much further, to a point where additional compacting — like that of forcing a cutting tool and tube through it — requires significant applications of force. *And, like Armstrong, Aldrin also had noticed that peculiar smell when he returned to Eagle and removed his helmet in the repressurized cabin. . . .*

Odor is very subjective, but to me there was a distinct smell to the lunar material — pungent, like gunpowder or spent cap-pistol caps. We carted a fair amount of lunar dust back inside the vehicle with us, either on our suits and boots or on the conveyor system we used to get boxes and equipment back inside. We did notice the odor right away.

It was a unique, almost mystical environment up there. »

One hundred eighteen hours and fifty minutes into the flight, nearly seven-thirty in the morning, July 21, Houston time. . . . All systems normal. . . . Mike Collins sleeping soundly in Columbia. . . . Neil Armstrong sleeping fitfully, perhaps not at all. . . .

Aldrin had the better sleeping position, lying on Eagle's floor. Armstrong had rigged up a strap around a vertical bar so that it formed a hammock for his feet. Then he found that the earth was peering at him through the telescope. The telescope was in such a position that it had the earth in its field of view, and it was like a big blue eyeball staring right at him. In weightlessness, it had been calculated, a man could sleep just as well standing up; but it did not quite work out that way. The natural way to sleep, on the moon as on the earth, was lying down. . . . And there was another problem which Buzz Aldrin described:

« The thing which really kept us awake was the temperature. It was very chilly in there. After about three hours it became unbearable. We had the liquid cooling system in operation in our suits, of course, and we tried to get comfortable by turning the water circulation down to minimum. That didn't help much. We turned the temperature control on our oxygen system up to maximum. That didn't have much effect, either.[6] We could have raised the window shades and let the light in to warm us, but that would have destroyed any remaining possibility of sleeping.

The light was sometimes annoying because when it struck our helmets from a side angle it would enter the face plate and make a glare which reflected all over it. Then when we entered a shadow, we would see reflections of our own faces in the front of the helmet, and they obscured anything else that was to be seen. Once my face went into shadow; it took maybe twenty seconds before my pupils dilated out again and I could see details. »

So there was to be no sleep for the first men on the moon; they felt as if they were freezing to death. In Columbia, Mike Collins slept well, and he did not feel lonely at all: "I wasn't. I had been flying airplanes by myself for about seventeen years, and the idea of being in a flying vehicle alone was in no way alarming. In fact, sometimes I preferred to be by myself." For these few hours, he certainly was.

If Mike Collins did not know where Eagle was — and he had practically given up looking — one mystery was about to be solved: the fate of the Soviets' Luna 15. At 8:47 in the morning of July 21, Houston time (ten hours later in Moscow), something unfortunate happened to Luna 15. It crashed onto the lunar surface. On July 22 the official Soviet news agency Tass announced: "On July 21, 1969, the program of research in the space near the moon, and of checking the new systems of the automatic station Luna 15, was completed. At 18 hours 47 minutes Moscow time on July 21, a retro-rocket was switched on and the station left the orbit and reached the moon's surface in the preset area." [7]

Luna 15 had reached the moon's surface all right, but not under ideal circumstances. It was not manned. Suppose the Soviet equivalents of Neil Armstrong and Buzz Aldrin had been aboard? Would the landing not have had a better chance of success? After all, if Armstrong and Aldrin had not been aboard Eagle, the lunar module of Apollo 11 would undoubtedly have met the same fate as Luna 15: it would have crashed, guided by a computer, in a lunar boulder field.

We were not distracted by the question of whether the ascent engine would light, but we were surely thinking about it.

NEIL ARMSTRONG

Spacecraft 107, alias Apollo 11, alias Columbia. The best ship to come down the line. God bless her.

MICHAEL COLLINS

13

"Open up the LRL doors, Charlie"

So MUCH redundancy had been built into the whole Apollo program that few spacecraft systems lacked a backup alternative. But there was no real backup for Eagle's ascent engine — its 3,500 pounds of thrust had to get Armstrong and Aldrin off the moon. It had to work. Emergency pickups had been rehearsed, of course, but it was not possible for Mike Collins to swoop down much below ten miles off the lunar surface. For steering purposes, Eagle did have a reaction control system which included four clusters of four engines each, but these sixteen RCS engines had only one hundred pounds of the thrust apiece, and a total of only four hundred pounds in the upward direction. That was not nearly enough if the ascent engine failed to work. Dr. John Houbolt, the author of the lunar orbit concept, had every confidence that the engine would work: "That was one of the best-tested engines in the universe." He felt that the hypergolic, or self-igniting, fuel burned by the ascent engine was an added safety factor; he had pressed for the use of that type of fuel from the beginning. And the engine itself, built, assembled and tested by Bell Aerosystems and the Rocketdyne Division of North American Rockwell, had been designed with maximum simplicity in mind. It had no frills. It could not be throttled. It had only two operating modes — on and off. Its only moving parts were the ball valves that flipped open to let propellants into the injector and thrust chamber — a primary valve and a backup valve opening into a bypass line. In fact, the engine looked like a tiny and insignificant piece of machinery. It stood less than four and one-half feet high, and it weighed only 172 pounds; for the Apollo 11 mission, thirty-four pounds had been trimmed from the engine's weight by replacing the thrust chamber's asbestos lining with a light, glass-reinforced ablator. The engine had passed its acceptance test firing, and similar engines had been ground-tested hundreds of times without a hitch. Predecessors had performed flawlessly on the flights of Apollo 9 and Apollo 10. But this engine had to fire once more if Armstrong and Aldrin were to return safely to the earth.

One hundred twenty hours and fifty-nine minutes into the mission. . . . Eight-thirty Monday morning, July 21, Houston time.

HOUSTON (Evans): Columbia, Columbia. Good morning from Houston.

COLUMBIA (Collins): Good morning.

HOUSTON (Evans): Hey, Mike. How's it going this morning?

COLUMBIA (Collins): I don't know yet. How's it going with you?

HOUSTON (Evans): Real fine here. Columbia, request Poo and ACCEPT. We'll shove the state vector in for you right away. . . . Okay, it's coming up now, Columbia. We're going to keep you a little busy here. As soon as we get the state vector in we'd like you to go ahead and do a P52 option 3 on this night pass, and then when you come on around the other side there we'll give you some landmark tracking information on prime 130. . . . We're working on the grid squares and will get them shortly. . . . Tango 2, 122 plus 21 plus 11 and six miles north of track.

COLUMBIA (Collins): You've updated your information as to the LM's position and this is your best estimate of where the LM is. Is that correct?

HOUSTON (Evans): Columbia, that's negative. This 130 is the little bitty crater there that you tracked — John Young's crater — that you tracked prior to descent.

COLUMBIA (Collins): Fine, okay. You've given up looking for the LM, right?

HOUSTON (Evans): Affirmative. We want this for one last fix on your plane. [*Collins simply was not destined to spot Eagle on the lunar surface. . . .*]

« I knew there were a number of things that could go wrong with the LM and that some of them would require a good deal of rescue work on my part, but I really wasn't apprehensive about it. In Columbia I had a happy home. Its construction is almost like that of a miniature cathedral, the bell tower being the tunnel which goes up into the LM. Because we had to prepare for a possible extravehicular transfer in case we were unable to dock the two spacecraft properly, I had removed the center couch, folded it up and stowed it underneath the left couch. This created a center aisle that gave me more room than I needed, and I rattled around in my minicathedral, bumping into the nave and transept when I wasn't careful.

Neil and Buzz could always talk to the ground. But I was whizzing around and around, and out of each two-hour revolution I was on the back side where I couldn't talk to anybody, anywhere, for more than forty minutes. Then as soon as I came within sight of the earth, I could talk to the earth. But the LM was still over the lunar horizon. So on any one pass on the front side of the moon I had roughly an hour and fifteen minutes in which I could talk to the world, but only six or seven minutes in which I could talk directly to the LM. Every time I would come around after a silent period I would be just like everybody down on the ground, wanting to know — 'What did they say? What did they say?' »

Loss of signal with Collins, over the hill again. . . . Armstrong and Aldrin now awake. . . . Liftoff less than three hours away. . . .

HOUSTON (Evans): How is the resting, standing up there, or did you get a chance to curl up on the engine can?

EAGLE (Aldrin): Roger. Neil has rigged himself a really good hammock with a waste tether and he's been lying on the ascent engine cover, and I curled up on the floor. Over. [*They could discuss the sleep problem further when they got back. . . .*]

HOUSTON (Evans): Roger. Copy, Buzz. We've got a couple of changes . . . the main one being that we do not want the rendezvous radar on during the ascent, and we think that this will take care of some of the overflow of program alarms that we were getting during descent.

EAGLE (Aldrin): Okay. We had the rendezvous radar in SLEW during descent, though.[1]

HOUSTON (Evans): Roger, we copy that. But there's a greater duty cycle on — there's a good fifteen percent duty cycle on the ascent program there, so just go ahead and leave it off.

EAGLE (Aldrin): All right, go ahead. I've got it out. [*More instructions from Mission Control. . . . After launch guidance system recommendation, add descent propulsion fuel vent open, add descent propulsion oxidizer vent open. . . . If 503 alarm occurs, designate fail. . . . A half-dozen small changes to keep the computer from having another case of indigestion. . . . And press on. . . .*]

EAGLE (Aldrin): Roger, I think I have that. [*Acquisition with Columbia. . . .*]

COLUMBIA (Collins): This is Columbia. . . . I'd like to know about this P52 coming up. Is that the one I just completed or do you want a pair of them back to back?

HOUSTON (Evans): Columbia, Houston. You do not need to do another P52 unless you want to. . . . You're looking good to us, Columbia.

COLUMBIA (Collins): Yes, sir. Keep it that way. . . . And a crew status report from Columbia. I figure I got about five hours' good sleep, although you guys probably know better than I do. [*Eagle's liftoff time now fixed at 124 hours GET, shortly before one in the afternoon Houston time, a little more than an hour and a quarter away. . . .*]

COLUMBIA (Collins): Whoever figured those hydrogens and oxygens out a couple of days ago must have known what he was doing.

HOUSTON (Evans): Okay, I think I read that oxygen wrong. It's a plus seventeen pounds.

COLUMBIA (Collins): Roger, still close.

HOUSTON (astronaut Jim Lovell): Eagle and Columbia, this is the backup crew. Our congratulations on yesterday's performance, and our prayers are with you for the rendezvous. Over.

COLUMBIA (Collins): Glad to have all you beautiful people looking over our shoulder.

EAGLE (Armstrong): We had a lot of help down there, Jim.

Loss of signal with Collins, liftoff during Columbia's next pass on the front side of the moon. . . .

North Amityville, Long Island

Herman Clark, the Grumman quality control inspector who had checked out lunar module No. 5 for the flight of Apollo 11, felt intensely emotional and a little apprehensive. "Here we were working on a piece of hardware," he said, "but when I knew they had landed the reaction really started." He enjoyed watching Armstrong and Aldrin bounce around on the moon ("They were really having a ball"), but he wished that the television reception could have been better. He badly wanted to have a good look at the LM. The astronauts said that it was in good condition, but Clark wished that he could see for himself. He thought of the things that had happened during the two years he had been working on LM No. 5, the arguments he had had, things that could have gone wrong, mistakes that had been caught; and he desperately hoped that he and his men had caught them all. There was the time the wrong kind of primer paint nearly got used on some machine parts which were going inside the cabin: "Somehow this engineering order hadn't gone the right route. Then my group leader started questioning and I said yeah, could be. So we sent a report over to the laboratory requesting the lab to run a test, and we found out that stuff was highly toxic and it would have an outgassing effect out in space where there's no atmosphere, or where you have almost a complete oxygen atmosphere. If it had got into the vehicle it could have killed the guys. Maybe this would have been picked up by somebody else; we don't know. But we did pick it up." As he waited for the ascent engine to fire, Herman Clark thought it was just as well to do a little praying.

EAGLE (Armstrong): Going to give you a few comments with regards to the geology question of last night. We are landed in a relatively smooth crater field of elongated secondary — circular secondary craters, most of which have raised rims, irrespective of their size. . . . The ground mass throughout the area is a very fine sand to a silt. I'd say the thing that would be most like it on earth is powdered graphite. Immersed in this ground mass are a wide variety of rock shapes, sizes, textures, rounded and angular. . . . The boulders range generally up to two feet with a few larger than that. Now some of the boulders are lying on top of the surface, some are partially exposed, and some are just barely exposed. . . .

HOUSTON (Evans): Tranquility, Houston. Roger, very fine description.

EAGLE (Armstrong): Now I suspect this boulder field may have some of its origin with this large sharp-edged, blocky-rim crater that we passed over in final descent. Now yesterday I said that was about the size of a football field, and I have to admit that it was a little hard to measure coming in, but I thought that it might just fit in the [Houston] Astrodome as we came by it. . . . The blocks seem to run out in rows and irregular patterns, and then there are paths between them where there is considerably less surface evidence of hard rocks. Over. *Liftoff an hour away.* . . .

Monday morning in the homes of the astronauts was quiet and sleepy. Everyone knew that the ascent engine had to fire, of course. But the tension of touchdown on Sunday, and the following moon walk, had been superseded by a feeling of confidence. Everything had worked up to now; why should not the ascent engine also work? In Nassau Bay, Pat Collins, who had stayed up until five in the morning to collect her thoughts — she strolled outside the house to look at the moon — was awakened at 7:30 A.M. by a telephone call from Capetown, South Africa. The caller said he was a journalist, and Pat Collins was incredulous. "What kind of a joker are you?" she asked. "Do you know where Capetown, South Africa, is?" The man on the other end of the phone explained that people in Capetown identified with Mike Collins because, like the South Africans, he had had no access to television. Then he proceeded to ask questions which Pat answered.

Joan Aldrin had swallowed too much coffee the night before and slept badly. She had been up early to take her son Andy to the doctor; he had an earache — "swimmer's ear," the doctor decided; no infection. Audrey Moon stretched out on a couch in the family room, still suffering from a mild degree of shock from a bad fall she had had the night before. Running out of the house to say goodbye to her son and grandson, who had been unable to get hotel accommodations nearer than Houston, twenty-five miles away, she went sprawling. She had bad cuts on one ankle and a nasty bump on her knee, but she was trying to keep Joan Aldrin from knowing about the incident. Bob Moon was busy emptying the ashtrays and taking out last night's garbage.

HOUSTON (PAO Terry White): Here in Mission Control center flight director Glynn Lunney is polling the various positions. . . . We're some fifty-three minutes now away from ascent. Meanwhile back at the scientific experiment situation, another attempt is scheduled today to shoot another laser beam up to the laser retroreflector. . . .

HOUSTON (Evans): Eagle's looking real fine to us down here. We have a fairly high confidence that we know the position of the LM. However, it is possible that we may have a plans change. But in the worst case it would be up to thirty feet per second, and of course we don't expect that at all. [*Three minutes to acquisition time for Columbia, coming around on its twenty-fifth lunar revolution. . . . Twenty-six minutes away from ignition burn on the ascent burn. . . . Here we go. . . .*]

HOUSTON (Evans): Columbia, Houston. . . . Loud and clear. If you would like to take it down, we have the latest position of Tranquility Base. Over. . . . [*Still trying. . . .*]

COLUMBIA (Collins): Go ahead.

HOUSTON (Evans): Roger. It's just west of West Crater, Juliette .5/7.7. [*This was the estimate of the geologists, passed up at last to Mike Collins, who had run out of time to look; but it was very nearly on the mark. . . .*]

COLUMBIA (Collins): Columbia is holding inertially at liftoff attitude. . . . I'm using B and D roll.

HOUSTON (Evans): Columbia and Tranquility, I'll give you a MARK at twenty minutes to go, and that's in about twenty seconds. . . . Stand by. . . . MARK twenty minutes. . . . Tranquility Base. . . . Just a reminder here. We want to make sure you leave the rendezvous radar circuit breakers pulled; however we want the rendezvous radar mode switch in LGC just as it is on surface 59. . . . Our guidance recommendation is PGNCS, and you're cleared for takeoff.

EAGLE (Aldrin): Roger, understand. We're Number One on the runway.

El Lago, Texas

Twenty minutes to liftoff. . . . Jan Armstrong sat on her bed eating a late breakfast of fried chicken which a neighbor had brought the day before. Her son Ricky joined her, and she called to Mark: "Come in and let's hear Daddy take off." Mark wanted to know, "Why can't we watch him?" . . . *Ascent propulsion system propellant tanks pressurized. . . . Starting in fourteen minutes, nearly five thousand pounds of propellant would be run through the ascent engine to boost the lunar module's upper stage to a velocity of 6,068 feet per second to achieve lunar orbit about sixty thousand feet above the surface. . . . Thirteen minutes and twenty-three seconds from ascent ignition. . . .*

HOUSTON (Evans): Tranquility Base, little less than ten minutes here. Everything looks good and we assume the steerable is in track mode AUTO. . . .

And the voices of Tranquility Base:

"Rate scale 25." "25." "ATT translation four jets. Balance couple on." "Balance couple on." "Stop, push-button reset, abort to abort stage reset." "Reset." "Reset." "Deadband minimum, ATT control, mode control, mode control AUTO." "AUTO, AUTO." [*Two minutes to go. . . .*] "Got your guidance steering in the AGS." "Okay, master arm on." "9, 8, 7, 6, 5 — abort stage, engine arm ascent, proceed. That was beautiful!" "26, 36 feet per second up. Be advised of the pitchover. Very smooth. . . . Very quiet ride. . . ."

They were up!

The time in Houston was 12:55 P.M., give or take a few milliseconds. In El Lago Jan Armstrong heard Buzz Aldrin mention "a quiet ride," stretched out on the bed, leaned on her elbow and said: "As long as that thing is lit, Mike will come and get them, wherever they are. Wherever they are, he would come. Nobody told me that. Nobody ever had to. He would."

In Nassau Bay Joan Aldrin sat cross-legged on the floor, watching the television screen. Her father asked her to take a chair. "No, Dad, I can't," she said. "Please no." Then Gerald Carr said, "It's firing and they're on their way." Joan kicked her legs in the air and fell backward. She sat up and smiled, then stood and propped herself against the wall. "I just think it's kind of funny," she said. "Well, I won't tell you what I think is funny, but it's strange . . . from the beginning I've worried more about the liftoff than the touchdown." Gerald Carr said, "Buzz says it's beautiful, just beautiful." Joan answered, "Bless him. Bless the baby." Jan snuggled against her grandfather and said, with an extra emotional tone in her voice, "They're on their way home." Joan Aldrin heard and said, "They're on their way back to mother. That's a good boy. . . . Just think, tonight we'll have that beautiful big burn and then — look out, world, here we come! . . . I don't have any more tears. . . . I think I've just cried for the past two weeks."

On Long Island, Grumman's Herman Clark had literally been holding his breath. With liftoff confirmed, he said to himself: "Okay. Job well done QC-wise!" Now it was time to go back to work and see if he and his men could do the job again. Perhaps, Clark reflected, they would feel more confident after the third or fourth lunar landing; each lunar module had been "tighter" and more "leakproof" than its predecessor. . . .

Eagle's ascent engine had a burn time of 465 seconds, using propellant at the rate of 11.3 pounds per second. . . . *One thousand feet off the surface, eighty feet per second vertical rise. . . . Three thousand up, fifteen hundred FPS horizontal velocity, 185 FPS vertical. . . . Go at three minutes, everything looking good. . . .*

EAGLE (Armstrong): We're going right down U.S. One.

Height approaching thirty-two thousand feet, right on track; horizontal velocity approaching twenty-five hundred FPS. . . . Data sources agreeing. . . . Horizontal velocity up to 4,482 FPS, one minute to go in the burn. . . .

When he and Neil Armstrong returned to Eagle's cabin after their EVA on the moon, Buzz Aldrin noticed the lunar particles they had brought back inside the cabin. . . .

« The particles started finding little homes for themselves on the flooring or the suits, rubbing up against things. Once we lifted off again and were in zero gravity we expected to see these particles emerge and float around. We didn't exactly expect a dust storm, but we certainly expected a considerable amount of it floating up from the floor and out of nooks and crannies. Surprisingly, it never did. We were able to go ahead and take off our helmets and gloves without worrying about getting dust in our eyes. »

EAGLE (Aldrin): Okay, I'm going to open up the main shutoffs. Ascent feed closed. Pressure's holding good. Crossfeed on. . . . Stand by on the engine arm. . . . Shutdown. [*Sixty thousand feet off the moon; rendezvous and docking with Columbia still about five hours away, but they had started the long trip home. . . .*]

Flight Operations director Chris Kraft thought one reason Eagle had got off the moon was the fact that half a billion people all around the world were helping push. In El Lago Jan Armstrong said, "I never had any doubts. Once they got off with that seven-minute burn, Mike could come and get them. Oh, I consider what might happen. I consider it. You just take life one day at a time because that's all you have. What happens is fate — yes, I suppose I am a fatalist. Why is one person killed in an automobile accident and not another?" In Nassau Bay Gerald Carr professed to have had complete confidence in the LM's hardware, and he had another observation to fuel the "manned" side of the manned vs. unmanned argument. He felt that the descent had been "the finesse and epitome of all our training. The computer put Neil down in the middle of a crater, and he just picked the thing up and moved it over a bit."

Strictly speaking, the astronauts were not yet "safe," but there was some justification for the sudden enormous feeling of safety. The two steps in the Apollo 11 mission for which there could have been no true rehearsal — land on the moon and lift off again — had been taken, and they had been fabulously successful. The things that remained to be done had been done before, and there was every reason to believe that they could and would be done again. Eagle and Columbia still had to catch up with each other, but on the flights of Apollo 9 and Apollo 10 the lunar module had rendezvoused and docked successfully with the command module. Then the precious rock boxes had to be transferred to Columbia by Armstrong and Aldrin and the ascent stage of the LM had to be jettisoned, but those operations posed no great problems. And finally, on the command module's thirty-first lunar revolution, its single engine had to fire one more time — to get TEI (transearth injection), freeing the spacecraft from the bondage of lunar gravity to send it on the long, long coast home.

EAGLE (Aldrin): Houston, Eagle. Did you copy our star angle difference and torquing angle?

HOUSTON (Evans): Eagle, Houston. We didn't have them on the downlink, but we copied them on the VOX.

EAGLE (Aldrin): Okay, it was zero for star angle distance minus 06 plus 64 and plus 1.37. Over.

COLUMBIA (Collins): Eagle, Columbia. Your Y-dot is minus 1.0. Over.

EAGLE (Armstrong): Mike, you're breaking up.

COLUMBIA (Collins): Eagle, do you read Columbia? Over. Eagle, this is Columbia, over. . . . Houston, Columbia. Over.

HOUSTON (Evans): Roger, Columbia, loud and clear now. This is Houston.

COLUMBIA (Collins): Roger, would you tell Eagle his Y-dot is minus 1.0. Over.

EAGLE (Armstrong): Roger, Houston. We got that, thank you. [*One minute until loss of signal with Columbia, about two and one-half minutes to LOS with Eagle. . . .*]

HOUSTON (Evans): Eagle, Houston. Recommend aft OMNI, and are you go for CSI so we can let Columbia know? Over.

EAGLE (Aldrin): Roger, we're go for CSI.

HOUSTON (Evans): Columbia, Houston, did you copy? Eagle is go for CSI.

COLUMBIA (Collins): No, I did not copy. . . . But thank you.

HOUSTON (Evans): Eagle, Houston. We'll see you coming around the other side. Your AOS time is one minute ahead of the flight plan. *Loss of signal with Eagle. . . .*

The conversation was getting to be a little sobersided. . . . *Ascent burn just completed, went quite well, on time at 124 hours 22 minutes zero seconds GET. . . . Eagle in an orbit of 9.4 by 46.7 nautical miles. . . . Next, circularize the orbit. . . .*

The first step was to execute, on Eagle's first pass on the back side of the moon following liftoff, the maneuver called CSI: concentric sequence initiation. Since Eagle's initial orbit, after liftoff, was elliptic or slightly elongated, like a football, CSI would take place when Eagle was more or less at apolune (46.7 nautical miles from the lunar surface). Eagle's reaction control engines, four quads of four engines each, would now have their big moment. One engine on each quad would be fired to get Eagle's perilune from 9.4 to about forty-five nautical miles off the moon — and achieve a nearly circular orbit. Then would come maneuvers called CDH, TPI and TPF — "constant Delta height," "terminal phase initiation" and "terminal phase finalization." CDH involved a maneuver to align Eagle's orbit with Columbia's. Forty minutes later or thereabouts, Armstrong and Aldrin would use the TPI maneuver to thrust visually along the line of sight toward Columbia when the line of sight was some twenty-seven degrees above the lunar horizontal. In TPF, still another forty minutes later, they would fire their thrusters in a series of small burns designed to align Eagle with Columbia for a short period of photography, then move on in for docking at approximately 128 hours GET — about six-thirty Monday evening, Houston time. . . .

. . . It all sounded so clinical and matter-of-fact when it was put into words; and yet there was nothing really matter-of-fact, or easy, about it; this was still the most difficult business in the world. . . . *Both spacecraft now over the hill. . . . Eagle's velocity measured at a little more than a statute mile per second, weight down to less than six thousand pounds, less than twenty percent of its original-loaded weight. . . . Time for the CSI burn, behind the moon . . . Reacquisition of signal a little more than half an hour away. . . . Less than one minute. . . .*

HOUSTON (Evans): Columbia, Houston. Heard you talking to Eagle. Do you have COMM with Eagle now? . . . You're very weak. Say again.

COLUMBIA (Collins): Roger, Houston, Columbia CSI nominal, no plane change. Everything's going beautifully, and the LM seems to be . . . [*static*] . . .

HOUSTON (Evans): Eagle, Houston. Stand by. . . . Can you give us a burn report?

EAGLE (Aldrin): Roger. Stand by. Okay, the CSI burn was on scheduled time. . . .

HOUSTON (Evans): Eagle, Houston. We copy. Any plane change? Over.

EAGLE (Aldrin): No, there was no plane change on CSI, and CSM had a 2.3 foot per second burn. We had a 2.9 and we elected to postpone that. Over.

HOUSTON (Evans): Roger. We copy, Eagle. Thank you.

COLUMBIA (Collins): Houston, are you reading Columbia now? Eagle, how about asking them for a high gain angle for me? Will you please?

HOUSTON (Evans): Columbia, Houston. You can go ahead and go to REACQ in the high gain. We should get you then. [*Coming up to the CDH maneuver, and Eagle seemed to have an onboard* CO_2 *sensor problem. . . .*]

EAGLE (Armstrong): And our water separator apparently isn't working too well. We're getting a lot of water through the suit loop, and we've changed water separators, but it doesn't seem to have improved the situation any.

HOUSTON (Evans): Eagle, Houston. Roger, we copy.

COLUMBIA (Collins): Eagle, Columbia standing by to back you up on the burn. Just let me know how it's going.

EAGLE (Aldrin): Okay. . . . I think you've already got the burn time, minus 8.1, minus 1.8, minus 18.2.

COLUMBIA (Collins): Okay that's a pretty close agreement. For burn time, I still have 126:17:46.

HOUSTON (Evans): For a warm feeling, we are agreeing with your CDH.

EAGLE (Aldrin): Congratulations. . . . Unfortunately the chart doesn't agree with us, because the range rate at 36 minutes was off the chart. [*Less than a minute until the CDH burn. . . .*]

COLUMBIA (Collins): Yes, I'm ready. Go ahead.

EAGLE (Armstrong): Burn complete.

COLUMBIA (Collins): Thank you. [*The two spacecraft now 91.3 nautical miles apart, closing at 119 feet per second. . . .*]

COLUMBIA (Collins): It sure is great to look down there and not see you.

EAGLE (Aldrin): Sure thing.

HOUSTON (Evans): On the water problem there, we can't add anything more to it except the fact that it looks like the water accumulators are up to speed to us down here.

EAGLE (Armstrong): Okay, it's not going to be too much trouble.

HOUSTON (Evans): Columbia, Houston. Our COMM problem was traced to a ground station here. . . . You're mighty fine now. [*Eagle and Columbia 67.5 nautical miles apart, closing at 121 feet per second. . . .*]

COLUMBIA (Collins): I have a TPI check when you guys want to compare them.

EAGLE (Aldrin): . . . [*static*] . . . 19.8 . . . Go ahead, Mike. What have you got?

COLUMBIA (Collins): 127:02:34.50.

EAGLE (Aldrin): You're about thirty-two seconds later than we are.

COLUMBIA (Collins): Okay, fine. [*Twenty-five minutes until the terminal phase initiation burn, then loss of signal again. . . . Mission Control's function now primarily advisory; in the final analysis the onboard computations by the crew of Columbia and Eagle had to bring about the rendezvous. . . .*]

COLUMBIA (Collins): Just as soon as you know what TPI TIG [time of ignition] is going to be I would appreciate a call.

EAGLE (Aldrin): How late can you take a revision?

COLUMBIA (Collins): Well, to stay on my timeline, I should have it in the next couple of minutes.

EAGLE (Aldrin): Okay. Latest estimate 127:03:39.

COLUMBIA (Collins): Thank you kindly.

HOUSTON (Evans): Eagle, Houston. . . . In the event of the possibility that we may have had some water channeling in those hydroxide canisters, we recommend you stay in the cabin mode from now on. Over. . . . Helmets and gloves on are your option and we really have no concern with the CO_2. Over.

HOUSTON (Evans): Columbia, Houston.

COLUMBIA (Collins): Columbia. Go ahead.

HOUSTON (Evans): Roger, Mike. You can go ahead and arm your logic any time you want to and we'll give you a go so you can hit your pyro arm at your convenience.

COLUMBIA (Collins): That's a good idea, babe. . . . MARK logic one, MARK logic two.

HOUSTON (Evans): Columbia, Houston. We need the SEC arm circuit breakers closed.

COLUMBIA (Collins): Okay. Going in, SEC arm BATT A and BATT B.

HOUSTON (Evans): Columbia, Houston. Logic looks good. You can arm your pyros at your convenience.

COLUMBIA (Collins): Thank you. [*Six and one-half minutes until the line-of-sight TPI maneuver. . . . Eagle and Columbia some 38.6 nautical miles apart, closing at 110 feet per second. . . .*]

EAGLE (Aldrin): And Mike, if you want our target Delta-V [velocity rate change], I'll give it to you.

COLUMBIA (Collins): Ready to copy.

EAGLE (Aldrin): 127:03:30.82, plus 22.7, plus 1.7, minus 10.6. Over.

EAGLE (Armstrong): I'm showing a good bit of out-of-plane velocity on my cross pointers, Mike.

COLUMBIA (Collins): Rog, I have no indication of it. . . . Coming up on one minute to TIG, Neil. How's it looking?

EAGLE (Armstrong): Pretty good. . . . That out of plane was in the AGS, not in the radar. . . . We're burning.

COLUMBIA (Collins): That-a-boy! . . . Burn complete?

EAGLE (Armstrong): Read, burn complete.

COLUMBIA (Collins): Roger, thank you.

HOUSTON (Evans): Eagle, Houston, aft OMNI, low bit rate and we'll see you at 127 plus 51 [GET].

Loss of signal again. . . . Terminal phase initiation due to take place with Eagle and Columbia behind the moon. . . . They should be within a few feet of each other when Mission Control reacquires signal. . . . Docking estimated at 128 hours GET, about 4:30 P.M. Houston time, Monday, July 21. . . . One hundred twenty-seven hours and fifty minutes into the mission, 4:22 P.M. in Houston, less than a minute from AOS. . . . Standing by. . . .

At 4:05 P.M. in Houston, when both spacecraft were behind the moon, the synchronized swimming team arrived at the Armstrong home to congratulate Jan Armstrong on her husband's successful lunar landing and liftoff. The leader of the group was Jan's friend and fellow instructor Jeanette Chase. The girls were going off to Toledo, Ohio, the next day to compete in the national Amateur Athletic Union synchronous swimming championships. It was hot in El Lago; the whole area was swarming with American and foreign press; and the team had made seventy dollars selling lemonade to newsmen to help finance the Toledo trip. Jan Armstrong wished them good luck: "I wanted to get up there yesterday and see you all. It's surprising how well you're doing without me. I've really missed you." Nurse Dee O'Hara was there, her dark hair cropped short as usual, her blue eyes a little bigger than usual. . . . *Static on the squawk box — and a peculiar buzzing sound which reminded Jan Armstrong of the noise she had heard during liftoff at the Cape.* "I bet you they've docked," she said. "Couldn't that noise be the thrusters, or powering up the thrusters?"

In Nassau Bay, the talk in the Collins home had been about how much weight astronauts' wives tended to gain during a mission because of all the food that neighbors sent in, and about the troubles of United States Senator Edward Kennedy in Massachusetts — the drowning of Mary Jo Kopechne at Chappaquiddick Bridge. Pat Collins's sister Ellie complained about humidity and fatigue, then fell asleep in a chair just before it was time for docking. Young Michael Collins burst in to ask, "Where is my Silly Putty?" The television screen showed a simulation of docking, and Ann Collins asked her mother, "Is that real?" Pat replied, "It's about as real as anything about this whole thing is." There was another complaint on television about the laconic quality of the astronauts' talk, or the absence of talk; but Pat defended them: "They may not be much on show biz, but the deed speaks for itself." Clare Schweickart agreed: "Astronauts get along so well because they don't talk." Somebody decided to start calling CBS's Walter Cronkite "Crankcase": "He always likes to get you psyched up for tragedy, and then when it doesn't happen everything is all right."

In the Aldrin home Joan said, with rendezvous and docking coming up, "I just keep ticking them off one by one." There had been a small family crisis earlier in the afternoon with Jan Aldrin, something involving a broken lamp, and Joan had forbidden Jan to go horseback riding the next day; so Jan was crying it out upstairs. Once she came down, leaned over her mother and tried to interrupt a conversation. Joan said, sharply, "Jan, not now — later. If you can't stop crying, go to your room." Jan ran out of the room and a few moments later had slipped out the gate in the backyard. But she came back, and she and her mother kissed and made it up. When Gerald Carr, who was listening to the squawk box, announced that the two crews were nearly four minutes ahead of schedule, Joan said: "They're stationkeeping already? That

Mike — bless his heart!" Carr reported again: "Okay, they're all lined up. Joan looked up briefly from her conversation with NASA's Bill Der Bing and said, "That's a *nice* Mike! . . . As long as they've found them, the docking itself is just mechanical."

Acquisition of signal, past terminal phase finalization, the two spacecraft within a few feet of each other, ready to dock. . . .

EAGLE (Armstrong): Okay, Mike. I'll get — try to get in position here, then you got it. How does the roll attitude look? I'll stop. Matter of fact, I can stop right here if you like that. [*The astronauts were now on their own; they had to hack it, and Mission Control could not help much. . . .*]

HOUSTON (Evans): Eagle, Houston. Middle gimbal. And you might pass to Columbia, we don't have him yet.

EAGLE (Armstrong): That's right. . . . I'm not going to do a thing, Mike. I'm just letting her hold in attitude HOLD.

COLUMBIA (Collins): Okay.

EAGLE (Armstrong): Okay. We're all yours, Columbia.

COLUMBIA (Collins): Okay. . . . I'm pumping up cabin pressures. . . . That was a funny one. You know, I didn't feel it strike and then I thought things were pretty steady. I went to retract there, and that's where all hell broke loose. For you guys, did it appear to you to be that you were jerking around quite a bit during the retract cycle?

EAGLE (Armstrong): Yeah. It seemed to happen at the time I put the plus thrust to it, and apparently it wasn't centered because somehow or other I accidentally got off in attitude and then the attitude HOLD system started firing.

COLUMBIA (Collins): Yeah, I was sure busy there for a couple of seconds. Are you hearing me all right? I've got a horrible squeal.

EAGLE (Armstrong): Yes, I agree with that, but we hear you okay. . . .

They were docked!

In the Armstrongs' El Lago home Mike Collins's remark about all hell breaking loose caused some confusion but did not shake Jan Armstrong's confidence. She said, "I guess they came all the way in and didn't feel it and then started backing out again — they had really docked." Shortly thereafter NASA administrator Dr. Thomas Paine came to see Jan. Dr. Paine talked about the smoothness of the flight ("Gee, the EVA — I was just overcome with admiration for those guys") and tried to pinpoint for Jan the exact landing site on a lunar map. (Nobody yet had that exactly right.) "I still can't believe that they've been there," she said. "I didn't until halfway through the EVA." "I was afraid when he dropped that camera [it was a film pack]," Jan said. "It's so hard to get up. I know he was afraid of stumbling, because he could tear his suit. . . ."

In Nassau Bay Gerald Carr told Joan Aldrin, "I think they're docked but I think we missed Neil's report." Bob Moon chimed in: "He said he didn't even feel the jerk . . ." Carr and Moon discussed what would happen next. Carr said, "After they've docked Buzz and Neil will take off their helmets, open the hatch, clear the tunnel. . . . Now they'll spend about an hour cleaning up, putting their things in order and passing them through. They have to vacuum each other, stow the vacuum. It'll be about an hour and a half from now before they move in. Neil goes in about an hour from now, then Buzz. It'll be about two and one-half hours before they're through [and back in the command module]. They've got all that cleaning up to do first." Meanwhile Joan Aldrin was examining a bottle of cologne she had received. She passed the accompanying note to her father and said, "It's from Turkey; I can't even read it." Michael Archer thought it smelled like bay rum, and the women in the room sniffed it with interest. The note was signed "Erkin Turkey Alasehir Turkey Made These Colognes from Flowers of a Tree called Philadelphia. Name of the Cologne is Luna." Finally Jeannie Bassett put the cologne in the bedroom. Joan Aldrin said, "I'm so selfish; I hate to contemplate that LRL . . ." *The Lunar Receiving Laboratory, twenty-one days of quarantine from the moment of lunar liftoff to release in Houston. . . .*

Pat Collins was doubtful about the first report on docking; she thought the dock had been "soft on, not locked on, and then the LM has to pressurize again. . . . See what I mean. . . . They're just filling air time on television trying to make the flight plan match their simulation." When docking was confirmed Clare Schweickart sighed and said, "Well, now we have only one more thing to worry about." (She meant transearth injection — the big burn that would let the world know "Here we come!") "No," Pat countered. "After TEI we'll worry about the flight home. Then we'll remember prop wash and I'll even worry about dropping the mobile trailer and sinking the raft." Dave Scott's wife Lurton arrived with some news: "Guess what, you're going to parades in New York, Chicago and Los Angeles, and then dinner at the White House." Pat Collins looked shaken and said: "The minute I hear a drop of martial music or see any people who look as though they're with you, I cry. Yesterday in church I almost didn't get through 'America the Beautiful.'" Someone told her she had better get used to it. "Oh no," she said. "It'll all be over by Christmas." No, she was told, it would never be over. She kept saying, "I'm not prepared for this. It's not going to make a difference" — but Pat knew, along with everyone else there, that it *would* make a difference.

Everyone listening to the squawk box heard Mike Collins say "all hell broke loose." Collins did not remember saying that — "but if it's on the transcript of communications, then I guess I did say it." . . .

« It was a nice, clear, crisp lunar day, if there is such a thing. The moon didn't look sinister or forbidding, as it can at very low sun angles. But that day, with a high sun angle, it was a happy place. It also was a happy situation, because here was the LM getting larger and larger, brighter and shinier, and right smack dab where it should have been. All the tricky parts of the rendezvous were over, and now all we had to do was dock and get home.

The computer, of course, had been telling me that everything was going well, but that's a rather impersonal message. It's not any substitute for being able to look out the window and really see Eagle fixed in the reticle pattern as if riding on railroad tracks.

The docking process begins when the two vehicles touch and the probe slides into the drogue. Neil made the first maneuvers to point his drogue at my probe, and then I took over — probably at a distance of twenty or thirty feet. And then I did the final maneuvers. It's easier to do it from my side. He's looking up over his head out through this little window, and I'm looking through a larger window straight ahead. It's just easier and more straightforward for me to do it. The probe and the drogue are held together by three tiny capture latches, and it's almost like tiny little paper clips holding together two vehicles, one of which weighs thirty thousand pounds and the other five thousand pounds. It's a tenuous grasp. To make the combination rigid you fire a little gas bottle that activates a plunger which literally pulls the two vehicles together. At this point the twelve capture latches fire mechanically and you are held together very strongly. That's the 'hard dock.'

Just as I fired the charge on the gas bottle we got a quite abnormal oscillation in the yaw axis. We had eight or ten rather dubious seconds then, when I really thought we were outside the boundaries for a successful retract and that I was going to have to release the LM and go back and dock all over again. I guess that's what my remark in the transcript was all about. But the two vehicles did not separate. If you had a go-go dancer and her topside were the command module and her bottom side were the LM — well, she has a certain degree of flexibility there, and she can change the angle of her hips by plus or minus fifteen degrees with no trouble at all. At any rate I instantly took action to correct the angle, and so did Neil in Eagle. Together we returned the two vehicles to an in-line position. I had plenty of fuel. All this time the automatic retract cycle was in fact taking place, and we heard a loud bang, which is characteristic of those twelve big latches slamming home. And lo and behold — we were docked, and it was all over.

The first thing I had to do then was get the tunnel cleared to remove the hatch and the probe and the drogue and to stow them. Then I floated up into the tunnel to greet Neil and Buzz. I could see them both, those beady little eyes, up there in the LM, and it's terrible — but I can't remember now which one of them was first to get back into Columbia with me. I met them both in the tunnel and we shook hands, hard, and that was it. I was glad to see them and they allowed as how they were happy to be back. They passed the rock boxes through to me, and I handled them as if they were absolutely jampacked with rare jewels, which in a sense they were. »

With the two spacecraft reunited, the next step was for Armstrong and Aldrin to transfer into Columbia with their precious lunar material, then get ready to jettison the ascent stage of the LM, which had done its job. . . . *Communications somewhat scratchy. . . .*

"Columbia has completed the leak check. . . . The leak check is complete, and I'm proceeding with opening the hatch up now." "Eagle, Columbia. . . . My hatch is removed. You can open yours and I'll start passing stuff up to you. . . . Standby." "Mike, you still there?" "Yeah, everything is going fine. Be with you in just a sec. . . ."

COLUMBIA (Collins): Houston, Columbia. You want me to roll over and get high gain or anything like that?

HOUSTON (Evans): Columbia, Houston. I can give you some REACQ angles for the high gain on the LM jettison attitude. And then you can go there whenever you want to. The angles are pitch minus 50 and yaw zero.

COLUMBIA (Collins): Houston, Columbia. Say the jettison roll, pitch and yaw, please.

HOUSTON (Evans): Columbia, Houston. Roll zero, pitch 025, and yaw zero.

HOUSTON (Evans): There's little noise there, Buzz, say again. . . . Eagle, Houston.

EAGLE (Aldrin): Roger, go ahead.

HOUSTON (Evans): Roger. Just a reminder to be sure and zero the AGS errors before you enable the AGS attitude hold there after you get in burn attitude.

EAGLE (Aldrin): Roger. You mean sep attitude. [*Separation attitude had to be aligned before the LM could be jettisoned. . . .*]

COLUMBIA (Collins): Houston, how do you read on Columbia high gain now?

HOUSTON (Evans): Hey, Columbia, Houston. Mighty fine, loud and clear.

COLUMBIA (Collins): Yeah, same here, Ron. Thanks. . . . Houston, do you have any preferences as to what you want us to do with the probe? . . . Over. . . . Okay, Eagle says they've got a place for it inside there, so no problem.

HOUSTON (Evans): Eagle, Houston.

EAGLE (Armstrong): Go ahead, Houston.

HOUSTON (Evans): Roger, Neil. Just a reminder again — the ACA out of detent to zero the AGS error there just in case you go to attitude hold, shortly.

COLUMBIA (Collins): Did you say you wanted the probe, Neil?

HOUSTON (Evans): Eagle, Houston. . . . That ACA out of detent didn't quite do it. . . . Request guidance control to PNGCS and then back to AGS, and that will zero the AGS errors, over.

EAGLE (Aldrin): Okay, we still have both mode control switches off. Over. [*Five minutes to loss of signal, LM jettison time 131:52 GET during the next pass over the near side of the moon. . . .*]

HOUSTON (Evans): Apollo 11, Houston. About a minute and a half to LOS. You're looking great. It's been a mighty fine day.

COLUMBIA (Collins): Boy, you're not kidding. [*Over the hill again on Columbia's twenty-seventh revolution. . . . Reacquisition of signal, a little past six in the evening, Houston time. . . .*]

HOUSTON (Charles Duke): Hello, Eagle, Houston. Do you read? Over.

COLUMBIA (Collins): Houston, this is Columbia, reading you loud and clear. We're all three back inside. The hatch is installed. Everything's going well.

HOUSTON (Duke): You guys are speedy. You beat us to the punch here. . . . We've got a state vector for you.

COLUMBIA (Collins): Buzz says the CO_2 sensor circuit breaker is in. . . . The RCS quantity was approximately 60 at A and 45 percent at B . . . and we're going P00 and ACCEPT.

HOUSTON (Duke): Columbia, Houston. Your friendly white team's going to be on till we get you on the way home, and we'd like to congratulate everybody on a successful rendezvous and a beautiful EVA. It was a great show for everybody. Over.

COLUMBIA (Collins): Thank you, sir. I'll tell Neil and Buzz. Houston, the hatch passes its integrity check. I'm going to go to "LM tunnel vent" now and leave it there.

HOUSTON (Duke): Roger, Columbia, we copy. That's good. . . . Columbia, Houston. It looks like you guys are so speedy on us that we're thinking about moving up jettison time to about a GET of 130 plus 30 if that's okay with you all. Over.

COLUMBIA (Collins): That's fine. I've still got to get a P30 pad from you.

HOUSTON (Duke): Mike, we can, for your druthers — we can do it either way. We can either let you do the jettison in P30 — correction, P47 — or we can send you a P30 target load up and then you — let you call P41, whichever you want to do. Over.

COLUMBIA (Collins): Yeah, I see Ron was going to give me a P30 pad and the flight plan says P47. Out of the two, I prefer to go the P30, P41 route. [*In Computer Program 47, Mike had to determine his own attitude angles and feed the information to the computer; in the Program 30–Program*

41 combination, the ground transmitted the computer data to set up for LM jettison. . . .]

HOUSTON (Duke): Rog, beautiful. . . . Columbia, we'd like you to terminate direct O_2 flow and stand by on your Poo and ACCEPT. We'll have to generate a new load due to the move up on time. Over. . . . Mike, it looks like if we move up this jettison time and give you a new load it would require a new attitude, and we can't do that due to the LM already closed out. And it would fight us all the way around and we'd lose COMM with it. We're thinking about separating in P47 in about ten minutes. We're looking at trajectories and we'll be with you momentarily.

COLUMBIA (Collins): Okay, it's no big thing with me either way. [*Mission Control was glad to get the jettisoning job done ahead of schedule; it was unwise to have the LM docked with the command and service module longer than necessary because its primary guidance system was not cooled and that could cause trouble. . . .*]

HOUSTON (Duke): Hello, Columbia, Houston. We'd like you to start down your jettison check list. . . . We'd like to jettison in ten minutes. That'll be 130:14:45. Over.

In Nassau Bay Gerald Carr was a little concerned about the advance in the jettison time; he wondered whether this would also mean advancing the time of transearth injection, and that would complicate the rest of the flight plan. He checked by telephone and informed the Aldrin family that TEI would be undertaken at the scheduled time. Dave Scott dropped into the Collins home to give an encouraging report; the spacecraft was in good shape: "They had a quick checkout and they are all ready to go."

HOUSTON (Duke): You can undock at your convenience — correction, jettison at your convenience. We would like you to jettison Eagle and stationkeep. . . . We'll have another attitude and a maneuver for you, so we'll be okay for TEI. Over.

COLUMBIA (Collins): Roger, that. Okay, I'm standing by to go P47 just as soon as you give me a go for pyro arm.

HOUSTON (Duke): Rog, I thought we gave you that. Mike, you're go for pyro arm and you're go for jettison.

COLUMBIA (Collins): Okay, letting her go in ten seconds. [*Unsnap the latches, get rid of the last of the LM, but don't bump into it either. . . .*]

COLLINS: . . . A fairly loud noise, and it appears to be departing — oh, I would guess several feet per second.

HOUSTON (Duke): Roger. Can you kind of stationkeep with it, Mike? Just stand by now. . . . Don't try to chase it, just hold what you've got.

ALDRIN: . . . Cabin pressure this time?

HOUSTON (Duke): Say again, Buzz. Over.

COLLINS: I'm now reading NOUN 83 plus 4 balls 4 minus 4 balls 8 and you want me to kill average G. Right?

HOUSTON (Duke): Stand by. . . . That's affirmative. You can exit P47.

COLLINS: There she goes. It was a good one.

HOUSTON (Duke): Roger-dodger! We got Eagle looking good. It's holding cabin pressure, and it picked up about two feet per second from that jettison.

ALDRIN: I believe that. I can see some cracks on the outer coating around the tunnel. . . . I don't think it has anything to do with structure.

HOUSTON (Duke): Hello, Columbia, Houston. We'll have an attitude and a little blip burn for you in about 130:30 so we can separate from Eagle. [*Eagle was still too close for comfort; the mother spacecraft, Columbia, had to move farther away* . . .] Columbia, Houston. Would you start a maneuver to about pitch of 230 for this little tweak burn? Over.

COLLINS: Roger. Pitch 230.

HOUSTON (Duke): Roger, Mike, and verify track mode in AUTO for the high gain. . . . We have a load for you. Over.

COLLINS: You got it.

HOUSTON (Duke): Roger, Mike. And our pitch attitude's a little wrong here. If you're ready to copy, I'll give you the SEP pad. Over.

COLLINS: Go ahead. Ready to copy.

HOUSTON (Duke): Roger. Starting with NOUN 33. . . .

COLLINS: Houston, Apollo 11. How about coming up with a good communications attitude for us to go to between now and the time we've maneuvered to TEI attitude.

HOUSTON (Duke): Roger, sir. Stand by. . . . Okay, Mike, you can maneuver to your preliminary TEI attitude as shown on page 398 of the flight plan, and the high gain angles are good as shown in the flight plan, and we'd like you to dump the waste water at 131:05 down to ten percent. [*Four minutes to the separation maneuver. . . . A two foot per second retrograde burn of seven seconds' duration. . . . Burn complete. . . . Transearth injection about five hours away, when the command and service module would be about twenty miles ahead of the lunar module and about one mile below it. . . .*]

COLLINS: You're loud and clear, Charlie. What numbers are you looking at for TEI TIG preliminary — 135 hours 23 minutes? Something like that.

HOUSTON (Duke): That's affirmative, Mike. We're looking at about nominal time. We've considered kicking it up rev, but we don't think this rev track is going to be any good since we had the RCS burn and so we need some more tracking to get you a good TEI. Over.

COLLINS: That's what we're looking for.

HOUSTON (Duke): Eleven, Houston. Looks like it's going to be a pretty relaxed time for the next couple of hours. We'll have you a pad, of course, the next rev or so, and we'll keep you posted on TEI. . . . And your little maneuver back here a moment ago will put you about twenty miles ahead of the LM at TEI.

COLLINS: Imagine that place [Mission Control] has cleared out a little bit after that rendezvous. You can find a place to sit down, almost, huh?

HOUSTON (Duke): Rog. Our MOCR's about empty right now. We're taking it a little easy. How does it feel up there to have some company?

COLLINS: Damn good, I'll tell you.

HOUSTON (Duke): I'll bet. I bet you'd almost be talking to yourself up there after ten revs or so.

COLLINS: No, no. . . . It'd be nice to have company. As a matter of fact, it'd be nice to have a couple of hundred million Americans up here. . . . Let them see what they're getting for their money.

HOUSTON (Duke): Rog. Well, they were with you in spirit anyway. At least that many. We heard on the news today, eleven, that last night — yesterday when you made your landing, the *New York Times* came out with a — the largest headlines they've ever used in the history of the newspaper.

COLLINS: Save us a copy. I'm glad to hear it was fit to print.

HOUSTON (Duke): It was great.

HOUSTON (Deke Slayton): That's why we didn't read you up any newscasts today. There really wasn't anything to talk about.

COLLINS: Hello there, boss. . . . [*Ten minutes to loss of signal. . . . Four and one-half minutes. . . . Over the hill again, the command and service module in a lunar orbit with a high point of 62.6 nautical miles and a low point of 54.9, traveling at a speed of 5,355 feet per second, about thirty-six hundred fifty statute mph. . . . One hundred thirty-one hours and forty-eight minutes into the mission, standing by to reacquire signal. . . . Five seconds. . . .*]

HOUSTON (Duke): Hello, Apollo 11, Houston. We are standing by. Everything is looking great here. Over.

ALDRIN: Apollo 11, Roger.

HOUSTON (PAO Douglas Ward): We suspect the crew is having something to eat at this time. . . . We will continue to monitor the LM and observe the performance of the primary guidance systems. All systems on the command and service module continue to function very well. . . .

HOUSTON (Duke): Got some news here we can read up. Over.

ARMSTRONG: Be pleased to have it.

HOUSTON (Duke): Roger, Neil. Starting: congratulatory messages on the Apollo 11 mission have been pouring into the White House from world leaders in a steady stream all day. Among the latest are telegrams from Prime Minister Harold Wilson of Great Britain and the King of Belgium[2] . . . And Premier Alexei Kosygin has sent congratulations to you and President

Nixon through former Vice-President Humphrey, who is visiting Russia. The cosmonauts have also issued a statement of congratulations. . . . And Mrs. Robert Goddard said today that her husband would have been so happy: "He wouldn't have shouted or anything. He would just have glowed. That was his dream, sending a rocket to the moon." . . . The Italian police reported that Sunday night was the most crime-free night of the year. And in London a boy who had the faith to bet five dollars with a bookie that a man would reach the moon before 1970 collected twenty-four thousand dollars. That's pretty good odds. You're probably interested in the comments your wives have made. Neil, Jan said about yesterday's activities: "The evening was unbelievably perfect. It is an honor and a privilege to share with my husband, the crew, the Manned Spacecraft Center, the American public and all mankind, the magnificent experience of the beginning of lunar exploration". . . . And Mike, Pat said simply, "It was fantastically marvelous." Buzz, Joan said: "It was hard to think it was real until the men actually moved. After the moon touchdown I wept because I was so happy." But she added, "The best part of the mission will be the splashdown." In other news, and there was a little bit, another explorer, Thor Heyerdahl, had to give up his attempt to sail a papyrus boat across the Atlantic. . . . While you were busy the other day, Joe Namath and football commissioner Pete Rozelle made the announcement that Broadway Joe had agreed to sell his interest in the Bachelors III restaurant and report to the New York Jets. . . . And in baseball, the west division of the National League remains a tight race. LA and San Francisco are one game behind league-leading Atlanta. The Astros have a record of forty-eight wins and forty-eight losses and are now in fifth place, seven games out. . . . The Chicago Cubs are still in first place in the east divison. They lead the New York Mets by four and one-half games. . . . Looking ahead, the all-star baseball game is scheduled for tomorrow. And President Nixon was scheduled to see the game and then leave immediately after the game for the Pacific splashdown area, before going on his tour of Europe.[3] . . . And that about covers the news this day. You guys have been making most of it. . . .

HOUSTON (Duke): Eleven, Houston. We've got a preliminary TEI 30 pad if you're ready to copy.

COLLINS: Ready to copy.

HOUSTON (Duke): Roger. Coming at you . . . [*One hundred thirty-two hours and twenty-nine minutes into the mission. . . . The crew of Apollo 11 collecting a string of numbers needed to start them on their way back to the earth in a little less than three hours, about midnight in Houston. . . . Time of ignition for TEI 135:23:42 GET. . . . Burn the 20,500-pound-thrust service propulsion system engine for two minutes and twenty-eight seconds. . . .*

Spacecraft weight at the time of the TEI burn 36,691 pounds. . . . About half an hour from loss of signal. . . .]

HOUSTON (Duke): Hello, Apollo 11, Houston. Seven minutes to LOS. Next AOS 133:46. You're looking good going over the hill. Out. [*Over the hill on the next-to-last full revolution in lunar orbit. . . . One hundred thirty-three hours and forty-five minutes into the mission, fifty seconds from reacquisition of signal on the last front-side pass before TEI. . . .*]

HOUSTON (Duke): Hello, Apollo 11, Houston. We're standing by.

ALDRIN: Roger.

HOUSTON (Duke): We'd like you, sometime at your convenience, to stir up the cryos on this pass, and we're wondering if you got the fuel cell purge. Over.

ALDRIN: Roger, the 02 fuel cell purge. . . .

HOUSTON (Duke): Say again. You're breaking up.

ALDRIN: Roger, the 02 fuel cell purge is complete.

HOUSTON (Duke): And eleven, Houston, for your information, Eagle, we had an ISS fail light come on . . . and about this time at AOS it looks like we're about to lose the platform. [*The lunar module's platform was about to go. . . . This was part of a preplanned test to find out just how long the LM's primary guidance system would continue to function without cooling. . . . It had functioned for a little more than four hours. . . . TEI less than an hour and a half away. . . .*]

HOUSTON (Duke): Apollo 11, Houston. We got the load in. You can have the computer.

ALDRIN: Eleven. Are you through with the computer?

HOUSTON (Duke): That's affirmative, Buzz.

ALDRIN: All right, that's timing for you.

HOUSTON (Duke): Apollo 11, Houston. Your friendly white team has your coming home information if you are ready to copy. . . . TEI 30 and then a TEI 31. . . . [*Final figures on the SPS engine's burn: two minutes twenty-eight seconds, consuming about ten thousand pounds of propellant. . . .*]

HOUSTON (Duke): Hello, Apollo 11, Houston. After the burn we'd like you to trim X and Z. Over.

COLLINS: Okay, Charlie.

HOUSTON (Duke): Rog, and that's two-tenths of a foot per second, as shown in the flight plan.

COLLINS: Sounds like there's a story behind that one, too.

HOUSTON (Duke): We'll tell you when you get back. [*Flight controllers reviewing their data, taking a good look at the spacecraft to make a go/no go decision on the TEI burn. . . .*]

HOUSTON (Duke): Apollo 11, Houston. You are go for TEI.

ARMSTRONG: Apollo 11. Thank you. [*Nine minutes until loss of signal, the TEI burn thirty-five minutes away. . . .*]

HOUSTON (Duke): Apollo 11, Houston. One minute to LOS. Go sic 'em.
COLLINS: Thank you, sir. We'll do it. [*The word was go. . . . Over the hill for the last time. . . . Burn time, but no data for another ten minutes or so. . . . Burn should be complete, and Apollo 11 should be traveling at a speed of about 8,660 feet per second, about fifty-nine hundred statute mph. . . . Less than one minute from reacquisition of signal. . . .*]

Midnight in El Lago and Nassau Bay. . . . Dick Gordon, command module pilot for the next lunar landing mission, Apollo 12, visited the Armstrong home. "It's just like the old fighter pilot's life," he told Jan Armstrong. "Long periods of boredom punctuated by moments of stark terror." When signal was lost, Jim Lovell, who had been named to command Apollo 13 in the spring of 1970, told Pat Collins: "You forget you are on the back side, so you start talking to Mission Control anyway. You're so interested in getting the burn going that you end up talking to yourself." There was some speculation about the first words they would hear when the spacecraft came around again. In the Aldrin home Joan Aldrin remarked, somewhat irrelevantly, "I don't know what we're waiting up for — but here we are." Her father settled back in his chair, closed his eyes and said, "Just wake me up when we get to Paris." Joan was sleepy. "Just tell me you're all right, Buzz boy, and we can go to bed." *Then . . . acquisition of signal. . . .*

HOUSTON (Duke): Hello, Apollo 11, Houston. How did it go?
ARMSTRONG: Tell them to open up the LRL doors, Charlie.
HOUSTON (Duke): Roger. We got you coming home.

When she heard her husband's voice, Pat Collins said, "Smile, honey! Put a smile in your voice." Everyone tried to guess how Neil and Buzz would describe to Mike what they had seen on the moon: "You want to see my contingency sample?" "If you play your cards right I got a small sample for you — one for you, one for you, one for me, and one for the scientists. . . ." Joan Aldrin nodded, smiled and said "Ummmm. . . ." The telephone rang and Audrey Moon answered it. Someone said "Congratulations," then hung up. It was time to go to bed. . . . *Apollo 11 traveling at a speed of 7,603 feet per second, less than five thousand statute mph, already slackening off. . . . Next order of business: get the crew to bed too. . . .*

Early in 1875 Cakobau, king of the Fiji Islands, visited Australia and returned home with a case of measles — a disease rarely fatal elsewhere but totally unknown at that time in the Fijis. In a matter of weeks Cakobau's illness caused an epidemic. Not having any natural resistance to the alien contagion, forty thousand Fijians died of measles. Admittedly it was a remote possibility that the Apollo astronauts might return from the moon carrying, either in their rock boxes or on their persons, a lethal contaminant which might cause a proportional epidemic on the earth. But the possibility had to be considered; indeed it had been popularized by Michael Crichton's current best-selling novel, *The Andromeda Strain* (Knopf, New York, 1969), which had as its theme the contamination of earthlings and the deaths of all but two people in a small Arizona town as the result of a deadly organism which an unmanned satellite had brought back to earth. Therefore the LRL — the lunar receiving laboratory — came into being. The decision was taken to quarantine Armstrong, Collins and Aldrin for twenty-one days. Some people thought the quarantine unnecessarily long, since Persa R. Bell, then manager of the LRL, thought the chance of anything harmful coming back with the astronauts was "probably one in a hundred billion." [4] Some ultra-conservative scientists would have preferred to keep the men isolated for months. So the time span of twenty-one days represented a sort of compromise, which covered the known incubation period of most earthly germs. The process of isolation would begin as soon as the astronauts left the spacecraft after splashdown in the Pacific and got into a raft; there they would don sacklike biological isolation garments (BIGs). Once aboard the carrier, they would step directly into a boxlike structure, thirty-five feet long, which looked like a large travel trailer and had in fact been adapted by the contracting companies (American Standard and Airstream) from an existing travel trailer design. This was the MQF (mobile quarantine facility), and it would be occupied by the astronauts, along with a doctor and an engineer, as carrier *Hornet* took them to Hawaii. Aboard *Hornet* the men of Apollo 11 could see the President of the United States, through a glass window, and talk with him, but they could not shake his hand. The facility was comfortable enough — a lounge with six aircraft seats and a table, a galley, and a sleeping and bath area. Once *Hornet* reached Hawaii, the MQF, astronauts and all, would be put aboard a C-141 jet aircraft for the nonstop flight to Ellington Air Force Base near Houston, then pulled by truck ten miles to the Manned Spacecraft Center, where a small army of technicians in the LRL awaited the MQF and the men in it. Some of the technicians would already be at work analyzing the contents of the rock boxes; the lunar rocks had a higher transportation priority than the astronauts themselves and were to precede them to Texas in a separate plane.[5]

None of the astronauts looked forward to the experience of spending all that time in the three-story building officially called Building No. 37; it had all the makings of a crashing bore. Some of the intimacy of the familiar homecoming scene inevitably had to be lost, although the excitement that attached to this particular mission could be counted upon to compensate. The men could talk to their families, face-to-face through glass or over the telephone, but the ground rules were simple: whoever or whatever entered the biologically sealed areas of the LRL stayed inside for the duration of quarantine. That rule would apply to Apollo 11's command module itself, although the astronauts had left their outer shoe coverings and their backpacks on the moon, and the raft into which they stepped after splashdown would be scrubbed down with disinfectant and sunk. And, theoretically at least, there was the possibility that the quarantine would have to be extended. The biologists planned to pulverize separate sets of tiny lunar samples and feed them to carefully cultivated (and germ-free) colonies of mice, birds, fish, insects and plants. It was not easy to accumulate a colony of germ-free mice; the mice had to be delivered by Caesarian section, and they were hard to raise. Three batches of mice were lost during trial runs and tests of the LRL systems. If the animals became strangely ill, or if an oyster turned into a man-eater after gulping a lunar lunch, that meant trouble — and a longer quarantine. Mike Collins had joked about such grim possibilities even before the flight. "Let us hope," he said mildly, "that none of those mice die."

Apollo 11 a little more than four thousand nautical miles from the moon, traveling earthward at precisely one statute mile per second. . . .

HOUSTON (Deke Slayton): This is the original CAPCOM. Congratulations on an outstanding job. You guys have really put on a great show up there. I think it's about time you power down and get a little rest. . . . I look forward to seeing you when you get back here. Don't fraternize with any of those bugs en route, except for the *Hornet.*

ARMSTRONG: Okay. Thank you, boss. We're looking forward to a little rest and a restful trip back, and see you when we get there.

HOUSTON (Slayton): Rog. You've earned it.

But still more questions:

HOUSTON (Duke): A couple of questions for the moon walkers, if you've got a second. Over.

ARMSTRONG: Go ahead.

HOUSTON (Duke): Roger, Neil. We're seeing some temperature rises on the passive seismic experiment that are a little higher than normal and we're wondering if you could verify the deployed position. We understand it is about forty feet from the LM in the eleven o'clock position. Over.

ARMSTRONG: No, it's about in the nine or nine-thirty position, and I'd say it's about fifty or sixty feet.

HOUSTON (Duke): Roger. Copy. Also did you notice — was there any indication of any dust cloud as you lifted off? Over.

ARMSTRONG: Not very much. . . . Parts of the LM went out in all directions, and usually the same distance as far as I can tell, but I don't remember seeing anything of a dust cloud. [*One hundred thirty-seven hours and fifty-two minutes into the mission. . . . Apollo 11 7,045 nautical miles away from the moon, velocity down to 4,868 feet per second, about thirty-four hundred statute mph. . . . Nearly two-thirty in the morning, Houston time. . . .*]

HOUSTON (Garriott): Howdy there, Mike. We're ready to go ahead and have you put your OMNI position for your sleep period, and we would like the following high gain positions. . . . Apollo 11, Houston. How do you read me through Honeysuckle now? Over.

COLLINS: You're loud and clear.

HOUSTON (Garriott): Very good. Reading you better now, and did you copy we'd appreciate going S-band OMNI and OMNI B at this time? Over.

COLLINS: Roger. [*Finally, the end of a long day's work and everyone really could go to sleep. . . . No plans to awaken the crew as long as the spacecraft systems continued to function well. . . .*]

Not much was planned for this day, Tuesday, July 22, either by the Apollo 11 astronauts or their families, although Valerie Anders, wife of the astronaut Bill Anders, was giving a small and very private pre-splashdown party Tuesday evening. There was one disturbing note in the news report: a tropical storm called Claudia was roaming around the Pacific, but as of Tuesday morning it was twenty-three hundred miles east of the landing area; if it stayed that far away it would not unduly affect the seas at splashdown point — *if* it stayed that far away. . . .

By shortly past noon, Houston time, all three astronauts were awake; in about half an hour the spacecraft would cross an imaginary line and come into the earth's sphere of influence. . . .

Stand by for a MARK. . . .

HOUSTON (McCandless): MARK, you're leaving the lunar sphere of influence.

COLLINS: Roger. Is Phil Shaffer down there?

HOUSTON (McCandless): Negative.

COLLINS: Roger. I wanted to hear him explain it again at the press conference. . . . That's an old Apollo 8 joke, but tell him the spacecraft gave a little jump as it went through the sphere [of influence].

HOUSTON (McCandless): If you're not busy now, I can read you up the morning news. . . . Apollo 11 still dominates the news around the world. Only four nations, Communist China, North Korea, North Vietnam and Albania, have not yet informed their citizens of your flight and landing on the moon. One newsman said that he has run out of ways to describe your success. Tonight President Nixon . . . will depart for the Pacific recovery area. . . . Accompanying the President will be Secretary of State William Rogers and Frank Borman. . . . Luna 15 is believed to have crashed into the Sea of Crises yesterday after orbiting the moon fifty-two times. . . . Sir Bernard

Lovell at Jodrell Bank Observatory said that Luna 15 hit the surface of the moon at a speed of about three hundred miles per hour. . . . Last night the Baseball Writers Association of America named Babe Ruth the greatest ballplayer of all time. Joe DiMaggio was named the greatest living ballplayer. Frank Borman made the announcements. . . . Mario Andretti won the two hundred-mile Trenton auto race Sunday. . . . And that's about the summary of the morning news. . . .

COLLINS: Marvelous.

ARMSTRONG: Look up the Dow Jones industrials for us.

One hundred fifty hours into the mission, about two-forty Tuesday afternoon in Houston. . . . Time for a midcourse correction, involving a ten-second burn of the spacecraft's reaction control system thrusters, needed to control the flight path angle. . . . Thirty seconds from initiation of the burn. . . . Burning now. . . . Burn off. . . .

ARMSTRONG: Houston, do you copy our residuals?

HOUSTON (McCandless): Roger, we've got your residuals. Do you have a counter reading for us?

Apollo 11 now 168,843 nautical miles from the earth, traveling at 4,078 feet per second, just under twenty-eight hundred statute mph. . . . velocity slowly building up again. . . .

HOUSTON (McCandless): For sixty-four thousand dollars we're still trying to work out the location of your landing site, Tranquility Base. We think it is located on LAM 2 chart at Juliette .5 and 7.8. . . .

The exchange was inconclusive, but a few minutes later the mystery was solved:

ARMSTRONG: . . . I took a stroll back to a crater behind us that was maybe seventy or eighty feet in diameter and fifteen or twenty feet deep. And took some pictures of it. It had rocks in the bottom. . . .

That did it; the geologists had the answer. The crater Armstrong mentioned had no name. But it was on the lunar maps, and Eagle's landing site could be pinpointed beyond a doubt. The geologists had been fairly sure a few hours after touchdown. Later Dr. Eugene Shoemaker said, "Had Neil told us about the small crater behind the LM, we could have pinpointed them right then within twelve to twenty meters." [6]

About eight in the evening, Houston time, Apollo 11 roughly one hundred sixty thousand miles from the earth.. . . .

HOUSTON (Duke): Hello Apollo 11, Houston. Your white team is now on. We're standing by for an exciting evening of TV and a presleep report. Over. . . . Apollo 11, Houston. You sure you don't have anybody else in there with you? [*No wonder Duke was puzzled; weird noises were being heard — sirens, whistles and screeches. . . .*]

COLLINS: Where . . . where did the white team go during their off hours, anyway?

HOUSTON (PAO Terry White): This is Apollo Control. Still no explanation of the weird noises emanating from Apollo 11. If indeed it *is* from Apollo 11, and it's reported from network that it's being received on the downlink at two different stations in the Manned Space Flight Network. Perhaps it will all shake out later. . . . [It "shook out" much later, when the Apollo astronauts held their post-quarantine press conference.[7]]

Nearly time for another television pass. . . .

HOUSTON (White): While we're waiting for the television pictures to come in, we have in the control room here a vase full of long-stemmed red roses, the card saying, "To one and all concerned. Job superbly done. From a moonstruck Canadian." [*Television pictures coming in. . . .*]

HOUSTON (Duke): The focus is a little bit out. We see earth in the center of the screen. . . .

ALDRIN: I believe that's where we just came from.

HOUSTON (Duke): It is, huh? . . . Hey, you're right.

HOUSTON (Deke Slayton): It's not bad enough, not finding the right landing spot, but when you haven't got the right planet. . . .

HOUSTON (Duke): I'll never live that one down.

COLLINS: We're making it get smaller and smaller here to make sure that it really is the one we're leaving.

HOUSTON (Duke): That's enough, you guys.

Knowing that scientists from many nations were watching, Neil Armstrong put the rock boxes on television: "They're vacuum-packed containers that were closed in a vacuum on the lunar surface, sealed and then brought inside the LM and then put inside these fiber glass bags." Buzz Aldrin used a few minutes to bring viewers up-to-date in the food department: "These bite-size objects were designed to remove the problem of having so many crumbs floating around in the cabin. . . . I think we've discovered that we could progress a good bit further than that — to some of the type meals that we have on earth. . . . I'll show you, I hope, how easy it is to spread some ham while I'm in zero G." And Mike Collins had "a little demonstration for the kids," using a spoonful of water. He said: "If I'm not careful I'm going to spill water right over the sides. Can you see the water slopping around on the top of the spoon, kids?"

Then he prankishly tipped the spoon upside down; the water stayed in, demonstrating to pop-eyed young viewers the facts of weightless life. "That's really not the way we drink. We really have a water gun which I'll show you. Here's the water gun. This cylindrical thing on the end of it is a filter with several membranes. One allows water to pass but not any gas; the other allows gas to pass out but not any water. So by routing the gaseous water which comes from our tank through the filter we're able to drink purified water without the gas in it. And of course all we do to get it started is pull the trigger. It's sort of messy. I haven't been at this very long. It's the same system that the Spaniards use to drink at a wine stand at the bullfights, only I think [that] would be more fun.

HOUSTON (Duke): Thank you from all us kids in the world, here in the MOCR, who can't tell the earth from the moon. . . . Looks like you need a wineskin up there, Mike.

That was about that for Tuesday, but a couple of hours later, about ten-thirty in the evening, Houston time, there was some more unscheduled "noise" . . .

ARMSTRONG: Charlie, could you copy our music down there?

HOUSTON (Duke): Did we copy what, Neil?

ARMSTRONG: Did you copy our music down there?

HOUSTON (Duke): Roger. We sure did. We're wondering who selected — made your selection?

ARMSTRONG: That's an old favorite of mine. It's an album made about twenty years ago, called *Music Out of the Moon.*

HOUSTON (Duke): Roger. It sounded a little scratchy to us, Neil. Either that or your tape was a little slow.

ALDRIN: It's supposed to sound that way.

COLLINS: It sounds a little scratchy to us, too, but the Czar likes it.

. . . Armstrong had really played the music for one person alone: his wife Jan. After he got home, she asked, "Where on earth did you find that? We used to listen to it all the time." *One hundred sixty-three hours and twenty-eight minutes into the mission, splashdown a little more than thirty hours away. . . . Apollo 11 135,920 nautical miles from earth, velocity up to 4,758 feet per second, over thirty-one hundred statute mph. . . . The crew asleep. . . .*

But these are deeds that should not pass away,
And names that must not wither.

LORD BYRON

Back to earth, the dear green earth.

WILLIAM WORDSWORTH

14

"It seems appropriate to share with you . . ."

IT HAS always been difficult for rational men to know how deeply they felt about anything until there was some event, some incident, some surprise — whether pleasant or unpleasant — to force them into an eyeball-to-eyeball confrontation with their inner emotions. Hardly any American, in 1861, would ever forget exactly where he was and what he was doing when he heard that Fort Sumter had been fired upon; the same was true when the news of Pearl Harbor came through in 1941. These were acts of war, and they unleashed emotions which for the moment were beyond taming, suggesting that man is not always so rational an animal as he thinks. No warlike — or lethal — weapon had been taken into space by Apollo 11, or for that matter by any American space flight, but the really singular and distinctive quality in the achievement of Apollo 11 was something the three men in the crew could not be fully aware of while they were doing their jobs. Despite Mission Control's conscientious relaying of the world news reports, which Apollo 11 was dominating; despite the fact that they knew that heads of state by the dozen were cabling messages of congratulation to the astronauts themselves, and to their country's President, they still could not get the full impact by way of CAPCOMs like Charlie Duke and Bruce McCandless. They could not, a quarter of a million miles away, have a proper gauge on the emotions they had tapped in the hearts of "all mankind" for whom they had journeyed in peace to the moon. They could not see the tears on the faces of grown men and women at critical points in the mission like liftoff at Cape Kennedy. They could not share the awed silence in New York's Central Park and London's Trafalgar Square as they were making their powered descent to the moon, and they could not hear the cheers and shouts of "Thank God!" that followed Neil Armstrong's announcement, in his normal matter-of-fact voice: "Tranquility Base here. The Eagle has landed." They could not share the champagne toasts at Harry's Bar in Paris, and they had to be above the inner tensions which temporarily destroyed the function of a million stomachs. These men had the professionalism to master and discipline legitimate concerns about risk and safety. But the vast majority of terrestrial mortals did not, and for a few marvelous days they did not mind admitting it to each other. Some strange worldwide process was going on which had in it the ingredients of wonder, terror, catharsis and thanksgiving — in no particular order. Whatever the common denominator in this process was, it was not political; people in Marxist countries (except the four which deliberately withheld the news) were cheering on Apollo 11 as fervently as the most chauvinistic American. And there was that congratulatory telegram from the Soviet cosmonauts. Perhaps Robert Ardrey had something when he suggested that the development of the American and Soviet space programs in the 1960's was coming close to "ritualizing" the cold war, and possibly was becoming a substitute for war between the two great superpowers.[1] Armstrong and Aldrin had walked on the moon, and other men would follow them on more ambitious lunar missions. But no succeeding lunar mission would ever create the same sense of shared excitement felt on the earth during the flight of Apollo 11. Amerigo Vespucci and Giovanni da Verrazzano contributed more to geographical knowledge of the western hemisphere than Christopher Columbus did, and following Apollo missions would vastly expand our knowledge of the moon. But no matter how many times we went there three names — Armstrong, Collins and Aldrin — would stand larger than life in the history books. Well before the flight, Neil Armstrong had tried to put his thoughts into some kind of focus. . . .

« It would be presumptuous for me to pick out a single thing that history will identify as a result of this mission. But I would say that it will enlighten the human race and help us all to comprehend that we are an important part of a much bigger universe than we can normally see from the front porch. I would hope that it will help individuals, the world over, to think in a proper perspective about the various endeavors of mankind as a whole. Perhaps going to the moon and back in itself isn't all that important. But it is a big enough step to give people a new dimension in their thinking — a sort of enlightenment.

After all, the earth itself is a spacecraft. It's an odd kind of spacecraft, since it carries its crew on the outside instead of inside. But it's pretty small. And it's cruising in an orbit around the sun. It's cruising in an orbit around the center of a galaxy that's cruising in some unknown orbit, in some unknown direction and at some unspecified velocity, but with a tremendous rate of change, position and environment. It's hard for us to get far enough away from this scene to see what's happening. If you're in the middle of a crowd, the crowd appears to extend in every direction as far as you can see. You have to step back and look down from the Washington Monument or something like that to see that you're really pretty close to the edge of the crowd, and that the whole picture is quite a bit different from the way it looks when you are in the middle of all those people. From our position on the earth it is difficult to observe where the earth is and where it's going, or what its future course might be. Hopefully, by getting a little farther away, both in the real sense and the figurative sense, we'll be able to make some people step back and reconsider their mission in the universe, to think of themselves as a group of people who constitute the crew of a spaceship going through the universe. If you're going to run a spaceship you've got to be pretty cautious about how you use your resources, how you use your crew, and how you treat your spacecraft.

Hopefully the trips that we will be making in the next couple of decades will open up our eyes a little. Jim Lovell, I know, pointed out that when you are looking at the earth from the lunar distance, its atmosphere is just unobservable. The atmosphere is so thin, and such a minute part of the earth, that it can't be sensed at all. This impressed me. The atmosphere of the earth is a small and valuable resource. We're going to have to face the fact that we have to learn how to conserve it and use it wisely. Of course, Jim was talking about pollution and that sort of thing, but down here in the crowd you are aware of the atmosphere and it seems adequate, so you don't worry about it too much. But from a different vantage point, perhaps it is possible to understand more easily why we should be worrying. »

Wednesday, July 23, eleven in the morning, Houston time, splashdown a little more than twenty-four hours away. . . . One hundred seventy hours and twenty-eight minutes into the mission, distance from the earth 115,470 nautical miles, roughly halfway home. . . . Velocity up to 5,317 feet per second, about 3,625 statute mph. . . . Eleven-thirty in the morning in Houston, crew now awake in the spacecraft after a ten-hour sleep period. . . .

HOUSTON (Garriott): Eleven, Houston. Got your signals loud and clear here. How are things this morning? Over.

ARMSTRONG: Do you read us, Owen?

HOUSTON (Garriott): Roger. Loud and clear, eleven.

ARMSTRONG: Okay. Everything seems to be all right here. So far we haven't been looking in the cockpit yet. We have been spending our time looking outside the cockpit.

HOUSTON (Garriott): Roger, eleven. You're breaking up just a little bit there, Neil. Your signals are loud, but breaking up occasionally. Your spacecraft all looks good here from the ground. . . . We noticed you stirring around the cockpit and thought we'd give you a call. . . . We do have a few items for you here — entry pads, consumables and so forth.

ALDRIN: Go ahead, Owen, I've got the books out. Ready to copy.

HOUSTON (Garriott): Okay, Buzz. On your flight plan items a few updates first of all. We have canceled the midcourse No. 6. Just remain in PTC. [*Stay in the barbecue mode, midcourse correction unnecessary.*] The PTC is a little bit ragged, and we'd like to make the water dump at a time which we think will hold it in its proper configuration, so it looks like we will have a desirable opportunity coming along in between fifteen and twenty minutes. . . .

ALDRIN: I got 'em. Midcourse correction canceled. Battery B charge and water dump on your call.

HOUSTON (Garriott): That's right, Buzz, and the last item here — we do request that we do a P52 even though we are not doing the midcourse correction, and we suggest you get to that after the waste water dump has been completed. We also have a state vector update for you, if you can give us P00 and ACCEPT. Over.

ALDRIN: Okay. You have the DSKY now.

HOUSTON (Garriott): Eleven, Houston. Present forecast shows acceptable conditions in your recovery area. Two thousand foot scattered, high scattered. Wind from 070 degrees, one to three knots, visibility ten miles, and sea-state about four feet. The forecast yesterday showed a tropical storm, Claudia, some five hundred to one thousand miles east of Hawaii. . . . Yesterday there was also a report of a tropical storm, Viola, further to the west. Its present location is some thousand miles east of the Philippines and moving northwest. Tropical storm Viola has been intensifying and should be transferred to the typhoon category within the next twelve hours or so. However, that will be far to your west. . . .

ARMSTRONG: Okay, sounds pretty good. . . . [*But now there were two tropical storms, not one, in the recovery area, and there was also the possibility of convergence. . . . One-thirty Wednesday afternoon, Houston time, Apollo 11 106,482 nautical miles from the earth, velocity 5,607 feet per second, 3,822.85 statute mph. . . . Spacecraft weight twenty-six thousand pounds, as compared to the Saturn V's six and one-half million pounds at the time of liftoff from Cape Kennedy.*]

HOUSTON (Garriott): Apollo 11, Houston. . . . I just wanted to make sure you fellows hadn't gone back to sleep again. And I also have a little bit of late news here. . . . We find South Korea's first superhighway linking Seoul with the port of Inchon has been named the Apollo Highway to commemorate your trip. . . . President Nixon has already started on his round-the-world trip, and today he is in San Francisco on his first stop — which will take him to the USS *Hornet,* from which he'll watch the return of your spacecraft. He plans to visit seven nations, including Rumania, during this trip. . . . The West Coast residents in Seattle, Washington; Portland, Oregon; Vancouver, British Columbia; San Francisco — all plan to make their areas visible to the three of you by lighting their lights between nine P.M. and midnight. We do have clear weather predicted there, so you may be able to see the Christmas lights, porch lights, store lights and whatever may be turned on. . . . [*Turn on the lights; the residents of Perth, Australia, had been doing that to identify their city as a landmark since the Mercury flight of John Glenn, and it was entitled to be called "The City of Lights."*. . .] A little closer to home here, back in Memphis, Tennessee, a young lady who is presently tipping the scales at eight pounds two ounces was named "Module" by her parents, Mr. and Mrs. Eddie Lee McGhee. "It wasn't my idea," said Mrs. McGhee. "It was my husband's." She said she had balked at the name Lunar Module McGhee, because it didn't sound too good. . . . The present score at the end of the fourth inning has the National League [all stars] leading the American League by 9 to 3. So the hitters are having a good day, as you can tell. And rain clouds are over the MSC area at the moment. It began raining here just about ten minutes ago.

ARMSTRONG: Thank you very much, Owen. I think my yard could use some water.

HOUSTON (Garriott): That's very true. I've forgotten exactly how many days it did go, Buzz. Something like thirty days without rain. . . .

ALDRIN: That was Neil. This is Buzz here. I wonder if you could find out when was the last time my lawn was cut.

HOUSTON (Garriott): That might be a little more difficult to find out. I'm not sure whether the — whether Mike [Aldrin] is ready to admit when he last did the job, but I'll look into that for you.

ALDRIN: He'll tell you. He's got a new mower.

COLLINS: Hey, ask my chinch bugs how they're doing.

HOUSTON (Garriott): Well, I'm not sure about yours. I can let you know about my own, and the report isn't very good. [*Apollo 11 now 115,165 nautical miles from the earth. . . .*]

HOUSTON (Garriott): Eleven, Houston, Over.

ARMSTRONG: Go ahead.

HOUSTON (Garriott): Joan wasn't home right now, Buzz, but Jan [Aldrin] reports the grass is getting pretty high, and I would estimate it's going to be close to your knees by the time you get out of quarantine. . . . And no reports from the chinch bugs there, Mike.

COLLINS: Well, they're sort of taciturn little fellows. They don't say much. They just chomp away.

HOUSTON (Garriott): Concur on that.

COLLINS: Which is about what we're doing up here.

HOUSTON (Garriott): We concur on that, too.

COLLINS: Houston, Apollo 11. We've been doing a little flight planning for Apollo 12 up here. . . . We're trying to calculate how much spaghetti and meatballs we can get on board for Al Bean.

HOUSTON (Garriott): I'm not sure the spacecraft will take that much extra weight. Have you made any estimates?

COLLINS: It'll be close.

HOUSTON (Garriott): Eleven, Houston. The medics at the next console report that the shrew is one animal that can eat six times its own body weight every twenty-four hours. This may be a satisfactory baseline for your spaghetti calculations on Al Bean. Over.

COLLINS: Okay, thank you. That's in the works.

COLLINS: Houston, Apollo 11. It was slightly colder in here last night than it has been on any previous night. Does EECOM notice any change in his data or any explanation for that? . . . Up until last night it was — if anything a little on the warm side at night. Last night it was on the chilly side.

HOUSTON (Garriott): Roger there. We'll run down the temperature for the two nights.

COLLINS: Oh, it's no big thing. Just as a matter of interest.

COLLINS: The peculiar thing, Owen, now, on the platform alignment, is that when I really take my time and do a very slow, careful, precise job of marking, I'm getting about the same star angle difference as when I'm doing it in PTC and have to do a hurried rush job with relatively poor tracking. . . . It almost made me believe there's a very small bias error somewhere in the sextant.

ALDRIN: Well, he's really trying to explain why he can't get all zeros.

COLLINS: I think Buzz is probably right. Matter of fact, one time I made a MARK which I thought was a little bit in error. But I thought — well, heck, I'll go ahead and see how it works out anyway, and I got five zeros that time. And when I have thought everything was exactly precisely on, I have consistently been getting .01.

HOUSTON (Garriott): Roger. Apparently it pays to hurry.

COLLINS: I usually do. The visibility through the telescope has been very poor. It's — I would say, even worse than the simulator is, right now. It requires long periods of dark adaptation which most times are most inconvenient, so it's really a tremendous asset to keep the platform powered up at all times. . . .

HOUSTON (McCandless): Eleven, this is Houston. With reference to your subjective evaluation that it felt cooler inside the spacecraft last night . . .

we did indeed see a drop of about three degrees. . . . Looking back, it appears that the crew of Apollo 10 reported similar feelings during the translunar and transearth coast phases. We're wondering if you could give us any indication of the relative amount of free or condensed water in the cabin last night and the night before from which we could infer humidity. Over.

COLLINS: There's more moisture in the tunnel now than there has been at any previous time. Subjectively we have been unable to determine any change in — any buildup in humidity. There appears to be no moisture any other place in the spacecraft — for example, the windows are not fogging, and various other cool spots around the spacecraft . . . all of them appear to be completely dry.

HOUSTON (McCandless): If you've got the entry operations checklist handy then I'll pass it up to you. Over.

ALDRIN: How can you make changes after liftoff? . . . Go ahead.

HOUSTON (McCandless): Okay. . . . Down towards the bottom we have three additional steps we'd like you to accomplish. The intent of this is to reduce the oxygen pressure in your manifold and to eliminate the oxygen bleed flow through the potable and waste water tanks during descent. . . . You've got 10,000 feet: main parachute deploy, main deploy push button, push within one second. And after that step we'd like you to insert surge tank 02 valve OFF, repress package valve OFF and direct 02 valve OPEN. . . . And then down at the very bottom of 6–2 where you see direct 2 OFF verify, delete that step completely.

ALDRIN: Okay, we've got it.

HOUSTON (McCandless): Apollo 11, this is Houston. For your information the all-star game has just ended with the National League winning 9–3 over the American. . . . And I have a message here for Mike that says, "All the chinch bugs are gone."

COLLINS: Having done their job, I guess.

HOUSTON (McCandless): Along with one tree, it turns out.

COLLINS: Yes. I heard about that. That was right before the flight.

HOUSTON (McCandless): Right. That big storm [the night before Apollo 11 lifted off at Cape Kennedy].

HOUSTON (McCandless): Apollo 11, this is Houston. Are you still up there? Over.

ARMSTRONG: Yes, we are, but not quite so far as we were a while ago.

HOUSTON (McCandless): Roger. We concur. We just wanted to make sure that we had good COMM with you. . . . For general information, eleven, you are now 95,970 miles out from the earth. Over.

ALDRIN: Right in our own backyard.

COLLINS: Starting to come downhill a little bit now. What's our velocity?

HOUSTON (McCandless): Your velocity is 5,991 feet per second. . . . And you are indeed coming downhill. [*Picking up speed now. . . . One hundred seventy-six hours and forty-four minutes into the mission, Apollo 11 just 94,961 nautical miles from the earth, velocity up to 6,029 feet per second, over forty-one hundred statute mph. . . .*]

HOUSTON (PAO Douglas Ward): The next item scheduled on the flight plan is a television transmission. That's scheduled to occur at Ground Elapsed Time of 177 hours 30 minutes, which would be 6:02 P.M. Central Daylight Time. . . . Reentry is scheduled to begin, based on no further midcourse corrections, at 195 hours 3 minutes 5 seconds. . . . *That would be shortly before noon Thursday, July 24, Houston time, less than twenty hours away. . . .*

Except for attending a luncheon given in Houston by North American Rockwell, the three Apollo 11 wives passed an uneventful day; tomorrow would be splashdown day, and as they waited for the parachutes to deploy some of the old tension would come back; but, like Scarlett O'Hara in *Gone With the Wind,* they could think about that tomorrow. One package that arrived at the Aldrin home did seem a little incongruous amid all the flowers and fruit baskets. It contained fireworks. Fireworks are legal in Texas only during the Fourth of July and Christmas seasons, but early in the mission Joan Aldrin had timidly inquired if there was any chance of getting a waiver; she felt like shooting off some fireworks at splashdown. With the assistance of a banker friend, Gene Lindquist, who had formerly worked for NASA, and the Nassau Bay fire chief, Chuck Miller, the waiver was obtained.

There was more than a faint suspicion that interest in Neil Armstrong's first words spoken from the surface of the moon had been created, somewhat artificially, by the pressures of journalistic curiosity. But there was nothing artificial about interest in what the astronauts would have to say this Wednesday evening on the telecast of the mission from the Apollo 11 spacecraft. The world was still moonstruck; the fact that two men had walked on the moon, and millions of people had *seen* them do it, was still too much for the finite mind to assimilate. Surely there was an explanation of what all this meant? It was too much for the journalists, too much for the theologians. Without being able to articulate their own feelings very well, men and women groped for some reassurance that they had not just dreamed the whole thing, for some kind of answer. . . .

"Goldstone COMM TECH Net One." "Goldstone Houston COMM TECH Net 1." "Goldstone COMM TECH." "Roger. Check for keying please." "Roger." "Goldstone, Houston COMM TECH." "Goldstone COMM TECH. One hundred percent keying."

ARMSTRONG: You got good S-band signal strength, Houston?

HOUSTON (McCandless): That's affirmative, eleven.

ARMSTRONG: You all set for TV?

HOUSTON (McCandless): Roger, we're all set whenever you're ready to send. . . . Okay, you're coming through loud and clear now, eleven, with your patch. [*The Apollo 11 insignia patch appeared on the screen, then the face of Neil Armstrong. His expression was friendly but serious; it was the expression of a man who had spent a few hours collecting his thoughts. . . .*]

Armstrong spoke in an even, measured voice; his tonal modulations had barely a suggestion — which was enough — of contained emotion.

"Good evening," Armstrong said. "This is the commander of Apollo 11. A hundred years ago, Jules Verne wrote a book about a voyage to the moon. His spaceship, Columbiad, took off from Florida and landed in the Pacific Ocean after completing a trip to the moon. It seems appropriate to us to share with you some of the reflections of the crew as the modern-day Columbia completes its rendezvous with the planet earth and the same Pacific Ocean tomorrow. First, Mike Collins."

HOUSTON (McCandless): Eleven, this is Houston. We have an LOS here.

ARMSTRONG: We'll be right back with you.

HOUSTON (McCandless): In the interim, you may be interested in knowing that Jan [Armstrong] and the children and Pat [Collins] and the youngsters and Andy Aldrin are down here in the viewing room watching this evening.

Mike Collins's face appeared. He had been growing a moustache, largely for his own amusement; it was now big, black and bushy, dominating his fine but small features. These were the last formal words he intended to speak from space, ever; well before the flight of Apollo 11 he had told his wife Pat that this would be his last mission. She advised him not to make up his mind until after the flight; his mind was made up, but he did not care to say so just now.

HOUSTON (McCandless): Okay, eleven, you're back on with Mike in the middle of the screen there.

"Roger," Collins said. "This trip of ours to the moon may have looked to you simple or easy. I'd like to assure you that that has not been the case. The Saturn V rocket which put us into orbit is an incredibly complicated piece of machinery, every piece of which worked flawlessly. This computer up above my head has a thirty-eight-thousand-word vocabulary, each word of which has been very carefully chosen to be of the utmost value to us, the crew. This switch which I have in my hand now has over three hundred counterparts in the command module alone. . . . The SPS engine, our large rocket engine on the aft end of our service module, must have performed flawlessly or we would have been stranded in lunar orbit. The parachutes up above my head must work perfectly tomorrow, or we will plummet into the ocean.

"We have always had confidence that all this equipment will work, and work properly, and we continue to have confidence that it will do so for the remainder of the flight. All this is possible only through the blood, sweat and tears of a number of people. First, the American workmen who put these pieces of machinery together in the factory. Second, the painstaking work done by the various test teams during the assembly and retest after assembly. And finally, the people at the Manned Spacecraft Center, both in management, in mission planning, in flight control, and last but not least, in crew training. This operation is somewhat like the periscope of a submarine. All you see is the three of us, but beneath the surface are thousands and thousands of others, and to all those I would like to say: thank you very much."

HOUSTON (McCandless): Eleven, this is Houston. We're getting a good picture of Buzz now, but no voice modulation. And would you open up the F stop on the TV camera? . . . That appears to be a lot better now. . . .

"Good evening," Aldrin said. "I'd like to discuss with you a few of the more symbolic aspects of the flight of our mission, Apollo 11. As we've been discussing the events that have taken place in the past two or three days here on board our spacecraft, we've come to the conclusion that this has been far more than three men on a voyage to the moon, more still than the efforts of a government and industry team — more, even, than the efforts of one nation. We feel that this stands as a symbol of the insatiable curiosity of all mankind to explore the unknown. Neil's statement the other day, upon first setting foot on the surface of the moon . . . I believe sums up those feelings very nicely.

"We accepted the challenge of going to the moon. The acceptance of this challenge was inevitable. The relative ease with which we carried out our mission, I believe, is a tribute to the timeliness of that acceptance. Today, I feel we're fully capable of accepting expanded roles in the exploration of space. In retrospect, we have all been particularly pleased with the call signs that we very laboriously chose for our spacecraft — Columbia and Eagle. We've been particularly pleased with the emblem of our flight, depicting the U.S. eagle, bringing the universal symbol of peace from the earth, from the planet earth to the moon, that symbol being the olive branch. It was our overall crew choice to deposit a replica of this symbol on the moon. Personally, in reflecting the events of the past several days, a verse from Psalms comes to mind to me: 'When I consider the heavens, the work of Thy fingers, the moon and the stars which Thou hast ordained, what is man that Thou art mindful of him?' "

The remarks were all in character, and they were saying just enough to give pieces of the answers to questions people on earth were asking themselves in the confusion that had replaced their shared exhilaration. Now it was Neil Armstrong's turn, and if he had been in doubt until the last minute about what he would first say on the moon, he knew exactly what he wanted to say now.

"The responsibility for this flight," Armstrong began, "lies first with history and with the giants of science who have preceded this effort. Next, with the American people, who have, through their will, indicated their desire. Next, to four administrations and their Congresses for implementing that will; and then to the agency and industry teams that built our spacecraft — the Saturn, the Columbia, the Eagle and the little EMU, the space suit and backpack that was our small spacecraft out on the lunar surface. We would like to give a special thanks to all those Americans who built those spacecraft, who did the construction, design, the tests and put their — their hearts and all their abilities into those craft. To those people, tonight, we give a special thank you, and to all the other people that are listening and watching tonight, God bless you. Good night from Apollo 11."

But not good night from Mission Control to Apollo 11. Ominous news began to come in from the Pacific about those two tropical storms which had seemed so harmless only a few hours earlier. Three and one-half hours after the conclusion of the telecast, at a little past ten o'clock in the evening Houston time, a vital piece of information had to be passed up to the crew.

HOUSTON (Duke): I got a couple of other things, Mike. We need to terminate battery B charge at this time — and also the weather is clobbering in at our targeted landing point due to scattered thunderstorms. We don't want to tangle with one of those, so we are going to move the — your aim point uprange. Correction, it will be downrange, to target for fifteen hundred nautical mile entry so we can guarantee uplift control. The new coordinates are 13 degrees 19 minutes north, 169 degrees 10 minutes west. The weather in that area is super. We got two thousand scattered, eight thousand scattered with ten miles visibility and six-foot seas, and the *Hornet* is sitting in great position to get to that targeted position. Over.

COLLINS: Roger.

Collins sounded a little glum; he disliked having to change his aim point, particularly as all the simulations had presumed a short-range entry. But he understood the wisdom of not landing in a violent storm. *Charlie Duke was wrong about one thing, though: USS* Hornet *was not exactly "sitting"; she was steaming northeast in something of a hurry.*

Aboard USS Hornet

Carrier *Hornet* was, relatively speaking, an old but competent naval vessel. Launched in 1943, she displaced only 30,800 tons (standard) as compared to 59,650 tons for the nuclear-powered *Forrestal.* A peaceful mission represented a change of pace for *Hornet*; for much of her life she had been at war: Marianas, Leyte Gulf and Iwo Jima in the Pacific war when she was a new ship; with the Seventh Fleet in the South China Sea during the Korean war, and more recently, in the Vietnam war. In command of *Hornet* was Captain Carl J. Seiberlich, but the senior officer aboard bore a name and rank famous in the United States Navy — Admiral John Sidney McCain Jr. He was the Navy's Commander of Pacific Forces.[2] On Wednesday afternoon *Hornet* was cruising lazily in the area of 10.6 degrees north latitude, 172.4 degrees west longitude, the targeted area of Apollo 11's splashdown. This was 1,197 statute miles southwest of Pearl Harbor and 449 statute miles southwest of Johnston Island. It was not far from the watery peak of Apollo 11 Mountain, one of several undersea mountains discovered by *Hornet* while doing underwater soundings en route to the recovery site. The ship had requested that the underwater peak be named for the voyage.

President Nixon was on the way. He had flown to San Francisco, pausing to telephone the Apollo 11 wives while they were attending the North American Rockwell luncheon in Houston, then flown to Hawaii and Johnston Island in Air Force One. He was due to fly that Wednesday evening by helicopter from Johnston Island to the Navy's communications ship *Arlington,* then transfer Thursday morning, again by helicopter, to *Hornet.* A good many things had to dovetail on this particular splashdown, including the weather.

For days — almost since *Hornet* had sailed from Pearl Harbor on July 12 — the Navy's weathermen had predicted acceptable splashdown conditions in the primary landing area. The predictions were based in part on the weather history for that part of the world. But by the night of Tuesday, July 22, *Hornet*'s radarscopes were picking up a disturbing number of thunderstorms. On Wednesday, local Hawaii time, the situation was becoming worse, not better. The prediction for early Thursday morning finally became unacceptable: heavy cloud cover, winds of fifteen knots, one- to three-foot waves on five-foot swells, scattered showers.

Captain Seiberlich had a press briefing scheduled for four o'clock in the afternoon, Hawaii time (it was then nine o'clock in the evening, Houston time). He led off with an announcement that Mission Control was considering moving the target area. Even as he was speaking, a messenger brought word to him that the splashdown site had been changed; it was now official. The new position was 13 degrees 19 minutes north latitude, 169 degrees 10 minutes west longitude. This was 241 statute miles south of Johnston Island and 906 statute miles southwest of Pearl Harbor — about two hundred fifty miles from *Hornet*'s position on Wednesday afternoon. The primary danger which attached to bringing down Apollo 11 in the original recovery site was turbulence generated in the air by the thunderstorms; there was a risk of structural damage to the spacecraft, just as a jet airplane runs an extra risk by going through turbulence of unanticipated force. The weather forecast for the alternate splashdown site was scattered clouds, winds of sixteen to twenty-four knots, and two- to four-foot waves on five- to seven-foot swells. The seas would be about the same as forecast for the primary site, but there was a vast — and important — difference in turbulence and visibility. The three C-130 jet aircraft, cruising uprange and downrange, would have — as would the two "swim" helicopters — a much better chance to achieve visual contact when Apollo 11 deployed its parachutes. The two hundred fifty miles did not pose much of a problem; *Hornet* had a maximum speed of 33 knots, communications ship *Arlington* could keep up and rendezvous, and the whole jigsaw could be made to come together, President of the United States and all. Captain Seiberlich did a quick mental calculation and estimated that *Hornet* would arrive, easily, at the new splashdown site several hours in advance. He was in a better situation than Captain Arthur H. Rostron of RMS *Carpathia,* dodging icebergs during a flank-speed dash through the night in April 1912 with the knowledge that *Titanic* would almost certainly be at the bottom of the ocean before he got to the right spot.[3] But there was one man on board *Hornet* who was a little concerned, and he would not sleep well that humid Wednesday night. This was Lieutenant Clancey Hatleberg, twenty-five years old and four years out of Dartmouth College, who had to deliver three biological isolation garments (the BIGs) to the Apollo 11 astronauts. He had practiced the recovery operation in the Gulf of Mexico with Armstrong, Collins and Aldrin personally, and he had made three trips to Pearl Harbor for rehearsal purposes yet he had never practiced in a sea as high as the one forecast for tomorrow morning. He did not consider these conditions quite "super," as they had been described to the crew of Apollo 11, and he allowed himself to worry a little. . . .

HOUSTON (Duke): Eleven, Houston. Mike, you get your chance at landing tomorrow. No go around.

COLLINS: Roger. You're going to let me land closer to Hawaii, too, aren't you?

HOUSTON (Duke): That's right, sir. [*One hundred eighty-two hours and six minutes into the mission, splashdown thirteen hours away, time for sleep. . . . Apollo 11 now 74,906 nautical miles from the earth, approaching at 6,954 feet per second, nearly five thousand statute mph. . . .*]

It was the second time that there had been a late change of signals on an Apollo splashdown. On the flight of Apollo 9 (Jim McDivitt, Dave Scott and Rusty Schweickart) in March 1969, carrier *Guadalcanal* had to turn southward at the last moment and steam to a new landing point. As in the case of Apollo 11, the weather report simply failed to hold up. There was one irony in the recovery operation: it was the one element in the space program in which the cost had been going down. The early Mercury flights required as many as fifteen recovery ships strategically posted, and the first Mercury recoveries cost about three million dollars apiece. As landing techniques were refined, NASA was able to get along with as few as five ships — three in the Atlantic and two in the Pacific — and the cost of an Apollo recovery came down to about one and one-half million dollars. Jerome Hammack, chief of NASA's landing and recovery team aboard *Hornet,* once told how that had happened: "You can draw a series of curves for the three programs. With Mercury you would peak up here for your first manned flight, and then the curve would go down. In the first manned Gemini flight the curve would peak up, but not as high as the Mercury peak, and then that curve would come down. And the same would be true of an Apollo curve. What you do when you gain experience is to cut your recovery costs. You add to your know-how, and you take advantage of more sophisticated equipment as it becomes available. This C-130 aircraft flies faster than some of the aircraft we originally used. It can cover a bigger area, so you need fewer aircraft. And it isn't just a matter of money. It's a question of how many resources you commit to cover a point where experience tells you there is hardly any chance of the spacecraft having to come down. When the decision was made, at the very beginning, to have water landings, the job naturally fell to the Navy, because the Navy had the ships. But the United States Navy is heavily committed these days. If we were requiring the same level of support that we had to have in the first Mercury flights — well, that would be quite a bind for the Navy."

If Captain Seiberlich was not bothered by the change in recovery plans, neither was the crew of Apollo 11. The last midcourse correction could still be skipped, and the lifting characteristics of the command module would be used to extend the entry range the proper distance. There did remain the problem of reentering the earth's atmosphere at the correct angle — about 6.5 degrees, with an error tolerance of no more than one degree. The earth's gravity was now causing the spacecraft to pick up speed sharply, and it would come screaming back into the atmosphere at nearly twenty-five thousand miles an hour. If it hit the atmosphere at too sharp an angle, it would burn up. If it hit too "flat," it would behave like a flat rock thrown to skip across a river. The spacecraft would skip up into orbit again, and it lacked the fuel and oxygen necessary to get home. Like everything else on the mission, reentry had to be done with precision; and, of course, there were those parachutes that had to open. . . .

One hundred eighty-nine hours and twenty-eight minutes into the mission, past six o'clock in Houston on the morning of splashdown day, Thursday, July 24. . . . Forty-one thousand nautical miles to go, velocity up to 9,671 feet per second, about sixty-six hundred statute mph. . . .

ARMSTRONG: What's the status on midcourse seven?

HOUSTON (Evans): We were going to let you sleep in until about 190 hours. Midcourse seven is not required. . . . And in the meantime, while you're eating your breakfast there, I've got the Maroon Bugle all standing by here to give you the morning news.

ARMSTRONG: Glad to hear it.

HOUSTON (Evans): Okay, Apollo 11 remains the prime story, with the world awaiting your landing today at about 11:49 A.M. Houston time. In Washington, House tax reformers have fashioned a provision which would make it impossible for wealthy individuals to avoid income tax entirely through tax-free investments or special allowances. . . . President Nixon surprised your wives with a phone call from San Francisco just before he boarded a plane to fly out to meet you. All of them were very touched by your television broadcast. . . . Air Canada says it has accepted 2,300 reservations for flights to the moon in the past five days. It might be noted that more than one hundred have been made by men for their mothers-in-law. And finally, it appears that rather than killing romantic songs about the moon, you have inspired hundreds of song writers. . . . Maroon Bugle, out. [*Less than thirty-five thousand miles to go, velocity 10,534 feet per second, over seventy-one hundred statute mph. . . .*]

HOUSTON (Evans): Apollo 11, Houston. Can you tell us where the visor assemblies ended up there?

ARMSTRONG: We're going to follow your suggestion and stow them under the right-hand couch.

HOUSTON (Evans): Roger, mighty fine. . . . The weather forecast in the landing area right now is two thousand scattered, high scattered, ten miles. The wind about 080 at eighteen knots. You'll have three- to six-foot waves. Your Delta H is plus ten feet. And it looks like you'll be landing about ten minutes before sunrise. Over.

ALDRIN: Houston, Apollo 11. Say Ron, I wonder if you could give us a good Navy explanation for this Delta H time? Over.

HOUSTON (Evans): Roger. Let me think about it, and I'll come back.

ALDRIN: Collins has got one, but I'm not sure I buy it.

HOUSTON (Evans): Apollo 11, Houston. . . . We don't have to worry about it anymore. The altimeter out there is now standard 29.92, but basically what it means is that if I give you a plus ten feet for instance — that means that you will hit the water with the altimeter reading ten feet. Over.

ARMSTRONG: All right.

ALDRIN: Ah, Collins was wrong. [*Past eight in the morning, Houston time, twenty-eight thousand nautical miles and less than four hours to go, velocity up to 11,689 feet per second, nearly eight thousand statute mph. . . .*]

HOUSTON (Evans): Apollo 11, Houston. I have your entry pad. Over.

ARMSTRONG: Okay, I'm ready to copy.

HOUSTON (Evans): Roger. Entry pad area is the mid-Pacific, roll 000 152 001. GET 194:46:06. . . . Sextant star 45-01-89-277. Boresight star, none available. Lift vector UP. . . . [*The UP instruction would apply to a much greater proportion of entry time than usual, extending the range of flight during reentry so that the spacecraft would rendezvous with* Hornet*'s new position. . . .*]

HOUSTON: This is Jim [Lovell], Mike. The backup crew is still standing by. I just want to remind you that the most difficult part of your mission is going to be after recovery.

COLLINS: Well, we're looking forward to all parts of it.

HOUSTON (Lovell): Please don't sneeze.

COLLINS: Keep the mice healthy.

COLLINS: The earth is really getting bigger up here, and of course we see a crescent. . . . We've been taking pictures and we have four exposures to go. We'll take those and then pack the camera. [*Seventeen thousand miles to go, velocity 14,633 feet per second, nearly ten thousand statute mph. . . . Recovery carrier on station, but far enough away to avoid getting hit. . . . Aircraft deployed. . . .*]

The wives of the Apollo 11 astronauts were prepared for what came next, or what would happen next; it had all been done before, and it could be done again. In Nassau Bay Joan Aldrin's father Michael Archer was up early, fussing with the champagne. Someone had brought three bottles of a French vintage which he recognized, and he put those three bottles aside: "That's for the family after Buzz comes home." He was concerned not only about cooling the champagne but also about cooling the glasses; he was not going to let anyone drink champagne out of an uncooled glass, not on this day. "Buzz is a perfectionist," he kept saying. "He'd want it this way." Joan sat down in that big new chair; she did not realize, until someone reminded her, that it was the first time she had sat in that chair since liftoff from Cape Kennedy. There were the usual supporting presences: the Reverend Dean Woodruff, Jeannie Bassett, John Young's wife Barbara, Gerald and Jo Ann Carr. . . .

Pat Collins was awake at eight in the morning in a house already full of kitchen sounds. There were three ten-pound bags of ice in the pantry, and someone had sent a case of champagne to supplement the solo bottles which had been collecting in a corner. The kitchen table was crammed with platters of deviled eggs, bean salad, homemade paté, cheese, cake, lasagna, shrimp casserole and dozens of other gifts of food — no wonder wives tended to gain weight during a mission. Pat hadn't cooked a thing: "All the neighbors come in with cakes and casseroles. It reminds me of a wake." But the atmosphere this morning was not that of a wake; this was a day of celebration. Sue Borman, Clare Schweickart, Mary Engle, Dottie Duke, Lurton Scott, Pat McDivitt, Barbara Young, Barbara Gordon and Sue Bean dropped in. Everyone was talking about reentry, and Kate Collins, wearing her mother's pink hostess apron, was passing coffee and sweet rolls. . . .

Everyone desperately wanted to *see* the parachutes open; they would be symbols of safety, signals for everyone to relax and be convinced that everything was all right. Everyone remembered the glorious moment during the splashdown of Apollo 9 when the parachutes of the McDivitt-Scott-Schweickart spacecraft burst onto the television screen in blazing color; but that had been daylight in the Atlantic. This splashdown, in spite of the range change, would occur in dawn's near darkness in weather which was not ideal for visual communication. *An hour and a half and 14,374 nautical miles to go, velocity 15,788 feet per second, nearly eleven thousand statute mph, entry velocity calculated to be 36,194 feet per second, 24,677 statute mph. . . .*

HOUSTON (Evans): Our faces are red here. We lost data with you there for a while. Did you do the P52? Over.

COLLINS: That's affirmative. We completed the P52. We'll give you the data from it in just a second. We passed our sextant star check at entry attitude and right now we're maneuvering to our first horizon check pitch attitude of 298 degrees. [*Velocity 17,322 feet per second, picking up speed fast now. . . . An hour away from separating the service module from the command module. . . .*]

HOUSTON (Evans): We don't want to jettison the hydrogen tank that is stratified, so could you cycle the fans in tank two, please? Hydrogen tank two.

COLLINS: You better believe. That old service module has taken good care of us. You better take good care of it.

HOUSTON (Evans): It sure has, hasn't it?

COLLINS: It's been a champ. [*But the champ had fought his last fight, and the service module, the mechanical brain of the CSM, would be jettisoned to burn up in the earth's atmosphere. . . . In fact it now had about thirty minutes to live. . . .*]

In El Lago Jan Armstrong said, as she had said many times before, "All you can do is take life one day at a time, because that's all you have." Once again, as at lunar touchdown and lunar liftoff, her face looked a bit more taut than usual. *Less than nine thousand nautical miles from the earth, velocity 19,512 feet per second, more than thirteen thousand statute mph. . . . Recovery team fully deployed. . . .*

Aboard USS Hornet

Lieutenant Clancey Hatleberg had spent a restless night. It was hot aboard *Hornet,* which was only partly air-conditioned. Lieutenant Hatleberg was a muscular young man, with a touch of baby fat in his face, but he was a frank young man too. He was a member of one of the Navy's five Underwater Demolition Teams. UDT men learn everything from parachuting to handling explosives, but mostly they learn to work and survive in the water. "I wanted to join the Foreign Legion, but it went out of business," he once said. "So I joined the UDT's." There were sharks in the Pacific, in that part of the Pacific where as a BIG frogman he had to swim, but Hatleberg said he was not worried: "The probability of getting bitten by a shark is less than getting hit by lightning. I read that somewhere." Hatleberg was also a man who was committed to what he was doing: "I really believe in this shot. It's probably the most significant development of the twentieth century. Just think. Man has looked forward to this during his whole existence, just sitting there, looking up at the moon — and now he's making the big leap." When he had rehearsed the recovery operation in the Gulf of Mexico with Armstrong, Collins and Aldrin, he didn't know what to expect: "Maybe I expected them to be radiating light or something. But they came aboard and I didn't even know it. Collins came over to me and we talked for about five minutes. He seemed really to be interested in what was going on around him. He was interested in my job, and he was saying it must be a lot of fun to be a swimmer, and I was saying I sure would like to see the moon." That sultry afternoon everything was business until after the rehearsal, when they were all suddenly scrubbing each other down with disinfectant. "They all got over on top of me," Hatleberg said, "and started laughing and washing me down and cutting up, and I wasn't about to say anything. Then they were gone." And Clancey Hatleberg worried about what could go wrong: "My job is based on three factors. First, they get to the moon. Second, they step out. And third, they get back. It's like riding an airplane, getting ready to jump. Anything could go wrong. Something could happen to the airplane. The wind could be too high. The equipment, the jumpmaster. . . . But when that light goes on, you know you're going to jump. . . . I worry that part of my job will go wrong. If one of the astronauts gets sick, what to do? If the raft gets punctured, what to do? If the prime swim team isn't there. . . . Will I make the right decision? When you're out there in the water, you never think like you do on land. Because in the back of your mind all the time is survival. You realize that if you do something wrong you could die. It's not like walking down the street. . . ."

. . . So on the night of July 23, 1969, Clancey Hatleberg lay on his bunk and talked with his three swimmer roommates — Wesley Chesser, John McLachlan and Robert R. Rorhback. It was too hot to sleep. They talked about general things — their career hopes and their past, their UDT training, about the funny things that had happened to them. They did not talk about tomorrow. When he finally dozed off on his sweat-drenched bunk, Hatleberg had a dream. He dreamed that the astronauts had splashed down and were out of the spacecraft — *and he couldn't close the hatch. An open hatch would destroy the quarantine.* Hatleberg was still struggling with the problem in his dream when the 2 A.M. alarm sounded. He got up, showered and went down to toy with his eggs, juice and coffee. He told a Navy chief about his dream. The chief laughed. "If you have a dream and tell about it before breakfast, it will come true," he said. Hatleberg hoped that the chief was wrong; he hoped that this dream would not come true. . . .

HOUSTON (PAO Jack Riley): Lieutenant Clancey Hatleberg of Chippewa Falls, Wisconsin, will deploy from Recovery One wearing a biological isolation garment and he will hand to the crew, through the hatch, their biological isolation garments. At 194 hours 18 minutes Apollo 11 is 7,512 nautical miles from earth, velocity 20,304 feet per second. [*Nearly fourteen thousand statute mph . . . about an hour to go. . . .*]

HOUSTON (Evans): Apollo 11, Houston. Command module RCS looks fine to us.

COLLINS: Same here, Ron. Looks very good. Doesn't make as much noise as we thought.

HOUSTON (Evans): Roger. And eleven, Houston. Weather still holding real fine in the recovery area. Looks like it's about fifteen hundred scattered, high scattered. And it's still three- to six-foot waves.

COLLINS: The air part of it sounds good. . . . And the sun's going down on schedule. It's getting real dark in here.

HOUSTON (Evans): Fine, Apollo 11. Houston, copy. [*Distance 3,896 nautical miles, velocity 24,915 feet per second, nearly seventeen thousand statute mph. . . . Ten minutes from time to separate the service module. . . . Picking up speed faster, faster. . . .*]

HOUSTON (Evans): Apollo 11, Houston. We see you getting ready for SEP. Everything looks mighty fine down here.

COLLINS: Same here, Ron. Thank you.

HOUSTON (Riley): We are awaiting confirmation of separation. . . . We confirm separation now from on-the-ground readings from telemetry. We can confirm separation. [*Nothing left of the old Saturn V stack now but the command module itself. . . . Look out world, here it comes — back. . . .*]

HOUSTON (Evans): Apollo 11, Houston. You're still looking mighty fine down here. You're cleared for landing.

COLLINS: We appreciate that, Ron.

ALDRIN: Rog. Gear is down and locked.

Cleared for landing! *Altitude 1,288 nautical miles, velocity 31,232 feet per second, over twenty-one thousand statute mph. . . .*

. . . But there were still left some painful minutes. Wives and friends, along with the other millions of well wishers, had to experience them. For a little more than three minutes, after the command module of Apollo 11 hit the earth's atmosphere, the crew would be out of communication, as absolutely alone as if they were on the back side of the moon. When the command module entered the atmosphere at its high speed, the generated heat of five thousand degrees Fahrenheit would cause a temporary communications blackout. The heat shield of the command module was built to withstand this, but the CM's radio would not function, and men and women everywhere could only wait, hope — and listen. . . . *Eight hundred nautical miles high, velocity 33,000 feet per second, 22,499 statute mph. . . . Thirty-five thousand feet per second, nearly twenty-four thousand statute mph, three minutes from reentry. . . . Thirty-six thousand feet per second, 24,544 statute mph. . . .*

ALDRIN: Houston, Apollo 11. I'm going to go to command reset and turn the tape on.

HOUSTON (Evans): Eleven, Houston. Recommend negative on that. That will put us in low bit rate.

ALDRIN: Okay, I already put it to command reset, but they still have barber pole on the tape. And now my switch is high bit rate.

HOUSTON (Evans): Okay, that will be fine. . . . And eleven, Houston. Don't mess around with that 225 there. . . . You're going over the hill there shortly. You're looking mighty fine to us.

ARMSTRONG: See you later. [*Entry time, blackout coming up. . . . Blackout. . . . Velocity 36,237 feet per second, 24,705 statute mph. . . .*]

The President of the United States and his Secretary of State had arrived aboard carrier *Hornet* about an hour earlier; this looked and was important, but it was not the most important thing in Lieutenant Clancey Hatleberg's mind. He had had his final briefing on weather and sea conditions. He had donned his black rubber suit, and now he was circling over the Pacific Ocean in a helicopter. The roar of the engines made his eardrums throb. For weeks Hatleberg had lived with decontamination, with biological isolation garments, with disinfectants. He had practiced so much that he did not like to look at a glass of water anymore. Now, on this gray, dreary morning, he wondered if he had practiced enough: "Just sitting around like that, your imagination starts to take advantage of the situation." There were three things in particular that Hatleberg was afraid he might forget: make sure the air vent valves on the spacecraft were closed; remove the tape from the filters on the astronauts' BIGs; inflate the small red life preservers on the BIGs. As Hatleberg sat waiting in the noisy helicopter, he took out a red grease pencil and wrote across his face mask: "Vents, tape, inflate."

The spacecraft first penetrated the atmosphere nearly fourteen hundred miles southeast of its ultimate landing point. Its first contact with the thin but hostile layer of gases surrounding the earth took place about eighty miles above the Solomons; now Apollo 11 was flying over, or nearly over, places with names that evoked terrible wartime memories for a middle-aged American generation: Guadalcanal, Tarawa, the Gilbert Islands. The gravitational force inside the spacecraft had now built up to 6.3 G. The command module's RCS thrusters would fire automatically to hold the spacecraft in the proper attitude for reentry; friction from the atmosphere would do most of the braking until the parachutes took over to drop Columbia into the water at a splashdown speed of about twenty statute mph.

In El Lago during the communications blackout Jan Armstrong was easing the tension by playing a card game called "Fish" with her son Mark. "I forget how you play it," said Mark. "Do you have a king? Do you have a jack?" Seated on the floor of the Armstrong home beside them was Jim Lovell; Marilyn Lovell was in the room too, along with Jan's mother, Mrs. Louise Shearon, and Ron Evans's wife Jan. . . . *Visual contact from the recovery carrier, quickly broken off as the spacecraft disappeared behind clouds; splashdown about nine minutes away. . . . Six minutes. . . .* "Drogue chutes should be coming out right about now," Jim Lovell said. Mark Armstrong asked, "Do you have an eight?" and waved a card in his mother's face.

COLLINS: Drogues.

HOUSTON (Riley): We take that to mean that the drogues deployed on time.

"The three main chutes should be coming out now," Jim Lovell observed. . . . *Less than four minutes to splashdown, no more calls from Mission Control now; turn the communications over to* Hornet . . .

HORNET: Apollo 11, Apollo 11. This is *Hornet, Hornet,* over.

ARMSTRONG: Hello, *Hornet.* This is Apollo 11 reading you loud and clear. [*Voice contact, visual contact with three full chutes, but not on television. . . .*]

HORNET: Report on condition of crew. Over.

ARMSTRONG: The condition of crew is . . . [*static*] . . . four thousand — three thousand five hundred feet on the way down.

HORNET: Eleven, this is *Hornet.* Copy. Eleven, what's your splashdown error? Over.

ARMSTRONG: Okay. Our splashdown error is by latitude, longitude, 133016915. That's . . . [*static*] . . .

. . . Apollo 11 was coming down right on its retargeted splashdown point, but the final descent would not be seen in El Lago and Nassau Bay, because of darkness, cloud cover, and the fact that *Hornet* was about thirteen miles away. (Oddly enough, Apollo 11 landed within two miles of communications ship *Arlington.*)

ARMSTRONG: Okay, *Hornet.* Apollo 11 is now . . . [*static*] . . . Apollo 11 at fifteen hundred feet.

SWIM ONE: This is Swim One, Apollo 11. [*That was the helicopter piloted by Commander Donald G. Richmond. His swimmers were John McLachlan, Terry Muehlenbach and Mitchell L. Bucklew. The recovery plan called for them to attach the orange flotation collar which would keep the spacecraft stable in a heaving sea. . . .*]

ARMSTRONG: One hundred feet.

SWIM ONE: Roger, you're looking real good. . . . Splashdown. Apollo has splashdown. . . . [*One hundred ninety-five hours and seventeen minutes GET, 11:49* A.M. *in Houston, end of the flight. . . .*]

SWIM ONE: This is Swim One. The command module is Stable 2, Stable 2. Over. [*That meant the spacecraft was upside down, but that could be fixed. There were those three switches for Mike Collins to throw, three compressor switches which had gone all the way to the moon and back without being touched, and which would now inflate three flotation bags to right the spacecraft . . .*]

AIR BOSS: This is Air Boss One. . . . It is still in Stable 2. The bags are inflating. It is not absolutely inverted now. It's seventy degrees to the vertical axis. Still in Stable 2. [*The aircraft squadron commander, Colonel Robert W. Hoffman, reporting to* Hornet. . . .]

HORNET: Air Boss One, this is *Hornet* bridge. Say when it is Stable 1. Over.

AIR BOSS: . . . [*static*] . . . stability above the vertical axis is approximately thirty degrees.

HORNET: Air Boss, *Hornet.* Recovery One is ready to deploy swimmers in one minute.

AIR BOSS: Swim Two, start recovery at once. You are cleared. [*So helicopter Swim Two, piloted by Commander Donald S. Jones, would get its swimmers into the water first — John M. Wolfram, then Wesley T. Chesser and Michael G. Mallory. . . .*]

ARMSTRONG: Air Boss, Apollo 11, everyone okay inside. Our checklist is complete. Awaiting swimmers. [*Three swimmers now in the water. . . .*]

SWIM TWO: Flotation collar halfway around the command module.
HOUSTON (Riley): The flotation collar is attached now. . . . And the collar is inflated. *Two rafts and three swimmers now in the water. . . .*

In El Lago, as *Hornet* steamed within camera range, Jan Armstrong spotted a little dot in the middle of the television screen. It looked like a very small cork bobbing up and down in a very large and unfriendly sea. "There it is, Marky!" she told her son. "I don't see it," Mark complained. When she heard that the flotation collar had been attached, Jan clapped her hands. Jim Lovell said, "You know what that means — save those rocks!"

In Nassau Bay, two minutes before splashdown, Pat Collins called for the flight plan and had a final look. NASA protocol officer Chuck Bauer told her, "It's going to take forty seconds longer than originally planned." Then Sue Borman cried out, "They're down!" Ann Collins wanted to know, "Where's Daddy?" Her mother replied, "You'll see him when the ship goes to get him." Then a raft became visible, but after a moment Pat turned away from the screen. "This is making me seasick," she said.

Recovery One moving into position, standing by to deploy. . . . It was eight minutes past six in the morning, Hawaii time, when Clancey Hatleberg finally dropped into the water. Then, "The first thing that went wrong was I couldn't find the vent valves. There was so much junk up on top of the command module that I just couldn't find them. I climbed up on top of that thing and looked and looked and looked, and finally I found one of them. It was closed, and I figured if one was closed the other must be." Hatleberg now had on his biological isolation garment, and when the astronauts opened the hatch he passed three others inside for Armstrong, Collins and Aldrin. Then the hatch was closed. After the astronauts had donned their BIGs, they opened the spacecraft hatch again. Hatleberg was startled by the way they seemed to come bounding out. Then he started to close the hatch: "I closed it, but it wouldn't lock. I just couldn't get it locked. I remembered that dream and I thought, my God, everything is falling apart. I motioned to the astronauts. Armstrong came back up, and we still couldn't get it closed. Then Collins came back and recycled the door, and we finally got it locked. Things went better after that, although the astronauts started to scrub each other [with sodium hypochloride] using the wrong method — it had been changed since they had last practiced. I had to stop them and start them over again. And the face masks on the BIGs fogged up, making it a little harder to see." Then a basketlike contrivance dropped from the helicopter piloted by Commander Donald Jones and his copilot, Lieutenant (j.g.) Bruce A. Johnson. It was Jones's third Apollo pickup; he had handled the helicopter recovery at the end of Apollo 8 (Borman, Lovell and Anders), and Apollo

10 (Stafford, Young and Cernan). Armstrong went up first, then Collins, then Aldrin. Clancey Hatleberg had to stay in the water for another two hours, waiting for the spacecraft itself to be recovered by another helicopter, but as soon as he saw the last astronaut enter the door of Recovery One he felt as if somebody had taken a hundred pounds of pressure off his head. Back on board *Hornet,* he reported to a NASA quarantine official named John Stonesifer. Satisfied that there had been no breakdown in the quarantine procedure, Stonesifer shook Hatleberg's hand. "Good job, Clancey," he said. Hatleberg said, "That was what I was waiting for. All the time, out there in the water, I was wondering if I was going to do a good job. I didn't realize how much pressure there had been until it was gone." His bunkroom was still hot and stuffy, but Hatleberg would not mind sleeping there anymore.

In the Armstrong home they had been waiting for Neil to get into the raft. When he did, little American flags suddenly appeared; Jan Evans had kept them tucked away in her purse. Flags were waved as each astronaut was hoisted up to the helicopter. Father Eugene Cargill had dropped in with Beth Williams. He proposed a champagne toast: "Here's to Jan and Neil, whom we'll never be able to visit again because of fear of contamination."

In Nassau Bay Pat Collins said weakly, "Just don't fall out of the raft, men." Just past one in the afternoon, Houston time, the helicopter landed safely aboard *Hornet.* Handclaps, screams and cheers mingled in the Collins living room. When the three astronauts emerged, wearing their BIGs, Kate Collins wailed, "Which ones are which?" Talking to no one in particular, Pat Collins said, "Surely they can let us look through the window and see them smiling through their masks. . . . All right, Kate, *now* — the champagne."

In the Aldrin home, Joan's father popped a champagne cork through the open kitchen hatch and into the living room. This was a minute and one-half before splashdown. The noise made Joan jump nervously. "Are the drogues out?" she wanted to know. She declined a glass of champagne; she would worry until they were all safe on *Hornet.* Somebody asked if Buzz would get seasick in the raft. "I feel sick," Joan said. When the third astronaut was lifted into the helicopter she clapped her hands and said: "That's Buzz. . . . I just know!" Somebody threw a champagne glass off the kitchen counter to smash it, and Uncle Bob Moon and the Aldrin children began setting off those firecrackers in the backyard, although Andy Aldrin was insisting that they save some for the day his father got out of the lunar receiving laboratory. As the astronauts left the helicopter, Joan was talking to herself: "Thank God, thank God. . . . That's Buzz at the end. It's the way he did that . . . I know. . . . Thank God." Now she was ready for a glass of champagne. "Now I can go back and live my life the way it used to be," she said to the room at large. "Forget it Joan, forget it," Gerald Carr cautioned her.

"Well, I can dream," she said defensively. In the kitchen her father was blowing more champagne corks and singing snatches from Italian opera. . . .

. . . Yet the cast of characters was not complete. Some men should have been there to share the triumph of this particular homecoming, but the fates had willed otherwise. Holding a glass of champagne in the Aldrin home, Jeannie Bassett was suddenly wracked by uncontrollable sobs. She put down her champagne and headed for the Aldrins' front living room. NASA's Bill Der Bing crooked his finger at the nearest woman[4] and asked for help. She followed Jeannie Bassett and comforted her. Jeannie was crying because her husband did not live to see this day: "If only he could be watching from somewhere. . . ." Her friend countered: "How do you know he isn't?" Jeannie straightened up, dried her tears, returned to the group in the other room and picked up her champagne. . . .

There were eight of them in all, eight astronauts who had been killed. Everyone remembered the pad fire which cost the lives of Grissom, White and Chaffee. But they were not the first. The first was Theodore C. ("Ted") Freeman, thirty-four years old, an Air Force captain and a bird lover. On the perfectly clear afternoon of October 31, 1964, he undertook a routine training flight from Ellington Air Force Base in a T-38 jet to get his flying hours in. On his landing approach he ran into a flock of wild geese. One hit his windshield and smashed it; another was swallowed up by the plane's air intake. The engines flamed out at about fifteen hundred feet. Freeman tried to glide the plane in, but he did not have enough lift. He banked sharply away from the barracks on the base and ejected. Because of the banking maneuver, he went sideways instead of up. His parachute opened only partially, and his body was found near the wreckage of the plane one mile from Ellington.

Next were Jeannie Bassett's husband, Air Force Major Charles A. Bassett II, thirty-four, and civilian pilot Elliot M. See Jr., thirty-eight. They had been named as the prime crew for Gemini 9. On February 28, 1966, as they were trying to land a T-38 at Lambert Field, St. Louis, in rain and fog, a wing tip caught the roof of the very building in which the McDonnell Aircraft Corporation was building their Gemini capsule. The jet caught fire and crashed, killing Bassett and See instantly. (Tom Stafford and Eugene Cernan, the backup crew for Gemini 9, were behind them in another jet. They did succeed in landing, and were named to replace Bassett and See on Gemini 9.)

Then came the pad fire in January 1967. The following June 6, Air Force Major Edward G. Givens Jr., thirty-seven, was driving to El Lago from Houston, where he had attended a flyers' reunion. His car went out of control on a sharp curve and ended up in a ditch. Givens died of a crushed chest. Then, on October 5, 1967, Marine Corps Major Clifton Curtis

("C.C.") Williams, thirty-five, on his way home from Cape Kennedy, was scheduled to make a refueling stop at Brookley Air Force Base near Mobile, Alabama. He never made it. At 1:24 P.M., Eastern Daylight Time, control towers in Tallahassee, Florida, Panama City, Florida, and Valdosta, Georgia, heard a weak and garbled "Mayday, Mayday, am ejecting." Then there was silence. Major Williams did not succeed in ejecting; his aircraft had "gone ballistic" on him. It flew itself into the ground, and his body was found near the wreckage.[5]

Aboard Recovery One

When Armstrong, Collins and Aldrin were hoisted into the recovery helicopter, they ran into an old friend: Bill Carpentier. Dr. William R. Carpentier, a Canadian and a flying enthusiast (he had a pilot's license by the time he was nineteen years old), interned at Ohio State University, stayed on for a residency program in aerospace medicine and managed to get himself "farmed out" to the Manned Spacecraft Center in Houston. He had a formal title: Flight Support Officer, Medical. From the beginnings of Mercury, people had been wondering about the possibility of mutations in space: "The cardiovascular system was adapting to a new environment, so to speak. It was not required to perform as much work as it would when you are walking around in one G. So they were getting changes in the cardiovascular system probably due, theoretically at least, to pooling of blood in the lower extremities after a long exposure to weightlessness. We were also finding decreases in the ability to perform work, to do exercise, after the astronauts came back. Right in the beginning of Apollo we were collecting samples on all crew members to study the microflora for a particular crewman — the bacteria that a person normally carries with him. Did this microflora change with space flight, the confinement of the three people together?" Bill Carpentier had been down at Cape Kennedy to check out Armstrong, Collins and Aldrin thirty days prior to flight, fourteen days prior to flight, and also for the intensive check five days before the flight. Having given the word go for the mission, as far as medical responsibility went, he headed for Hawaii to get aboard *Hornet*. Aboard Recovery One, he was a little more than a highly specialized aerospace physician; he was also a frogman. He had been involved in the recovery operations for Gemini 5, 6, 7 and 9, and for Apollo 7 and 9. He had never had to jump in the water yet, but he had to be prepared to do so: "If there is anything wrong with a crewman requiring medical attention, then I'll go into the water with the swimmers to do whatever has to be done. We can't expect a swimmer to make a decision that has medical implications. Once a swimmer opening the hatch got hit on the head, and he was a little dizzy for a few seconds. He was thinking of calling for help, but he didn't. That was the closest I came to going into the water myself."

When his three prime patients came up, one by one, in the helicopter rescue basket following the splashdown of Apollo 11, Bill Carpentier was beside himself — "although not as much as I thought I would be, and looking back, not as much as I thought I should have been. Only later did I realize that — really, three guys went to the moon and brought back some rocks. It's a little like the time I took a trip to Europe. It took me six months to realize I had a marvelous time." Carpentier's first words to Armstrong, Collins and Aldrin were: "Don't take off your BIGs till I've got all the swabs." This took place aboard *Hornet*; conversation had been impossible during the helicopter ride. Each astronaut, as he entered the helicopter, gave Carpentier a thumbs-up sign to indicate that he felt fine, and that had to do until they were in the doorway of the mobile quarantine facility after landing on *Hornet.* Once on *Hornet* the Apollo 11 astronauts, encased in their BIGs, were theoretically harmless to earthlings; they might as well have paraded twice around the decks. However they stayed in the helicopter until it was lowered to the hangar bay area; then they walked a few steps to the MQF, with disinfectant sprayed behind them. Inside the MQF was John K. Hirasaki, whose grandfather had emigrated from Japan to a small Texas farm in the early 1900's. He was the recovery engineer in the MQF, and he had to take care of the Apollo 11 spacecraft when it was lifted aboard to be tied into the MQF by a collapsible tunnel. John Hirasaki and Bill Carpentier had played a lot of bridge together, and would play a lot more during the next few days; that was all there had been to do in leisure time when they rehearsed this operation at the end of the Apollo 9 mission. But they would not play cards right now; there was work for Bill Carpentier to do, and the President of the United States would have to wait a few minutes; in fact about forty minutes of "free time" had been programmed here as an interlude. Microbiology sampling took about ten minutes for each man. Carpentier recalled that Neil was first: "He had twenty minutes to take a shower. Mike was next and Buzz was last. He had the least time in the shower — only about six minutes."

And now, the ruffles and flourishes, "Hail to the Chief" . . . The President walked forward with a different spring in his step, and a different smile on his face; he brought with him the presidential presence, but he had no intention of invoking that presence very much. Indeed he seemed deliberately to shed some of the dignity of his office; this was a day when he could do that, and share the triumph of many years and many people, like many another American citizen. He *was* the President, of course; but this was Armstrong's day, and Collins's day, and Aldrin's day, and President Nixon wanted to keep it that way: "Neil, Buzz and Mike, I want you to know that

I think I'm the luckiest man in the world, and I say this not only because I have the honor to be President of the United States, but particularly because I have the privilege of speaking for so many in welcoming you back to earth. I could tell you about all the messages we have received in Washington. . . . They represent over two billion people on this earth, all of them who have had the opportunity through television to see what you have done. . . .

"But most important, I had a telephone call yesterday. The toll wasn't, incidentally, as great as the one I made to you fellows, incidentally, on the moon. I made that collect, in case you didn't know.

"But I called three of — in my view — three of the greatest ladies and most courageous ladies in the whole world today — your wives. And from Jan and Joan and Pat I bring their love and their congratulations. . . . And also I've got to let you in on a little secret. I made a date with them. I invited them to dinner on the thirteenth of August, right after you come out of quarantine. It will be a state dinner, held in Los Angeles. . . . And all I want to know: will you come? We want to honor you then."

The smiles on the faces looking through the rear window of the mobile quarantine facility were very wide, and Mike Collins's moustache looked very big; it itched and it would soon be off, as soon as he got out of quarantine. . . . But the President of the United States had asked a question, and the commander of Apollo 11 answered it for the crew. Neil Armstrong said, "We'll do anything you say, Mr. President." Still the President was reluctant to end the conversation; he had to go from *Hornet* back to Johnston Island, and then all the way around the world, but he wanted to talk more with these three men. . . .

"Frank Borman," he said, "says you're a little younger by reason of going into space. Is that right? You feel a little younger?"

"We're a lot younger than Frank Borman," Mike Collins said.

"Come on over, Frank," the President said. "You going to take that lying down? Come on, Frank."

"Mr. President," Borman said, "the one thing I wanted — you know we have a poet in Mike Collins. . . . In three minutes up there he used four fantastics and two beautifuls."

. . . Still the President did not want to leave; and it was just as well, for his very presence communicated to the astronauts some of the awe and exhilaration felt across the whole planet about what they had accomplished: "This is the greatest week in the history of the world since the Creation, because as a result of what happened, in this week, the world is bigger infinitely, and also, as I'm going to find on this trip around the world . . . as a result of what you've done, the world's never been closer together before. . . . Incidentally, the speeches that you have to make at this dinner can be very short. And if you want to say fantastic or beautiful that's all right with us. Don't try to think of any new adjectives. They've all been said."

It was time for the President to go; he would be flying westward, to Asia, Eastern Europe and then home, and the astronauts would return to Houston by way of Hawaii. But first the President thought it appropriate to have a prayer of thanksgiving, and he called on *Hornet*'s chaplain, the Reverend John Plirto. "Let us pray. Lord God, our Heavenly Father, our minds are staggered and our spirits exultant with the magnitude and precision of this entire Apollo 11 mission. We have spent the past week in communal anxiety and hope. . . . The reality is with us this morning in the persons of astronauts Armstrong, Aldrin and Collins. We applaud their splendid exploits, and we pour out our thanksgiving for their safe return to us, to their families, to all mankind." . . .

The astronauts had their heads bowed and their hands folded during the chaplain's prayer, but at one point Mike Collins looked up. There was a certain chair beside the fireplace in the Collins home in Nassau Bay, and he knew that Pat Collins would be sitting in it. He looked in a certain direction, at an angle calculated as carefully as he had calculated reentry. He knew that he was looking straight at Pat Collins, and the faintest suggestion of a smile crinkled the corners of his eyes.

In Nassau Bay Pat Collins's eyes met her husband's. She shook her head and chided him gently, "That's his casual Episcopal upbringing," she said, unsuccessfully attempting to frown as a couple of tears slid down the side of her nose.

Watching her husband, Joan Aldrin had been clasping her hands and blowing kisses: "That's Buzz. . . . Hello, Buzz. . . ." When the prayer of thanksgiving was offered she bowed her head, took off her glasses and cried a little. Then she thanked Gerald Carr for his help in interpreting what had been going on. Carr put his arm around her and said, "Oh Joan, we're just the luckiest people in the world."

Jan Armstrong was wearing a carnation lei sent by friends in Hawaii. She clapped when the astronauts came to the MQF window, then laughed at Mike Collins's moustache. Holding young Mark in her lap, she bowed her head during the prayer. When the ship's band struck up "The Star-Spangled Banner," she touched Mark on the shoulder so that he would stand up. Everyone else stood. At the end she wiped her eyes.

In the MQF on board *Hornet,* Bill Carpentier went back to work. He took the astronauts' temperatures: they were normal. He weighed them: "Neil lost five or six pounds, and then I got Buzz and Mike mixed on the first report. It was Mike who lost a few pounds, not Buzz. He stayed pretty much the same." Carpentier took blood samples — thirty-nine cc's from each man, to be divided up in five ways and analyzed for white cell count, differentials, hemoglobin, etc. The blood would go flying off to Houston from *Hornet* that same day. Carpentier found a little fluid in Armstrong's right ear; it had developed during reentry and it soon cleared. But that was all.

Newsmen on board *Hornet* kept asking him for quotes: "Somebody said, 'Did you ask them what did it feel like to be the first men on the moon?' Well, I didn't ask them that; I'm their doctor, not their storyteller."

. . . On Sunday, July 27, at one o'clock in the morning Houston time, an Air Force C-141 jet landed at Ellington Air Force Base. The MQF was aboard, and Armstrong, Collins and Aldrin — along with Bill Carpentier and John Hirasaki — were in the MQF. It seemed as if it would take forever to unload. But finally it was unloaded — and finally the three wives of Apollo 11 could greet their husbands again face-to-face, even if they still had to be shielded by a window, like pheasant under glass. Joan Aldrin did not care who knew about her emotions that night; wearing the new hunting pink dress she had bought on her shopping expedition early in the mission, she threw her arms in the air like a woman transported. The smiles on the faces of Jan Armstrong and Pat Collins also testified to relief, and to joy; the smiles on the faces of Neil Armstrong and Mike Collins were the smiles of triumphant men who really had touched the stars. Buzz Aldrin was biting his lower lip, as if he were determined not to put some inner emotion on display; but when the television camera zoomed in closely, some onlookers noticed that there were tears in *his* eyes, too. . . .

But on balance, there were many more smiles than tears. For Kenny Kleinknecht, manager of Apollo's command and service modules, splashdown day was his fiftieth birthday: "I got the biggest birthday present ever. George Mueller got his at launch." Jean Gilruth, the wife of MSC director Robert Gilruth, and Deke Slayton's wife reminisced about the old days when they had been down at the Cape for John Glenn's shot, standing on the beach near the Holiday Inn and crying their eyes out; and Marge Slayton said, "That's gone from it now." Jean Gilruth said, "It's deep, it's inside. I cry harder and easier now. But it's a funny thing — I'm crying but there's not a sound, and it comes quicker. Just certain voices or a certain scene or one statement will get me." At a post-splashdown press conference in Houston the Apollo spacecraft program manager George Low paid an unexpected, and unrehearsed, tribute to Bob Gilruth. The next morning Gilruth awakened and said to his wife, "I've got the sorest muscles *here*" — in his face. Jean Gilruth said, "It's that great big grin you've been walking around with." The Gilruths went out and walked alone to a nearby bayou. "We just sat there," Jean Gilruth said later. "We didn't say a word." They had done that once before, wordlessly, after the pad fire in January 1967. . . .

. . . And then there were the telegrams — hundreds of them, thousands of them. Somebody named Tom E. Slater, in Chanute, Kansas, addressed his telegram to "Commander Armstrong, Apollo 11, Splashdown, U.S.A." A family in Gomel, Byelorussia, U.S.S.R., cabled: "Dear Lunamen, we are overwhelmed and very proud of your remarkable achievement." It was signed "the family of Murtazin." From St. James's Palace in London came

"Warmest congratulations on your historic achievement." The cable was signed by one Colonel Armstrong Aldrin Collins of the Queen's Guard. . . .

. . . And the rituals: ticker tape parades in New York and Chicago and the President's state dinner in Los Angeles, all in one tumultuous day in which the Apollo 11 astronauts finally got a glimpse, jumbled and kaleidoscopic but accurate for all that, of the emotions their achievement had evoked in the American people. Then visits to hometowns,[6] and a formal appearance at a joint session of the Congress of the United States, a trip around the world (twenty-three countries in thirty-eight days), and for Neil Armstrong, another trip halfway around the world again to visit American troops in Vietnam at Christmastime. It was not all travel and public appearances; within a week after they were discharged from quarantine on August 10 (in perfect health, and the mice Mike Collins had worried about doing fine), Armstrong was back in the lunar module simulator and Aldrin was back in the one-sixth G simulator; for the benefit of succeeding missions, NASA wanted them to check out the rehearsal part while their memories of the real thing were still green. But there was one ritual which the astronauts and their wives could share only with the rest of the astronaut community, and that was the traditional "pin party" at which, following a successful mission, each member of the crew receives a gold lapel pin. This had been a very special mission, and Jim Lovell and Fred Haise, both of whom had been on Apollo 11's backup crew, wanted the party to be very special. There was a motion picture, shot on NASA's mock lunar surface at the Houston MSC, showing one of the "astronauts" slowly strangling from a tangled telephone cord but unable to hang up on the President of the United States. And the lonely hours of Mike Collins, along with his vain search for Eagle on the moon, made too good a subject to ignore. Suddenly a recorded voice boomed across the room: "Eagle, this is Columbia. Eagle, do you read me? . . . Houston, Mission Control, this is Columbia. . . . Houston, this is Columbia. Do you read?" Then silence, and the same voice, more nervous this time: "Ellington Tower? Ellington Tower, this is Columbia. . . ." More silence and finally, the voice now desperate: "Luna 15. . . . Luna 15, this is Columbia. . . . Luna 15. . . ."

. . . And there would be a happy ending for the boy who wanted a snake for a pet: young Mike Aldrin. He did receive a baby boa constrictor for his fourteenth birthday. Then he got another one — from Senator Mark Hatfield of Oregon, who somehow had heard about this matter. According to Joan Aldrin, "Seems the Senator's son was given one and Mrs. Hatfield said it would have to go. So, thanks to the Senator, for a short while we were inundated with crawly reptiles, and things that go squeak in the night — the mice, that is — to feed the snakes. But we managed to give away that one, so we are now down to three cats (the twelve kittens, two litters, were given away), one snake and two dogs."

It was the end of a mission, and the beginning, indeed the very beginning, of the exploration of space; yet because of Apollo 11's unique watershed quality some of the Apollo astronauts had concluded that the time had come to pass the torch of exploration to a new generation of astronauts. Mike Collins made known his decision not to fly again in space on August 17, 1969, in a television interview; the only reason he had not mentioned it earlier was that nobody had thought to ask him the question in public. He later explained: "Space flight is odd in that the whole is *not* equal to the sum of its parts. The flight itself is wonderful. The parts are tedious, annoying and frustrating drudgery. During the six months before our flight I had more than four hundred hours in simulators. I had no desire to best my own figure. I really thought that if I had to spend another two years locked up in those things I'd go bughouse. Don't mistake me — I did want to make the flight. If our flight had had to abort, I would have been back here knocking on Deke Slayton's door. But we did complete the flight, and I can only stay 'up' on those things for a certain amount of time. I've been more or less 'up' for five years now, and it's difficult to sustain that pace. Perhaps you could compare it to an athlete peaking for a performance, though it's not quite the same. If I thought I were leaving Deke, or the program, in the lurch I wouldn't quit. But Deke has got a whole lot of new guys, young, zealous guys who haven't flown and who want to fly. Let them have their crack at it. And then there is the matter of having a normal family life with these peculiar working schedules we have. You can do both jobs, space flight and family man, but I don't think you can do a really good job of both at the same time."

At the time he made his decision Collins expected to remain with NASA in some capacity, but that was not to be. Before the end of the year President Nixon appointed him Assistant Secretary of State for Public Affairs. Even before the launch of Apollo 11, Bill Anders had elected to become executive secretary of the National Aeronautics and Space Council. "A lot of little things add up to a decision of this kind," he said later. "Before my own space flight, Apollo 8, my strongest motivation was to get on a flight — see the moon up close. That overrode all other motivations. My personal choice after Apollo 8 would have been first to land on the moon, second to make an earth orbital flight — because I think the earth is a fascinating thing to study — and third to circle the moon again. These were steeply degrading choices from my point of view. I had to face the fact that on my next flight, when I got one, I probably would go as command module pilot again, and maybe on the flight after that I would get to land. Given the long intervals between flights, and the crew assignments already made, I felt that my chances of actually landing on the moon were getting rather remote. The position I accepted, to become the President's adviser on space and aeronautical matters, is quite a responsible one. The job is quite different from the work I have been doing, but it does meet some of my basic criteria: a challenge and an opportunity for involvement."

As in the case of Mike Collins, Frank Borman told his wife Sue well before the launch of Apollo 8 that this would be his last flight. She found the news almost too good to believe and asked him, "But what if for some reason you are unable to complete your mission? Wouldn't you then want one more flight?" Borman said no; and much later he said bluntly: "I gave up flight status because I got so I didn't like it anymore. I didn't like the long hours sitting in simulators; I didn't like not having enough time to read the things I want to read. I didn't like being away from my family for two hundred days a year. I had seven years of that. I'm interested in a lot of things other than technology. I just didn't think I could face up to it again, for another flight, and that's the truth." For the time being Borman would stay with NASA,[7] as would Jim McDivitt, who in September 1969 succeeded George Low as manager of the Apollo spacecraft program at the Houston MSC. But Wally Schirra had left the program for good to become president and chief executive officer of Regency Investors, Inc., a Denver-based leasing company. He had taken himself off flight status much earlier, having announced before the flight of Apollo 7 in October 1968 that this would be his last mission. As he was preparing to move to Denver, Schirra said, "I quit while I was ahead, and I'm glad I did. I had three truly memorable flights, one each in Mercury, Gemini and Apollo, and I enjoyed ten years association with NASA. As the space program matured, so did I. I matured so much that sometimes I was afraid I was losing my sense of humor. The space age is very hungry. I felt after Apollo 7 that I had been completely devoured by it. And I believe all the things I learned in the space program can be put to good use in my job. I'm getting to be a bear on pollution and earth resources. I want to run around the world trying to convince people to take better care of what we have right here. If I'm lucky, maybe we can help nudge NASA to cooperate with industry, and nudge industry to cooperate with NASA."

And Tom Stafford: in five years he had been on eight flight crews, prime or backup, and had flown three flights. Now he came to a fractionally different decision. When Al Shepard, the first American in suborbital space, was restored to flight status and given command of Apollo 14, due to land on the moon in the autumn of 1970 (Shepard's old inner ear trouble had been repaired), Stafford replaced him as chief of the astronaut office at the Houston MSC. It was widely assumed that Stafford had taken himself off flight status, but he had not. He was thinking far into the 1970's, about an orbiting space station and the "space taxi" which was on the drawing boards to take astronauts up there and back: "I'd really love to fly the space shuttle. It would be great to land on the moon. I got down to about eight miles, but the shuttle really is better. Then we can plan for a mass number of flights and significantly reduce the cost per pound in orbit. With the shuttle and a station, we can reduce this cost by a factor of ten. Pound for pound, the development of the Saturn V didn't cost much more than the development of the 707 aircraft or the DC-8. The big difference is that when you fly from Los Angeles to New York you don't dump the plane in Long Island Sound. If you did it this way, only Howard Hughes and a couple of other guys could afford to fly on the airlines. What the hell, I'm thirty-eight and healthy, and I can wait."

. . . Then there would be manned flights to other planets, probably beginning with Mars, but that was for the 1980's and younger men who had not yet been chosen as astronauts. It was predictable that, by then, memories of Gemini and the first year of Apollo manned space flight — Apollo 7, 8, 9, 10 and 11 — would have acquired the patina of wistfulness and nostalgia. Men would tell and retell the Gemini 6 story of Wally Schirra's harmonica and Tom Stafford's jingling bells, and the story of the weird noises that Neil Armstrong, Mike Collins and Buzz Aldrin had sent back to the earth from their Apollo 11 spacecraft because it would be fun to confuse Mission Control. In the process of retelling, these stories — like battlefield stories — would become exaggerated and would take on a dimension larger than literal truth, and some of them would wind up in the unforgettable library of apocrypha, along with the stories told by space veterans who, at the time of Apollo 11, found their minds wandering back to the cozy and less formalized early days of Mercury. . . . *But meanwhile* . . .

There were four stars on the patch of Apollo 12, one for each of the crewmen and one for C. C. Williams, who would have been the lunar module pilot on this flight had he lived. . . . Al Bean had replaced C.C. as LMP, and it had been his idea to put the extra star there. . . .

November 19, 1969: at 12:45 A.M. Houston time, Apollo 12's lunar module Intrepid touched down on the moon. The landing was made in the Ocean of Storms, 955 miles west of where Apollo 11's Eagle had landed. It was a pinpoint landing; it had to be if Commander Charles ("Pete") Conrad Jr. and Commander Alan Bean were to retrieve a piece of an unmanned Surveyor which had soft-landed on the moon at a known spot in April 1967. When Conrad walked around behind Intrepid he could *see* the Surveyor; they were in the right place. While Commander Richard F. Gordon Jr. orbited the moon in the command module Yankee Clipper, Conrad and Bean spent eight hours walking on the moon — two EVAs, this time, instead of one. Then, once again, the lunar module's ascent engine fired perfectly, the two spacecraft docked without incident, and the crew of Apollo 12 splashed down in the Pacific on November 24 at 2:58 P.M. Houston time.

Now the age of lunar exploration had truly begun. Now the dreams which had been dreamed so long ago — before Mercury, before Robert Goddard, before Leonardo and Galileo and the nameless starstruck inhabitants of a dozen ancient civilizations — now the dreams had *substance*. Modern man could and would dream on, but he would also know with awe that the time had come for his most ambitious dreams to be realized. This was the unique glory of the trail-blazing flight of Apollo 11.

EPILOGUE

Beyond Apollo

by ARTHUR C. CLARKE

1. *"Where the action is"*

SOME years ago, a *New Yorker* cartoonist made a profound and witty observation on the nature of life. He drew a primeval beach, lined with the giant ferns that have long since vanished from the earth. Crawling up out of the ocean was a clumsy, archaic fish — a coelacanth, or a close relative — while a few yards to the rear a companion lingered nervously in the deeper water. The intrepid adventurer, already half on dry land, was looking back at his anxious colleague. And he was saying, with a rather patronizing expression: "Because this is where the action is going to be, baby."

If such a prediction had really been made, half a billion years ago, it would have been remarkably accurate — but there would have been very little evidence to justify it. From the point of view of any fish, the land is a most unpleasant, hostile place. One would have to be crazy to explore it — and as for *living* there . . . !

Look at the disadvantages. In the sea, there is no such thing as weight; a hundred-foot whale floats as effortlessly as a one-inch jellyfish. On land, the relentless pull of gravity drags all creatures downwards from birth to death. Even when it fails to kill them outright, as it frequently does, it often cripples them through the physical defects it induces in bone and muscle.

The sea is also a far more benign environment than the land. It has none of the violent temperature extremes — ranging over more than two hundred and fifty degrees Fahrenheit — which occur between the tropics and the poles. There is no fierce source of ultraviolet radiation overhead, which can burn and even kill an unprotected organism. And apart from rare submarine earthquakes, it is always calm; only the uppermost skin of the sea is ever disturbed by storms.

Yet despite this, life came out of the sea — its ancient birthplace — and conquered the alien land. In so doing, it opened up whole new possibilities of existence which we now take for granted, but which would not have been at all obvious even to a fish of human intelligence — could one imagine such a thing — back in the Cambrian period.

Because it has many parallels with our present situation, it is worth pursuing this fantasy a little further. A genius-type coelacanth, peering up through the wavering surface of the water at the dimly seen world of trees, mountains, clouds, volcanoes, thunderstorms, could have made a good case for exploring this strange environment in the cause of science. He could have confidently, and correctly, pointed out that a wealth of new knowledge would come from such an investigation. He might have argued "How can we possibly understand our universe while we are restricted to one small portion of it?" And, slightly anticipating Marshall McLuhan, he might even have realized that he could not hope to study water — until he had left it behind him.

But not even the most brilliant and farsighted of fish could have imagined the ultimate consequences of the exploration — and colonization — of the land. He could not have anticipated the rise of new life forms, with much superior senses and greatly improved ability to manipulate the environment. Long-range vision, and the dexterity of human fingers, could never have evolved in the sea. Nor, in fact, could higher intelligence itself — simply because the benevolent sea does not provide the same challenges as the fierce and inhospitable continents. (Today's intelligent marine animals, the whales and dolphins, are of course all dropouts from the land.)

But, above all, our Paleozoic Leonardo could never have imagined the new technologies which would be discovered and exploited once life had escaped from the all-embracing sea. In particular, the very existence, and the infinite uses, of fire would have been utterly beyond his comprehension. The taming and control of fire is the essential breakthrough which leads to the working of metals, to prime movers, to electricity — to everything, in fact, upon which civilization depends. Though an underwater culture is not inconceivable, it would be forever trapped in the Stone Age.

There is no need to pursue the analogy further; the lesson is obvious. When we escape from the ocean of air, we will be moving out into a whole universe of new sensations, experiences, technologies — only a few of which we can foresee today. Zero gravity research, industry and medicine, which will be discussed in detail in section 4 of this chapter, open up such immense vistas that our descendants will find it impossible to believe that we ever managed without them. Yet the greatest boons from space, it is fairly certain, will come from discoveries still undreamed-of today. Waiting for us beyond the atmosphere are the equivalents, and perhaps the successors, of fire itself.

Prometheus stole the sacred flame from heaven and brought it down to earth as a gift to the human race; and on a pillar of flame man is now riding back into the abode of the gods. What other divine powers still remain there for us to discover — and to exploit?

When we regard space exploration from this point of view, we can see at once the ludicrous shortsightedness of those who have regarded it merely as a competition between two ephemeral political entities of the late second millennium. It is true that the "Space Race" was a fact of life in the 1960's, but the rivalry between Spain and Portugal was about as important — and as transient — in an earlier age of exploration.

There are some who may agree with these long-range historical, and evolutionary, implications of space travel, but would argue that they are so far in the future that they do not concern practical politics — that has to deal with the down-to-earth problems of transportation, housing, education, medical care, poverty, and so forth. Until these matters are settled, they ask, have we any right to throw billions of dollars into space?

There are so many answers to this question that it is difficult to know where to begin. At the most elementary level, it may be necessary to point out to some of the more naïve critics that the twenty-four billion dollars devoted to the space program was not spent on a kind of astronomical foreign aid program. It was all fed back into American industry, where it generated skills, hardware and technologies which in the long run will be worth far more than they cost — though it may be impossible to prove this beyond question for some years, just as no one could prove in 1903 that the Wright brothers had not wasted their entire investment. Moreover, much of the vast payroll of the space program has been injected into the more backward areas of the United States. The chambers of commerce concerned may not care to advertise the fact, but it could be claimed with some justice that the Apollo project did more to drag certain states into the twentieth century than a great many programs of direct social improvement.

There is a superficial reasonableness about another very common criticism of space expenditure. If one *is* determined to spend $24,000,000,000 — surely it could be better used to build schools, hospitals, homes, roads — or to pay higher salaries to teachers, policemen and other underprivileged public servants?

Yet indirectly, as we have pointed out, it has helped to do just this. How much more efficient the process would have been had the amount been devoted *entirely* to this purpose could be argued endlessly — but the discussion would be a complete waste of breath, for it would have no connection with the realities of political life. It has often been remarked that money "saved" from one worthwhile project cannot be switched to another, but has to be voted all over again — which in fact seldom happens.

The "spaceships or schools" argument is a particularly unfortunate example of this fallacious reasoning. Indeed, the space program is one of the best things that ever happened to the United States educational system, both financially and psychologically. The shock of Sputnik — America's technological Pearl Harbor — focused attention on schools and colleges in a way that nothing else could possibly have done.

Until October 4, 1957, even Admiral Rickover was a voice crying in the wilderness; after that date, education became a number-one priority, and suddenly the money became available. The very requirements of the space program created an unprecedented demand for highly qualified men, and NASA was to become a major supporter of university research. Far from space robbing the schools and colleges, it contributed both directly and indirectly to their financial well-being. One can sympathize with those irate scholars who see millions being devoted to space research, when they cannot get modest grants for their own pet projects; but they should realize that cutting space expenditure is much more likely to *reduce* their chances than to increase them. Budget-slicing is a contagious disease.

Moreover, the inspirational value of the space program is probably of far greater importance to education than any input of dollars. No one who has lectured extensively to young people can have any doubts of this, though it may not be realized by those elderly academics who, as is well known, are careful to have little contact with students. But a whole generation is growing up which has been attracted to the hard disciplines of science and engineering by the romance of space. This body of trained men will, in the years to come, be one of the world's greatest assets, and no price can be put upon it. The starry-eyed youngster of today, watching TV broadcasts from the moon, will be the inventor, discoverer or technical administrator of tomorrow. And it is worth pointing out, even at some risk of invoking the horrid specter of the "two cultures," that he will be among the students least likely to set fire to the dean's office, exchange fisticuffs with the campus police, or generally go to pot. . . .

Individuals, as well as societies, need goals to inspire them; otherwise their existence becomes pointless, and the realization of that fact (whether consciously or unconsciously) results in those psychological and social ills with which we are all too familiar. And however much that mythical creature, the hardheaded, practical man-in-the-street may resent the fact, the most inspiring goals often have no obvious connection with the problems of everyday life.

The supreme physical achievement of the age of Pericles is the Parthenon; the foundations might still be unlaid, had Phidias been compelled to justify his construction program by the improvements that would eventually result in Athenian housing. It is quite possible that such a "spin-off" did occur — but what does it matter? For more than two thousand years, the marble columns standing on the Acropolis have been among the most precious treasures of mankind — though doubtless the hoi polloi would have preferred better drains.

A few days after the first landing on the moon, this point was excellently made by the London *Economist* — a very down-to-earth journal, not noted for imaginative flights of fancy. In its editorial for July 26, 1969, the *Economist* remarked:

> And as the excitement dies and familiarity sets in, the voices that say the money could be better spent on ending wars and poverty on earth must gain converts.
>
> But this argument overlooks the factor in human make-up that sets us apart from the apes. When man first became a tool-maker, he ceased to be a monkey. The human race's way of sublimating its highest aspirations has been to build the greatest and grandest artifact that the technology of the time can achieve. Through the pyramids, the parthenons and the temples, built as they were on blood and bones, to the be-spired cathedrals conceived and constructed in ages of great poverty, the line runs unbroken to the launch pad of Apollo 11. Oddly — or perhaps not so oddly — the churchmen with their unstinting praise of the astronauts have recognised this where the liberally educated rationalists with their bored carping, and their ill-bred little jokes, have not. Spiralling to the planets expresses something in human nature that relieving poverty, however noble a cause that is, does not. And to the planets, sooner rather than later, man is now certain to go.

Unlike the pyramids and the cathedrals, the exploration of space will have so many practical justifications that our descendants will think us mad that we ever doubted its value. But they will remember us, when all the creations of our hands have passed away — because we were the first men to set our sign among the stars, and our feet upon the moon.

2. *The Price of Space*

HOWEVER great the benefits — scientific, cultural, philosophical — manned space flight may bring, it is obvious that at its present enormous cost it can never be much more than a very rare and exceptional form of activity. At the current level of spending, the United States might be able to afford a landing on the moon every couple of months, and the USSR could probably do the same. Any attempt to establish permanent bases on the moon would demand a greatly increased level of expenditure — and a manned expedition to Mars would cost several times as much as a lunar mission. Even a prosperous and scientifically oriented world state might be hard pressed to manage more than one planetary flight a year.

But this assumes that space technology remains at its present primitive level — which is absurd, though it is an absurdity constantly committed by scientists and engineers who should know better. The very first decade of the space age showed spectacular improvements in cost effectiveness; these will continue (though perhaps not at the same rate) and as the price per pound of payload in orbit steadily falls, so space operations will become more complex, more ambitious — and more commonplace.

Some actual figures from the recent past will give us a better perspective by which to judge the future. On January 31, 1958, the United States placed its first payload (Explorer 1) into orbit; it weighed only thirty-one pounds, yet was rightly regarded as an outstanding achievement. At that time, no one would have dared to predict that *less than ten years later* (November 9, 1967) a rocket capable of orbiting 140 *tons* would make a flawless maiden flight from Cape Kennedy, and would carry men around the moon by Christmas of the following year. . . .

Even more important than this amazing 10,000-fold increase in payload, however, was the reduction in the price tag. In 1958, it cost half a million dollars to put one pound in orbit. Ten years later, the Saturn V could do it for five hundred dollars — a mere one thousandth of the initial rate.

Now, it would be stupid to extrapolate these figures and to predict that by 1978 we will be orbiting million-ton payloads at a bargain rate of fifty cents per pound. The law of diminishing returns has already set in, and future improvements will become progressively more difficult. But they *will* occur, just as they have done in every other form of transportation, once the initial, experimental stage has been passed. As they do, the cost of space travel will continue to decline. A decade ago it was preposterous; today it is merely exorbitant. In a few years it will be extravagant — by the end of the century, no more than expensive. And in the early 2000's, flight between earth and moon will be an ordinary commercial operation.

It is important to realize that we require no hypothetical inventions or breakthroughs — no science-fictional "spacewarps" or gravity screens — to bring this about. Existing knowledge, materials and fuels are adequate; what is still required is the experience, skill and engineering competence which only time and millions of man-hours can provide. Once again, the history of aviation gives us some instructive parallels.

The first heavier-than-air machines could barely stagger off the ground supporting a pilot and zero payload. Even this had been roundly declared impossible by authoritative critics, who when confronted with a fait accompli retreated a few steps and declaimed "Well, the airplane's of no practical importance — it can never carry a passenger as well as a pilot." Although this prediction was also rather swiftly refuted, for its first decade the airplane was considered of use only for sport and, just possibly, military reconnaissance. It took the First World War to prove that it could carry useful payloads — needless to say, bombs.

Through the 1920's, imaginative writers produced a spate of now mercifully forgotten stories about sky pirates, aerial explorations of the poles and darkest Africa, and romances of transatlantic flight. The experts, of course, continued to know better. Listen to the astronomer William H. Pickering (no relation to Jet Propulsion Laboratory's William H. Pickering): "The popular mind often pictures gigantic flying machines speeding across the Atlantic and carrying innumerable passengers in a way analogous to our modern steamships. . . . It seems safe to say that such ideas must be wholly visionary, and even if a machine could get across with one or two passengers the expense would be prohibitive to any but the capitalist who could own his own yacht."

Well, there are not too many yacht-owning capitalists among the thousands who are in the air over the Atlantic at this very moment; in this case the "popular mind" (one can almost hear the professorial sneer) was completely right, and the authority was wholly wrong. In the case of "negative predictions," this is quite a common phenomenon. The experts can spot all the difficulties, and lack the imagination or vision to see how they may be overcome. The layman's ignorant optimism turns out, in the long run — and often in the short one — to be nearer the truth.

When we look from the first airplane — the 1903 Wright Flyer — to the last of the propeller-driven airliners, such as the DC-6 just fifty years later, we see a perfectly uniform line of development without any jumps or discontinuities. All the major components of the DC-6 existed in embryo in the Flyer — engine, propeller, airfoils, fuel. The fantastic improvement in performance, which changed aviation from a dangerous sport to one of the world's greatest industries, was the result of countless steady but undramatic advances. Their cumulative effect, during half a century of research and development, transformed the airplane from little more than a toy to a dominant factor in global transportation.

It is instructive to list these advances, to see if they have any relevance to astronautics. They include: improved materials — especially the replacement of wood by metal; streamlining; the retractable undercarriage, which reduced unnecessary drag in flight; the variable-pitch airscrew, which allowed maximum propulsive efficiency over the whole range of operations; slots and flaps, which improved lift and landing characteristics; and supercharging, which allowed engines to operate in the thin air at high altitude. Another major development, though it did not involve the airplane itself, was the concrete runway, which permitted greater landing and takeoff speeds and much higher weights.

Now, there is nothing in this list that could not have been foreseen in 1903 — but even the most farsighted prophet could scarcely have imagined the combined effect of all these improvements — and the many others that also occurred. For separate increases in efficiency do not merely add; they tend to multiply, and the total effect is far greater than could ever have been anticipated by considering them individually. And because the human mind finds it easier to add than to multiply, it always underestimates ultimate possibilities.

With this history lesson as a guide, we may be able to make a better job of forecasting the future of astronautics. And, indeed, it is not difficult to see many ways in which the economics of space transportation may be drastically improved from the present figure of five hundred dollars a pound.

The first, and most essential, step will involve a complete break with today's space booster philosophy. It is impossible to tolerate indefinitely a situation in which a gigantic, complex vehicle like a Saturn V is used for a single mission, and destroys itself during flight. The Cunard Line would not stay in business for long if the *Queen Elizabeth* carried three passengers — and sank after her maiden voyage. Yet project Apollo is even more extravagant than this, because each mission also jettisons two enormously expensive pieces of equipment — the lunar module, and the service module. And even the command module itself, which does return intact, is not intended for reuse.

This really fantastic state of affairs is the result of an accident of history. Today's manned spacecraft evolved from missiles, which from the nature of things are expendable. But they might have evolved from aircraft, and indeed the pioneer writers on the subject assumed that space travel would come about in this way. They envisaged airplanes flying faster and higher, switching from propellers to rockets, and finally leaping out into space like flying fish escaping from the ocean. This certainly seems a more logical — if less melodramatic — way of doing the job; it is also the way it will *have* to be done eventually, if space flight is to make commercial sense.

It may well be argued that only the competition of the Cold War forced the development of man-carrying missiles, and that the "rocket plane" approach was the more natural and rational one. However, it would probably have taken much longer — and might therefore have been more costly — because it requires a much higher level of technical sophistication. We are now approaching this level, as a result of the experience gained with our first-generation vehicles, and can consider the next step forward — the reusable spacecraft, or space shuttle.

Ideally, the space shuttle should be able to take off like a conventional airplane, climb up to orbit, deliver its payload, reenter the atmosphere, and land on an ordinary runway, ready for another mission as soon as it has been serviced and refueled. Such a "single-stage-to-orbit" space transporter is still a long way beyond present capabilities, but there are intermediate designs that could be developed by the middle 1970's.

One concept employs the idea of drop-tanks — already commonplace on military aircraft, but taken here to its ultimate. Each stage of today's liquid-propellant spacecraft consists of huge tanks to which are attached a relatively small amount of electronics, control equipment, and rocket engines, the whole assembly being discarded as soon as the fuel is exhausted.

This is a very wasteful procedure, because the electronics, engines and so forth are ten or twenty times more expensive than the propellant tanks, yet are dropped with them. It therefore makes good sense to design a "core" vehicle, carrying payload, rocket engines and all the associated flight systems, and to surround this with a cluster of tanks which can be jettisoned when empty. In this way, all the costly equipment can be reused; if it was felt worthwhile it might even be possible to recover the empty tanks by parachute or rocket braking, so that they too could be brought back to the launch site and used again. There is an obvious analogy with the procedure used to extend the range of high-performance fighters; for spacecraft, however, the drop-tanks would weigh several times as much as the basic vehicle.

Another concept is the two-stage launch vehicle, in which the lower or booster stage, after its propellants are exhausted, can turn around and fly back to base, landing either as a giant glider, or with the assistance of small jet engines to give it additional range. A very similar idea appeared briefly in the early days of transatlantic flight, with the Mayo Composite Aircraft. This consisted of a large flying boat which carried on its back a small, high performance airplane, too heavily overloaded with fuel to take off under its own power. The flying boat thus acted as an aerial launch pad or reusable booster, which could be employed for an unlimited number of missions. The upper stage, when it had been lightened by burning most of its fuel, could make a normal landing at its destination.

This may be yet another example of an old idea, by-passed by the advance of technology, which makes an unexpected reappearance in a more sophisticated form. But there are many other concepts for cheap earth-to-orbit transportation being investigated by NASA and the aerospace industry; only time will tell if any of them are viable, or whether the answer to economical space travel lies in something which is not yet even a doodle on the back of an envelope. Nevertheless, with any luck the DC-3 of the space age should begin its career in the 1970's.

It is perhaps hardly necessary to point out that we also have to get away as quickly as possible from today's primitive "splashdowns." The USSR has always brought its astronauts down on land, and has thus been spared the enormous expense of maintaining large fleets of communications and recovery ships. Sea-based operations, of course, also impose great restraints on mission timing. Launches can take place only when weather conditions in the recovery area are forecast to be good, and as flight durations increase this will become less and less predictable. The situation is bad enough for the eight-day trip to the moon and back — but suppose a returning Mars expedition was lost, after two years in space, because there were storms in the Atlantic *and* the Pacific! It is obviously essential to develop techniques for coming down on land — preferably at a modest rate of approach on normal runways.

The experts who have made a special study of space-transporter concepts — such as Dr. George Mueller, NASA's associate administrator for manned space flight — believe that when all these ideas have been perfected, we will be able to put payloads into orbit at prices approaching five dollars per pound. This represents a hundredfold reduction of the present figure, which, it will be remembered, is itself a thousandfold improvement on the rate of ten years ago.

Looking still further ahead, we can also be quite certain that another factor will enter the picture — the technological breakthrough, which always occurs sooner or later. To revert once more to the history of aviation, we have seen how the Wright Flyer evolved smoothly into the commercially successful propeller-driven aircraft of the 1950's. But then the jet engine took over, and within a few years speed, payload, economy and comfort jumped to new levels. Inevitably, something like this will happen in astronautics.

That this is not just wishful thinking is indicated by the fact that we can already identify possible areas for such breakthroughs. They include: revolutionary new structural materials, such as carbon or boron composites, which are far stronger than steel but lighter than aluminum; use of the oxygen in the atmosphere to support combustion during the early stages of flight, so that the weight of oxidizer necessary for a mission is reduced; and, most important of all, nuclear propulsion.

The atomic rocket is a goal that engineers and scientists have pursued for more than twenty years; indeed, speculations about the use of radioactivity for space flight may be found in Goddard's notebooks half a century ago. Weight for weight, nuclear reactions liberate millions of times more energy than chemical ones; the three thousand tons of propellants in a Saturn V could — in theory — be replaced by a couple of pounds of nuclear fuel.

However, pure energy is useless by itself. It can provide heat and light in abundance, but not propulsion. The only way in which the power of the atom can drive a spacecraft is by expelling matter — usually hydrogen gas — at high speed. This is still the thousand-year-old rocket principle, with nuclear instead of chemical energy driving the exhaust jet.

To put these ideas into practice has required an immense engineering effort on the part of NASA and the Atomic Energy Commission, but in December 1967 there was a full-power ground test of the NERVA (nuclear engine for rocket vehicle application) at the quaintly named Jackass Flats, Nevada. The engine operated at a thrust level of 55,000 pounds — which may seem a very modest figure when compared with the 7,500,000 pounds of the five F-1 engines which power the first stage of the Saturn V. However, NERVA ran for the fantastic time of one hour, compared with the F-1's mere two-and-a-half minutes, so the total impulse delivered by the two systems was almost the same. What is much more important is the fact that the nuclear rocket — though still at the very beginning of its development — is already about three times more efficient than the chemical one, and has an almost unlimited potential for growth.

Atomic spaceships could be flying in the late seventies, multiplying the payloads we can take to the planets. And during the coming century they will open up the solar system — as, a hundred years earlier, the internal combustion engine opened up the remotest places of the mother world.

3. *Expedition From Orbit*

IT is now no secret that when President Kennedy committed the United States to a landing on the moon, the experts were bitterly divided over the best way of achieving this goal. Eventually, the choice was between two approaches: earth orbit rendezvous and lunar orbit rendezvous. The latter was finally adopted; it had the great advantage that the mission could be performed with a single launch vehicle, and because the lunar landing would be carried out by a specialized shuttle craft which would not be carried back to earth, the launcher rocket could be much smaller than the one required for a direct earth-moon return flight.

Yet earth, rather than lunar, orbit rendezvous offered many advantages, and the possibility of carrying out the mission with a vehicle a good deal smaller than the 6,500,000-pound Saturn V — a Saturn IV, let us call it. However, *two* Saturn IV's would be needed, and they would have to be launched within a short time of each other. If only the first got off the ground and the second was delayed by technical problems, the whole mission would have to be aborted. After elaborate calculations and at least one furious debate in front of an embarrassed President, it was agreed that lunar orbit rendezvous offered the best chance of fulfilling the prophecy "in this decade." Today, there are few who would argue against the wisdom of this decision.

Nevertheless, there can be little doubt that eventually *both* techniques will have to be exploited for maximum efficiency in space flight. This was pointed out decades ago by the pioneer writers on astronautics, and their line of reasoning can best be understood if we try to envisage what might be called the "ultimate" earth-moon transportation system. (Ultimate meaning what may be achieved within the next fifty years; beyond that, all crystal balls are made of frosted glass.)

Some of the required characteristics of the system will be obvious when we consider what is wrong with the present one. The appalling waste involved in throwing away expensive spacecraft and launch vehicles — the lunar module, the service module, the Saturn V itself — has already been mentioned. But this is only part of the sad story.

Look at the command module itself — the only component of the 363-foot-high Apollo-Saturn assembly that does come safely back to earth. A substantial fraction of its weight consists of the massive heat shield that protects it during reentry into the atmosphere, and therefore serves no purpose at all except during the last hundred miles of the half-million mile round trip. Yet this dead weight has to be carried all the way to the moon — and back! — at a cost of, literally, hundreds of tons of propellants.

The same argument applies to the three huge parachutes which lower the command module gently into the ocean; they operate only during the final two miles of the descent. And to make matters worse, they have to be designed to cope with the weight of the now useless heat shield, which has completed its task many miles overhead. The heat shield and the parachute are, of course, essential for the safety of the mission; the pity is that they have to be carried all the way to the moon, when they are needed only on the very last leg of the return journey. It would clearly be much more efficient if they could be left parked in earth orbit, to be picked up when they were needed just before reentry.

This is the philosophy underlying what have been called "orbital techniques," of which the Apollo mission's lunar orbit rendezvous is but one example. The aim is that of any experienced traveler: don't carry things you won't need, but make sure they are available when you *do* require them. And the laws of celestial mechanics, which allow us to leave supplies or equipment parked in space so that they can be located with astronomical precision a thousand years hence, cooperate ideally in this respect.

Although the pioneering studies of Tsiolkovsky in Russia, Oberth in Rumania and Baron Guido von Pirquet in Austria had emphasized the value of orbital rendezvous as early as the 1920's, the fact that spacecraft could meet and dock routinely at 18,000 miles an hour was not demonstrated until the brilliantly successful Gemini flights of the mid-60's. These — and their Russian Soyuz counterparts — are perhaps a better guide to future space operations than the Apollo mission, which was determined by deadlines rather than long-term economy.

What we may see now is the development of vehicles that are specialized for the different types of space operation, and can therefore operate with maximum efficiency in their particular regimes. The first would be the earth-to-orbit space transporter, or "space shuttle" discussed in section 2, designed to place payloads in stable orbits at an altitude of a few hundred miles.

We are beginning to discover, somewhat to our surprise, that space is a remarkably benign environment — at least for machines. Once clear of the atmosphere and set circling at the correct speed, properly packaged supplies and equipment can remain as good as new for perhaps literally astronomical periods of time. For in space, neither moth nor rust corrupt — and it is relatively easy to guard against the minor hazards of cosmic radiation and micrometeorites. Even the first generation of space probes and satellites has shown astonishing durability, often far exceeding planned lifetimes. (Occasionally with embarrassing results, as when satellites' radio transmitters fail to shut up when their mission is completed.)

Men have also demonstrated that they can work in orbit, once they have grown accustomed to the peculiar conditions (and advantages) of weightlessness. It will therefore be possible to assemble, check-out, or refuel types of spacecraft designed to operate *only* outside the atmosphere, and which will not be handicapped by having to fulfill the very difficult requirements of withstanding the thrusts and vibrations of blast-off, and the searing heat and even higher stresses of reentry.

The Apollo lunar module was a first step in the direction of a true spacecraft, and fits rather well the description written some twenty years ago for *The Exploration of Space:* "It would be a very curious-looking contraption. Probably if we saw a photograph of one we would not realise we were looking at a spaceship at all, so alien might it be to our present-day [1951!] ideas . . . it would have no vestige of streamlining and could be of whatever shape engineering considerations indicated as best. . . . It would have about as much structural strength as a Chinese lantern." Apropos this last sentence, the lunar module crews have already commented on the "tin-foil and tissue paper" appearance of their vehicle, and the crinkling noises it makes when thrust is applied.

But even the lunar module has to be built to withstand, during the few minutes of the launch from earth, ten times the forces it will experience during the landing on the moon. It also has to be folded and packaged so that it will fit into its snug little garage in the Saturn third stage — a feat made quite difficult by its wide-spread undercarriage and the small forest of antennas sprouting from it. There is obviously a great deal to be said for assembling true spacecraft *in* space — their natural environment, where their form can be perfectly fitted to their function and no engineering compromises are necessary.

Before we achieve this rather sophisticated level of operations, refueling spacecraft in orbit will become commonplace. This will be particularly advantageous as soon as reusable space transporters are available, for one vehicle of modest size, shuttling propellants up to orbit in multiple flights, could do the work of a giant Saturn-type launcher costing many times as much — and, of course, capable of only a single mission.

There are several approaches to space refueling, and perhaps they will all be employed. One could launch spacecraft with empty propellant tanks, and then — in an exact analogy to aerial flight refueling — send up one or more tankers to rendezvous with it and pump propellants across through flexible pipelines or rigid couplings. Alternatively, complete modules full of propellants could be flown up and attached to the waiting spacecraft.

This last idea was demonstrated on two of the Gemini flights in 1966. Gemini 10 and 11 both made a rendezvous with target vehicles launched earlier by an Atlas-Agena, and still containing considerable amounts of propellant. Once the spacecraft were docked together (Gemini 11), the Agena motor was restarted, and the combined vehicle was boosted to then record heights. (Who would ever have dared to predict that in the *two* years between 1966 and 1968 the altitude record for manned flight would increase from 850 miles to 240,000 miles?)

Although the first space refuelings will be carried out by means of tankers designated for specific missions, the time should eventually come when it will be worthwhile building up fuel supplies in orbit — in other words, to establish satellite filling stations, perhaps at various distances from the earth. And going beyond this, one can envisage the construction of veritable space ports, providing all the facilities — servicing, communications, navigation, quarantine — that are found on their terrestrial counterparts. When one remembers that it was only fifty years from Kitty Hawk to Idlewild, another half century would be ample time for such a development, if there proves to be need for it.

We can imagine that at least three distinct types of vehicle will operate from such a space port. There will be the earth-to-orbit shuttles, which lift propellants, supplies, cargo and passengers up through the atmosphere and, after docking and unloading, reenter for another mission. They may resemble today's ultra-high-speed rocket aircraft, or they may have wings or rotors which unfold after they have lost most of their orbital speed against air resistance.

Then there will be the lunar landing ships — larger descendants of the Apollo lunar module — which will be totally unstreamlined and will have shock-absorbing undercarriages so that they can touch down on the moon. But unlike the lunar module, they will not abandon their landing-gear-plus-descent-stage on the moon; they will need it again for the next mission.

These lunar shuttles would first operate between earth orbit and the surface of the moon, but at a later stage they might become still more specialized. They might only ascend from lunar base to lunar orbit, and make a rendezvous there with a third type of vehicle — a pure-space ferry designed to operate between earth orbit and moon orbit.

Because it would never have to land on any celestial body — not even an airless, low-gravity one like the moon — the "deep-space" transporter could be a very efficient vehicle indeed. Heat shields, wings, landing gear — all would be unnecessary. Moreover, only very low-powered and hence lightweight rocket engines would be required to nudge it from orbit to orbit. It would probably employ some form of nuclear propulsion, and as a result of all these improvements its payload would be a very substantial fraction of its total mass — not the miserable one percent or so which is the best that surface-to-space launchers can achieve.

A full-fledged attempt to show these three types of vehicle was made in the movie *2001: A Space Odyssey*. The winged Pan American craft that deposited Dr. Floyd at the Orbital Hilton was an earth-to-space transporter. The spherical ship that descended onto the moon was an earth-orbit-to-lunar-surface shuttle. And the huge, segmented "Discovery" which carried HAL and his human colleagues to Jupiter was a pure space vessel — capable of planetary flybys, but never of landing.

Whatever advances may be made in spacecraft, however, the greatest hope for improvement — without which transportation beyond the earth will never be really economical — lies not in clever engineering, but in using the natural resources of the solar system. All our present manned missions labor under an inevitable, but quite crippling, handicap. We have to carry propellants for the homeward as well as the outward voyage. . . . This does not merely double the difficulties of a mission; it would be truer to say that it *squares* them.

Anyone who doubts this should consider the price of a transatlantic air ticket — if today's jets were unable to refuel when they landed, but had to carry sufficient reserve to get back to their base. In such circumstances, the gloomy prediction of Professor Pickering, quoted in section 2, to the effect that "even if a machine could get across with one or two passengers the expense would be prohibitive to any but the capitalist who could own his own yacht" would be very nearly true.

Obviously, refueling bases on the moon and planets are a very long way ahead, and will evolve only if there is a really substantial space traffic to justify them. But this involves a typical "chicken or the egg" situation — we will not have such traffic until we have developed the vehicles to make it feasible.

Though there will always be good reasons for scientific expeditions to other worlds — for the mysteries of the universe are infinite and inexhaustible — large-scale space flight will have to pay for itself. That it will someday do so is as yet an act of faith, made reasonable by the fact that *every* form of exploration man has ever attempted has eventually led to economic benefits.

But it is not an act of blind faith, or a mere argument from analogy. Within the first decade of the space age, overwhelming practical reasons for the exploitation of this new medium had been demonstrated. There were even strong indications that no land on earth — not even the gilt-edged sidewalks of Manhattan — was as valuable as the unmarked strip of sky exactly 22,300 miles above the equator, where the synchronous satellites hover effortlessly over the same fixed spot on the globe.

We will now take a closer look at the forthcoming business of space — the next Industrial Revolution. It will provide the answers to many of the questions raised here.

By the time we have built the spacecraft that will be needed to service the orbital laboratories, workshops, factories, observatories — even hospitals and hotels — of the next generation, we will as an inevitable by-product have developed most of the technology, and the hardware, for the exploration of the moon and planets. And having won that power, it is inconceivable that we will not use it.

4. *The Business of Space*

WHENEVER new territory has become available to mankind, it has sooner or later been developed, colonized, or otherwise exploited; there are no exceptions to this rule, if a sufficiently long time scale is adopted. From discovery to full exploitation however, may take anything from a century to several hundred thousand years.

For example: men (or their immediate ancestors) have been in Africa for at least a million years, and there is still plenty of room there. North America was discovered very much later — a mere ten or twenty thousand years ago — yet, as is all too obvious, much of it is hopelessly overexploited. In some parts of California, the sequence from first scout to second mortgage has taken only a couple of lifetimes.

The Antarctic was first glimpsed in 1820, and its population density even now is less than one man per thousand square miles. The continental shelf became generally accessible in the 1950's; though some parts are already overrun with tourists, its *permanent* human population is still zero. These are the last two frontiers left on earth; it can hardly be doubted that foreseeable technologies are perfectly capable to turning them both into industrial slums within a century — and will indeed do so if we are not careful.

In the sections that follow we shall be talking about the colonization of the moon and planets, but long before we embark seriously upon such projects there is work to be done much closer to home. For just beyond the fringes of the atmosphere lies an almost perfect vacuum, that will soon be more valuable to mankind than any piece of terrestrial real estate stuffed with dross like gold or gems.

Since the beginning of history, much human effort has been devoted to reaching high ground, for a very wide variety of purposes. These have included communication (by the use of semaphores and mountaintop beacon fires); reconnaissance (Moses viewing the Promised Land); defense (innumerable castles and hilltop fortresses); Revelation (Moses again, receiving the Laws); and exotic articles of commerce (ice from the Alps to cool Nero's brow in the hot Roman summer). All these examples have — or will have — strikingly close parallels in space.

We can now reach the highest possible ground by launching ourselves, with our tools and our instruments, beyond the atmosphere. Artificial satellites can be established in orbits at any angle to the earth's axis, and at any altitude. They can move in paths that are perfectly circular, or highly eccentric — swinging out beyond the moon, and dropping back to within a few hundred miles of the earth. There is an orbit for every taste; but by far the most valuable one is that at a height of 22,300 miles, directly above the equator.

For here, and only here, a satellite can be "geostationary" or synchronous — that is, it can hover motionless over the same spot on the earth. Though this seems like pure magic, it is an elementary consequence of the law of gravity. The moon takes twenty-seven days to go around the earth; if it was closer (as indeed it was in the remote past, and may one day be again); it would complete its circuit more swiftly. The law governing altitude and period was discovered by Kepler in 1617; thereafter, any astronomer could have calculated that a satellite at a height of 22,300 miles would take exactly one day to circle the earth. It could therefore remain apparently fixed forever in the sky, because it would be moving along its orbit at the same rate as the planet spinning beneath it.

In effect, therefore, the laws of celestial mechanics allow us to construct a line of invisible towers, 22,300 miles high, completely around the equator. If we placed them a mile apart, there would be room for 160,000 separate satellites or space stations in this band of sky — every one hovering motionless above the hemisphere beneath it, and thus able to watch over, or communicate with, half the planet.

The economic and political consequences of this will be incalculable, and we have seen merely their beginnings in the setting up of Intelsat, the global satellite communications system. It is already hard to remember that, only a few years ago, radio and telephone links across the great oceans were scarce and often unsatisfactory — while TV service was completely impossible. Yet now we take it for granted when we watch an Apollo splashdown, in full color, while it is actually happening in the central Pacific. This is entirely due to the still relatively primitive communications satellites of today.

Tomorrow, as "Comsats" become more powerful and can carry more circuits, they will trigger an accelerating revolution in human affairs, perhaps even exceeding that wrought by the printing press. *Direct* broadcasts into the home will eliminate the need for thousands of ground stations — and so will open up the remotest and most backward parts of this planet to modern communications, with all that this implies for education, culture, business and politics. A hundred years ago, the electric telegraph made possible — indeed, inevitable — the United States of America. The communications satellite will make equally inevitable a United Nations of Earth; let us hope that the transition period will not be equally bloody.

Comsats can also, if used properly, improve and immeasurably enrich the everyday life of the individual. Although the multiplicity of television and radio channels will simultaneously multiply the amount of airborne trash, they will also make possible high-quality information services which are totally uneconomic today. By the 1980's, every home could have a display console on whose screen could be flashed instantly any picture or text stored in any library on earth. "Orbital newspapers," updated every hour, could be available on a global basis. Doctors, lawyers, engineers, scientists — in fact, all professional men — could have their own information channels which could keep them up to date in a way which today's journals and abstracting systems are hopelessly failing to do. If necessary, coding systems could ensure that only authorized viewers could tune in to these specialized services. The greatest use of "pay TV" may be to worldwide professional organizations and hobby clubs.

But this merely hints at the ultimate consequences of the satellite communications revolution. The telephone transformed business and social life, at the beginning of this century; the forthcoming home console will have an even greater impact, because it will allow men to meet effectively face to face, to exchange any type of information, to converse with their computers and consult information banks — without ever leaving home, unless they wish to do so.

Consider the implications of this. It means the eventual end of our present office-oriented society, with its twice-a-day traffic jams. It may even mean the end of the city — for one of its main functions will cease to exist, when men anywhere on earth can be, in all but physical fact, in each other's presence at the touch of a button. The motto of the future will be "Don't commute — communicate." And thus the automobile, which has destroyed so much of our environment, will itself be superseded by a later technology — giving us, hopefully, a chance to repair the ravages it has wrought.

It will indeed be ironic if those who attack the space program because it diverts funds from urban improvements are opposing the only force that can solve the problem — by abolishing it. This shows the danger of short-range, unimaginative thinking; major crises often demand radical and unorthodox solutions, as Hercules demonstrated when he cleansed the Augean stables by diverting a river through them.

Much nearer to the present, but of equal economic and social importance, will be the impact of the meteorological and earth resources satellites. The first are already changing weather forecasting from an art to a science; the second will eventually detect mineral and oil deposits, crop diseases, regions of land and ocean fertility, forest fires, water and atmospheric pollution, and multitudes of other phenomena of vital concern to mankind. It has been conservatively estimated that, when they are fully operational, these satellites will save enough money, and create enough new wealth, to pay for any conceivable space program.

It is often argued that all these services can be provided by unmanned, automatic satellites like Early Bird, Tiros, Essa and Nimbus, the pioneers in the field of space utilization. This, however, completely misses the point. As satellites grow larger and more complex — and our global society comes to depend upon them more heavily — the stage will very soon be reached when space-borne installation, repair and maintenance crews will be absolutely essential. There have already been cases where satellites costing tens of millions of dollars have been rendered completely useless by the failure of some small component; as one embarrassing example, just before the launch of Apollo 11 a vital communications satellite was put out of action because the bearing of a rotating antenna seized. Today, all we can do is launch another satellite and write off the first one; tomorrow, as the expense and difficulty of space transportation decreases, orbital repairs will become routine. This will result in a double saving, for it will no longer be necessary to build satellites to the fantastic standards of reliability now demanded. When they are designed for maintenance, and not for indefinite life, the cost of satellites will drop from Rolls-Royce to Volkswagen levels.

As orbital operations become more and more commonplace, we will develop the most economical mix of manned space stations and unmanned satellites. There are many tasks, such as the collection of huge amounts of scientific or meteorological data, relaying of radio and TV, or the establishment of navigational grids, which do not require human intervention except at rare intervals. But there are other types of activity which would be very difficult, or virtually impossible, without the participation of men on the spot.

Perhaps the most important of these is scientific research. For years, astronomers have dreamed of establishing observatories above the murk and haze of the atmosphere, so that, for the first time, they can see the universe as it really is. Climbing a mere hundred miles can bring the stars and planets ten times nearer, for the performance of our instruments has long been limited by the air above us, not by optical considerations. In space, telescopes could at last function with one hundred percent efficiency.

The first robot astronomical observatories have already been launched, and the torrents of information which they are sending back have revealed a new cosmos; it is as if a fog has lifted, and we are starting to see an unsuspected landscape. But the great orbiting telescopes of the future — larger than any that will ever be built on earth — will require teams of highly trained specialists to install and operate them. Only routine observations can be mechanized; for active research on the frontiers of knowledge, the scientist must be there to adjust, modify — and often rebuild — his own instruments.

This is even truer in physics than in astronomy; the very idea of operating an experimental lab through remotely controlled puppets would be enough to make a Rutherford or a Fermi turn in his grave. Moreover, many of the greatest breakthroughs in science have come from the observation of some unexpected and often trifling anomaly — exactly the sort of thing no robot or telemetering system would be likely to detect.

Orbiting laboratories will soon give us an unprecedented opportunity of studying matter, and the laws of nature, under wholly new conditions. Never before have we had access to a virtually perfect vacuum, of unlimited extent. In the past, a major part of the effort and expense required to run any physics lab has gone into the production of high vacuua, so that materials and atomic particles can be observed without contamination. Vacuum chambers more than a few feet across are very expensive; a space lab need only open the door, and it has a vacuum reaching all the way to the stars.

However, it is the condition of weightlessness (often inaccurately called zero gravity) that gives us the most exciting opportunities for research in space. Weightlessness can never be truly simulated on earth, except for a few seconds in freely falling containers or diving aircraft. But in orbit — as the world knows from Apollo telecasts, not to mention countless science-fiction movies — it is the normal state of affairs.

In this novel regime, liquids form themselves into perfect spheres, smoke does not rise (because no direction is "up") and the behavior of objects is controlled by forces such as electrical and magnetic attraction, adhesion and surface tension, which on earth are usually masked by gravity. When all vestige of weight is removed, we will be able to study chemical and physical reactions at a new level of sophistication. It is even possible that we will at last be able to make fundamental advances in our understanding of gravitation itself, when we can neutralize its influence. Studying gravity here on the earth's surface, where we are wholly under its control, may be as futile as trying to unravel the behavior of sound in the high-decibel environment of a discotheque.

Among the most promising areas of weightless research are investigations of living organisms, because gravity is a major factor controlling growth and other aspects of physiology. Even before Laika orbited in Sputnik 2, medical scientists devoted a good deal of effort to launching mice, dogs, monkeys and other small animals into space. When the researchers can accompany their subjects, and indeed share some of their experiences, progress in this field should be greatly accelerated.

It is possible that orbiting space labs may give us fresh insights into molecular biology — the fundamental science of life. If we can study cells under weightless conditions, we may gain new information about the laws that control their growth and reproduction. Any knowledge in this area could contribute to an understanding of one form of cell growth that concerns us all — cancer. Although it is never possible to predict beforehand what may come from the opening up of a wholly new field of inquiry, or the discovery of a new technique, it never fails to produce advances that could not have been achieved in any other way — especially by a direct assault on the problem.

Those well-meaning people who say "Why not spend the money used for space on cancer research?" should remember the obscure German physicist of the 1890's who attempted to find what happens when electricity is passed through rarefied gases. Doubtless his friends occasionally suggested that he should do something that would benefit mankind; luckily, Röntgen continued with his experiments. And he brought about the greatest advance in the history of medicine — for he discovered X-rays.

Lest it may seem naïve to expect dramatic advances in medicine to come from space research, it is worth mentioning that Dr. Christiaan Barnard, who should know something about the subject, has proposed to NASA a series of experiments on the behavior of DNA molecules and immunological mechanisms (graft rejects) in the absence of weight. And looking very much further ahead, there are dazzling prospects for new forms of therapy in zero-G hospitals. The removal of all weight would be an immeasurable boon to patients suffering from severe burns, and recovering from many types of operation; it would also eliminate one of the commonest and most painful of complaints, the bedsore.

It will be quite a few years before we can transship sick and injured people up to orbit, though doubtless the concept of the air ambulance would have seemed equally incredible to the Wright brothers when they began their experiments. Before the end of the 1970's, however, we may well see the beginning of specialized manufacturing processes in space.

There are many items of regular commerce whose price-to-weight ratio is so high that even a $500-a-pound freight charge would be trivial; obvious examples are rare pharmaceuticals, electronic components, and the movements of high-quality watches, but there are countless others. Not only will many products be much cheaper when we can manufacture them in vacuum or weightless conditions; there will be others which can be made *only* in space, and can then be shipped back to earth.

There are already a few processes which depend upon the temporary absence of weight. One may be encountered in any kitchen — when the cook turns over a flapjack by flipping it in the air. Those who have tried other methods of inversion *not* utilizing free fall will confirm that they are tedious and messy. . . .

A less trivial use of weightlessness was invented about two hundred years ago for the manufacture of lead shot. Some unknown genius realized that if one pours molten lead through a sieve at the top of a high tower, by the time it reaches the bottom, surface tension will mold it into a perfect sphere. By varying the mesh of the sieve, shot of any size can thus be produced very cheaply in unlimited quantity. Any alternative method of manufacture would almost certainly be orders of magnitude more expensive.

Even these simple examples are enough to hint at what may be done when we can maintain the state of weightlessness indefinitely. The manufacture of perfectly spherical (and even hollow) ball bearings more cheaply and with a higher degree of tolerance than is possible on earth is one suggestion that has been made. Another is "foamed metal." Normally, if gas is blown into molten metal, it will rise to the top. But if there *is* no "top," it will stay where it finds itself; the result could be gas-and-metal foams which would be extremely light, yet might have all sorts of useful mechanical properties.

In much the same way, pseudo-alloys could be made of metals which do not mix, and normally separate into layers when poured together. Whole new areas of chemical, electrical, biological, and metallurgical experimenting and manufacturing will become possible. What the space industries of the future may do is limited only by our imaginations; and we know so little, as yet, of this strange new realm that we can be quite certain all our guesses will fall short of the truth. If today we are shooting billions of dollars into space — it is because we will get trillions back. This is no wild exaggeration; in a mere half century, mankind has already spent almost a trillion dollars on airplanes. And space is going to be with us for a long, long time to come. . . .

Finally, although the subject may seem frivolous, let us not overlook the use of space for recreation. This, after all, will be the greatest industry of the centuries that lie ahead, when the computer has taken over mankind's traditional chores.

As soon as the cost of earth-to-orbit transportation descends below ten or so dollars a pound, space tourism will become feasible. For at least twenty years various travel organizations have been issuing, as a publicity stunt, reservations for flights to the moon. They may have to honor them sooner than they expected.

The magnificent color photographs and movies already taken of the planet earth by our outward-bound astronauts provide one completely adequate reason for going into space; the view is superb — and inexhaustible. We here at the bottom of the atmosphere never look twice at the same sky, or the same pattern of land and sea. Yet how much more varied will be the perpetually changing vista of the entire globe, with its bands of clouds, its cyclonic storms, its white pencils of jet streams, its veiled glimpses of deserts and mountains and savannas!

Five years before the launching of the first satellite, when that event still seemed decades in the future, an imaginative editor commissioned me to write an article on an orbital vacation resort. I suggested that it would be in two portions — one with gravity, one without, so that the residents could get the best of both worlds. This could be achieved if it was built in the form of a central, fixed ball, surrounded by a slowly rotating ring, so that the whole structure would look rather like the planet Saturn.

Most vacationers [I wrote in *Holiday,* November 1953] go up there to enjoy the fun and games under zero-gee — but weightlessness is not so amusing when you want to eat a meal or take a bath, and some people find it impossible to sleep under free-fall conditions. Hence the dual-purpose design of the hotel. The central ball contains the gymnasiums and the fantastic swimming pool, while over in the ring are the bedrooms, lounges and restaurant. As the ring rotates, centrifugal force gives everyone inside it a feeling of weight which can't be distinguished from the real thing. . . .

. . . Because "Up" always points to the center of the ring — to the invisble axle on which it turns — all the floors are curved, like the inside of a drum. . . . When you're dining, your table seems to be at the bottom of a smoothly curving valley, while everyone else is sitting at improbable angles up the slope. . . .

A dozen years later, Stanley Kubrick persuaded MGM to construct this scene, at almost the cost of the real thing, in the "Hilton Space Station" sequence of *2001: A Space Odyssey.* Even before it had appeared on the screen, Barron Hilton had used stills to illustrate a lecture he delivered at a Dallas symposium on the commercial uses of space. And at the same meeting Dr. Krafft Ehricke had carefully worked out the economics of an orbiting hotel on a rate-per-day basis.

Meanwhile, back in 1953:

Ignoring such activities as poker and canasta, which are highly independent of gravity, there are two classes of recreation aboard the hotel. In the ring you can play most of the games that are found on Earth — with suitable modifications. The billiard tables, for example, have to be curved slightly; at first sight it looks as if they dip down in the middle, but in this radial gravity field, this makes them behave like flat surfaces. You very quickly get used to this sort of thing, though it may throw off your game for awhile when you return to Earth.

However, since there seems little point in going out into space to indulge in terrestrial-type sports, most of the excess energy in Sky Hotel is expended in the zero-gee rooms aboard the ball. The one thing that nobody misses is a chance to do some flying — *real* flying, of the kind we've all dreamed about at some time or another. You may feel a little foolish as you fasten the triangular wings between your ankles and wrists and secure the free ends to your belt. Certainly your first few strokes will start you turning helplessly over and over in the air. But in a few hours you'll be flying like a bird. . . .

When the first men on the moon were bounding about on the lunar surface, obviously having a good time, they were pathfinding for generations yet unborn. Our picture of space is not complete if we think of it only in terms of power and profit and knowledge; for it is also a playground whose infinite possibilities we shall not exhaust in all the ages that lie ahead.

5. *The Next New World*

ANYTHING written about the moon at the beginning of the 1970's will probably look silly in the 1980's, and hilarious in the 1990's — particularly to the increasingly numerous inhabitants of our first extraterrestrial colony. For the moon is now changing before our eyes, and will continue to change with every successive landing.

To the small handful of astronomers who paid any attention to it a generation ago, the moon seemed a completely lifeless, static world, where nothing had happened for millions of years, or would ever happen again. Although its cratered countenance gave unmistakable proof of catastrophic violence in the past, its history seemed essentially over. The main interest of observers — nearly all amateurs — lay in mapping its spectacular surface features, and arguing endlessly about their origin. The futility of those arguments can be judged by the fact that, even under the best viewing conditions, earth-based instruments cannot detect objects on the moon less than half a mile across. There was no way, therefore, of learning anything about the small-scale nature of the lunar surface; in this atmosphere of ignorance, the most fantastic theories could flourish without danger of refutation. One German astronomer maintained, with Teutonic fanaticism, that the moon's visible features are made of ice; another observer put forward the delightful suggestion that its characteristic circular walled plains are coral atolls, left high and dry by retreating oceans.

It is easy enough for us to laugh at these ideas now that we have photographs showing every boulder on the moon, and have brought its rocks and soil back to earth for chemical analysis. Yet we must realize that it will be decades — perhaps centuries — before we have uncovered all the secrets of the world next door. Even on our own planet, which we have been able to examine in intimate detail for thousands of years, there are still plenty of major mysteries which can be guaranteed to disrupt any geological convention. To imagine that we will have discovered all that there is to know about the moon after a few Apollo landings is ludicrous. They will merely whet our appetite; Neil Armstrong's remark that he and Aldrin felt like small boys in a candy shop, because there was so much to see and do, will be echoed for a long time to come.

The moon's total surface area is almost 15,000,000 square miles — roughly equal to that of Africa, or the two Americas combined. Until the coming of the space age, only one hemisphere had ever been seen by man; in the remote past, friction probably produced by tidal forces robbed the moon of its rotation, so that now it keeps the same side always turned toward us. This visible face, as is obvious even to the naked eye, is about equally divided into light and dark areas; the dark regions are relatively flat and uniform plains (the *maria* or seas) while the light areas are mountainous highlands.

It appeared reasonable to assume that conditions would be much the same on the invisible "far side" — but this has turned out *not* to be the case.

There are almost none of the great dark plains on the hidden face of the moon; it is virtually all hill country. Why this should be so, nobody knows; it is certainly strange that the earth has somehow affected the topography of the land on which it never shines. And it is also strange to think that the moon would be considerably brighter in our skies if it had come to rest the other way around. Obviously there is some good reason why it did not do so; like the Apollo command module when dunked into the ocean, it may have a Stable 1 and a Stable 2 position.

Another surprise was the nature of the lunar surface itself. As no telescope could reveal anything smaller than the Pentagon building, the detailed texture could only be a matter of speculation, and rival theories could make convincing cases for hard lava, finely powdered dust, and needles of glasslike rock. But the Luna and Surveyor automatic probes, and later the Apollo astronauts themselves, found that the surface layers consist of soft dirt, which behaves rather like the damp sand left at the edge of the retreating tide. It might almost have been designed for the safe reception of descending spacecraft.

These examples are merely the first of innumerable surprises — some of them perhaps less pleasant — which the moon will spring upon us as we explore it. For we must never forget that we are dealing with an entire world, where we may find a range of geological features and environments almost as varied as on earth. Besides the obvious plains and highlands, there are canyons, deep crater pits, impressive near-vertical cliffs, meandering valleys which appear to be the beds of dried-up rivers — though whether of water or lava remains to be discovered — low domes that may be the roofs of bubbles blown in molten rock, and immense sheets of once-fluid material (again, presumably, lava) which has flowed for hundreds of miles, drowning the hills and craters in its way.

Though they have not yet been discovered, we may also expect to find caves and those curious pipes — as much as a mile long and fifty feet wide — sometimes formed in the neighborhood of volcanoes by rivers of molten rock. Because of the low gravity, these "lava tubes" might be much larger on the moon than on earth, and could provide very useful shelter. It is also quite possible that a certain amount of igneous activity still takes place from time to time; the evidence for this comes from reported red glows, the "moonquake" tremors picked up by lunar seismographs, and the clearly marked trails of giant boulders which ground disturbances have sent rolling for hundreds of yards. Any residual volcanic activity would be of the greatest importance to explorers; it could provide sources of heat during the long lunar night, and supplies of useful chemicals brought up from the deep interior of the moon.

There are many lunar vents and craters which have dark stains around them, where gases appear to have emerged and produced some kind of fallout over areas of hundreds of square miles. The materials involved will probably be carbon and sulfur compounds, and as they will almost certainly be associated with water vapor, this raises a fascinating though admittedly very remote possibility. Even if ninety-nine percent of the moon is completely sterile, the existence of small "oases," with microclimates of their own, is not wholly out of the question. These little islands of life — isolated from each other by an impassable wilderness as hostile as space itself — might have undergone quite separate evolutionary sequences, and it would be a great mistake to assume that any life forms that have managed to survive on the moon would be primitive. They are much more likely to be very specialized indeed; it may be rash, therefore, to relax the elaborate precautions of the Lunar Receiving Laboratory completely, even if half-a-dozen Apollo landings show no trace of biological activity. Our early expeditions will all be in fairly accessible places; who can say what may be lurking in the rugged foothills of Tycho, among the 30,000-foot-high peaks of the Leibnitz Mountains, or in the vast, drowned crater of Tsiolkovsky, which dominates the far side?

Perhaps even more to the point — what may we find when we start to dig? It has often been suggested that there may be underground ice on the moon; if there are local sources of heat, there may also be underground *water*. In the finest and most poetic of all spatial romances, *The First Men in the Moon*, H. G. Wells envisioned a luminous Central Sea "in perpetual flow around the lunar axis" and winding through labyrinths in which lurk "terrible and dangerous creatures that all the science of the Moon has been unable to exterminate." He described the capture of one "many-tentaculate, evil-eyed black thing" concerning which: "Afterwards, when fever had hold of me, I dreamed again and again of that bitter, furious creature rising so vigorous and active out of the unknown sea. It was the most active and malignant thing of all the living creatures I have yet seen in this world inside the moon. . . ."

Fantasy, of course. But such fantasies — if they have a plausible scientific basis — can serve a useful purpose in preparing us for the strangeness we will encounter as we venture out into the universe. The moon, in all probability, has no Central Sea, no vestige of life. Yet it may have other things quite as interesting, and even more surprising.

It is one of the characteristics of life (whether intelligently directed or not) that it radiates out from some original source and eventually occupies every possible biological niche. We are now trying to prevent this happening on the moon at one level, while encouraging it at another.

The first effort is doomed to failure, if it has not failed already. Contamination of the moon by terrestrial microorganisms may have occurred as early as 1959, when Luna 2 impacted in the Mare Imbrium; it is even more likely to have begun with Apollo 11, despite the stringent precautions taken. A great deal of equipment (not to mention bags of body wastes) was left at Tranquility Base, and although the raw ultraviolet radiation of the unshielded sun is an excellent sterilizer, any agile microbe would be perfectly safe from that danger a couple of millimeters underground.

As manned lunar operations become more extensive, efforts to keep the moon biologically clean will become hopelessly impractical, and may in any case prove to be scientifically unnecessary. And eventually, of course, we will deliberately introduce plants (and animals) onto the moon, in an effort to set up balanced life-support systems. This will mark the real beginning of lunar colonization, after the hit-and-run raids of the 1970's.

The parallel has often been drawn between the exploration of the moon and of the Antarctic, which had no permanent inhabitants until very recently. Then, as part of the 1957–58 International Geophysical Year — which also ushered in the space age — the United States Navy built the Amundsen-Scott South Pole Station, and men learned to survive even in the depths of the fearsome Antarctic winter.

With the exception of water and air, everything needed for survival in the Antarctic has to be shipped and flown in from other continents. Yet if it was worth the effort, the polar base could be made almost self-sufficient. The largest single item on the list of supplies is fuel oil for power and heat; this could be readily replaced by a nuclear reactor — as indeed has been done in Antarctica. And food, the other major import to Latitude Ninety South, could be produced there if a closed cycle economy was developed. One can imagine large conservatories in which crops would be grown during the long hours of daylight, or under artificial illumination during the night. Some of the plants could be eaten directly, others converted into protein by being fed to livestock — or, more probably, treated by purely chemical methods. Of course, all organic waste products would be cycled back into the system — which, apart from the inevitable losses, would be self-sustaining. It would be a small working model of the earth itself.

Today, it is not economically worthwhile to go to this trouble; it is far cheaper to fly in supplies. But the situation will be very different on the moon, for it will always be much more expensive to transport goods from earth to moon than from New Zealand to the South Pole. It also appears, somewhat surprisingly, that the lunar environment may be in many ways more hospitable than the Antarctic; although the absence of an atmosphere greatly increases the complications of living, it is perhaps better to have no atmosphere at all than one at minus ninety degrees Fahrenheit, and moving at a hundred miles an hour. . . .

A vast amount of paper work has been devoted to the construction and maintenance of the lunar base, and the time is rapidly approaching when theory must be put into practice. It is generally assumed that we will start with inflatable shelters, pumped up with oxygen brought from earth, and continually regenerated by chemical purification systems. However, if large numbers of people are ever to exist on the moon, this can only be a short-term solution. It will be essential to obtain oxygen — and as many other expendables as possible — from local resources. This is where supplies of free water would be so important, as water is ninety percent oxygen.

If water is not available, we will have to use rock. Most people are quite surprised to discover that the soils and stones of the earth's crust contain approximately fifty percent of oxygen; as expected, this has also proved true of the moon. In principle, therefore, we can generate an atmosphere from the lunar rocks by chemical engineering techniques — though at the cost of a great deal of hardware and considerable amounts of power.

Perhaps we can persuade growing plants to do the work for us; after all, they originally produced the oxygen we parasitic animals breathe here on earth. They could certainly be used, inside a lunar base, to regenerate the atmosphere by removing the exhaled carbon dioxide. Most projects for permanent settlements on the moon feature enclosed gardens or farms, where swiftly maturing crops could be raised during the fourteen days (earth time) of continuous sunlight.

Once this technology has been perfected, and lunar settlements have become self-sufficient in the basic necessities of life, they could be made of virtually any size. It is perfectly reasonable to talk of roofing the smaller craters, and building cities inside them — *if* there is any occasion to do so.

At this point in time, no one can say whether such grandiose schemes will ever be required. We may well discover all that we wish to know about our giant natural satellite by a network of automatic recording instruments, plus occasional traverses by parties of scientists and technicians. In this case, the permanent population of the moon may never be much larger than that of the Antarctic — say, about one thousand.

On the other hand, it seems more likely that the moon offers such opportunities in almost every scientific discipline that it could keep armies of researchers busy for generations. Consider geology (or selenology) alone; there are secrets waiting on the moon which will reveal the past of our own planet — including, perhaps, the full explanation of the metal and mineral deposits upon which human civilization depends. It has been well said (by Michael Collins, among others) that the moon, which has not been scoured by billennia of rains and winds, may be a Rosetta stone carrying on its surface the story of the solar system. The pages of that story have long been lost on earth; their rediscovery would be of inestimable scientific and economic importance.

Many of the types of research and production (excepting only weightless manufacturing) that could be carried out in space stations could eventually be transferred to the moon — which has the enormous advantage that it could provide unlimited amounts of raw material in situ. The astronomers might find it a firmer base for their great telescopes; the physicists could build their giant particle accelerators on the lunar surface, where the necessary vacuum is provided automatically. On earth, we are already contemplating atom smashers a mile in circumference; the accelerators of the next century may be wrapped around the moon. . . .

All these ideas are doubtless laughably crude and unimaginative, compared with the fantastic realities that lie ahead. The ultimate uses of the moon are still hidden from us in the mists of time, as effectively as the value of Alaska was concealed from those congressmen who, in 1867, castigated Secretary Seward for squandering $7,200,000 on a worthless wilderness of snow and ice.

But perhaps its greatest service to our descendants will be as a launching platform from which they can reach still stranger and more marvelous worlds. Because of the low gravity, it is comparatively easy to take off from the moon; in terms of energy, it is more than twenty times harder to escape from the earth. It follows, therefore, that if we can locate supplies of rocket propellant there — and hydrogen is the best — the moon could serve as the ideal base for the exploration of the solar system.

We are indeed lucky to possess, so close at hand, a celestial neighbor which can serve us as a stepping-stone to the stars, and as a proving ground on which to test our techniques for the exploration of remoter worlds. From now onwards, the moon is irrevocably bound up with the future of mankind.

And well before the end of this century, the first human child will be born there. It would be interesting to know the nationality of its parents; but such fading symbols of the old world will not be long remembered, in the fierce and brilliant light of the lunar dawn.

6. *The Solar Century*

DURING the week of the first lunar landing, there was a sudden change of focus of the human imagination. Until then, the moon had been the limit of most men's thoughts; that we might soon walk on its surface — though accepted as a logical possibility — was something that the mind could not really grasp. Then, almost in a moment, dream became reality; Armstrong and Aldrin might have explored no more than a few square yards of the Sea of Tranquility, but no one could doubt that the entire moon had now come within reach. Its exploration, and perhaps its colonization, was only a matter of time.

In a few dazzling hours, a once mythical world had become real estate. And because the human spirit must always have fresh goals, the new frontiers of the imagination swept overnight out to the planets. Men began to speak of Mars as only months before they had spoken of the moon.

Some of this talk was uninformed speculation — post-Apollo euphoria

— by those who had no idea of the technical problems involved. It might almost be compared with the hopes ignorantly expressed, at the end of the eighteenth century, by the people who believed that now men had risen into the air by the newly invented balloon, they could easily fly on to the moon. It was not quite as simple as that; the lunar voyage had to wait for another two hundred years. . . .

This time, however, the popular reaction was somewhat more in accord with the facts. Although — even at its nearest — Mars is almost a hundred and fifty times further away than the moon, in frictionless space distance is no measure of difficulty. Once a rocket has escaped from the earth's gravitational field, which it has to do even to reach the neighboring moon, it requires very little extra fuel to travel on to Mars, Venus, or indeed any of the planets.

The chief problem in manned interplanetary travel is not propulsion, but life-support. If we employ the orbits which require the minimum fuel — as we will be compelled to do in the early days of exploration — the very shortest round trips will last about two years. This means that we must preserve life in space for a hundred times the duration of a lunar journey; thus vast improvements in our present systems are required. One of the most important functions of manned space stations will be to test and develop methods of air and temperature control, as well as food regeneration which can be relied on for periods of years. This can be done safely in an environment which is only a few hours' flight time away from earth; if anything goes wrong, one can come home in a hurry. This would not be true, halfway to Mars. . . .

Even when we are fairly confident of our techniques, it would be rash to set out for Mars or Venus in a single spacecraft. When he sailed for the New World, Columbus very wisely used three ships. That would be a reassuring number for our early interplanetary expeditions.

At a reasonable rate of development — assuming *no* crash programs — such expeditions should be possible in the 1990's. Even if we made no deliberate attempt to reach the planets, well before the end of this century we would have automatically evolved all the necessary techniques during the exploitation of earth-orbital and lunar space. Thereafter, the opening up of the solar system would be inevitable.

Before we go to the planets, we will know far more about them than we do today, when most of our ideas are still little better founded than the early speculations about the moon. Powerful telescopes in orbit, and TV cameras carried on space probes like the Mariners and Voyagers, will have dispelled much of our present ignorance. But there is no substitute for a man on the spot, and in the case of the planets there is another reason — often overlooked by those who advocate purely robot explorations — why we must eventually send our eyes and brains across the abyss.

We can, and probably will, examine much of the moon with the aid of unmanned "lunar rovers" under radio control from earth. But the moon is only *two and a half seconds* away by radio waves; there is time to react to emergencies (such as rocks or crevasses in the line of travel) or to take advantage of those unexpected opportunities which are the essence of successful exploration.

Such operation in real time will be quite impossible on the planets. Even when Mars is at its closest, a radio signal takes five minutes for a one-way journey. If one of our Martian rovers got into trouble, it would be ten minutes before our corrective instructions could reach it — fifteen before we could know if we had done the right thing — twenty before we could change our minds. And this is the *minimum* time lag; for the outer planets the delay would be several hours. Though we will be able to give our robot explorers a great deal of autonomy, so that they can look after themselves in most foreseeable situations, when it comes to really strange and hostile environments it will be essential to have human controllers only a second or so away — that is, in orbit around the planet being explored — if not actually on its surface.

A striking example of the limitations of robot scouts was given recently by Dr. Ernst Stuhlinger of the Marshall Space Flight Center. In January 1967, with Dr. Wernher von Braun and Dr. Robert Gilruth, he visited the U.S. Antarctic Base to get a better understanding of the problems of interplanetary exploration. They were shown around by one of the local biologists who "pointed out to us the various forms of algae and arthropods that live in that region, and he described their ingenious ways of adaptation to this unusual environment. Each rock sample which he selected with the trained eyes of the research biologist contained on its protected underside some specimens of algae, mites or small insects; the samples which we untrained space engineers picked up did not show any traces of life. 'No wonder,' said Dr. [Russell] Strandtmann. 'This is the difference between a live, alert, intelligent, highly trained and motivated scientist, and a lifeless robot. Do you see now,' he added, 'why we scientists believe that man should go in person to the Moon and to Mars?' "

When nuclear propulsion is perfected, it will be easier (and cheaper) to reach the nearer planets than it is now to travel to the moon. And as our technology continues to advance, the velocities we can attain will steadily rise. Flight times will drop from months to weeks — and ultimately to mere days. We will be able to travel anywhere we wish in the solar system, and visit all the children of the sun.

Some of those are very strange offspring indeed. Yet perhaps we had better get used to the fact that *our* planet is the anomaly, with its hard, cold rocks and its unique, oxygen-bearing atmosphere. Any explorers from interstellar space might overlook it completely during their first examination of the solar system; they might notice only the four "gas giants" Jupiter, Saturn, Uranus and Neptune. Compared with these monstrous worlds the so-called terrestrial planets earth, Mars, Venus and Mercury are insignificant dwarfs, huddling round the warmth of the central sun.

It is almost half a billion miles to Jupiter, and nearly three billion to Neptune, the outermost of the giants. Yet despite the unimaginable distances involved, we have an opportunity of examining *all four* of these worlds within the next two decades — using the launch vehicles we already possess. An automatic probe aimed toward Jupiter in 1977 could swing past that planet in 1979 and, if its timing was correct, could get a boost from the Jovian gravity field which would flick it on to Saturn. Here the same thing would be repeated, when it arrived in 1981, and four years later it would reach Uranus. Cannoning off a third gravitational field in this game of cosmic billiards, it would arrive at Neptune in 1989, after a total flight of twelve years. Although this may seem a long time — and it will pose some very severe problems to the electronics designers — the direct flight *to Neptune alone* would normally take twenty years.

Opportunities for this multiple mission (the "Grand Tour") — occur only in the period 1976–78. Thereafter, the planets will not get lined up properly again for a hundred and seventy-five years. It will be a pity if we miss the best chance to survey *all* the giant planets, this side of the year 2153.

Even by that date, it seems unlikely that we will have got very far in unraveling the mysteries of Jupiter, the lord of them all. Mere statistics cannot convey the sheer size of Jupiter; it means little to say that the planet has 11 times earth's diameter and 318 times its mass. There is one mental image, however, that does give a faint impression of Jovian magnitude. If some cosmic collector of worlds skinned our planet and pinned its pelt like a game trophy on the face of Jupiter, it would appear there about as large as India on a terrestrial globe. In other words, Jupiter is to earth as earth is to India; anyone who has ever flown over the endless plains of *that* country may well feel numb at the thought of exploring Jupiter.

However, it is by no means certain if Jupiter has a solid surface to be explored. All we can see through our telescopes are the outer layers of a colorful and extremely turbulent atmosphere which consists largely of hydrogen, methane and ammonia. The rapid spin of the planet — despite its size, its rotation period is only ten hours — produces trade winds that must make ours seem like gentle zephyrs; the resulting immense belts of cloud, parallel to the equator, give Jupiter a characteristic banded appearance.

The nature of these clouds, if that is the appropriate term for them, is a major mystery. Hydrogen, methane and ammonia are quite colorless — yet the Jovian atmosphere shows a wide range of pinks and blues and salmons. Its most conspicuous feature, the 30,000-mile-long oval known as the Great

Red Spot, has even been described as brick-red, though sometimes it fades (or sinks?) to invisibility.

Because it is at five times the earth's distance from the sun, Jupiter might be expected to be extremely cold; this is certainly true of its upper atmosphere. However, recent observations indicate that it is not quite as cold as it should be; like our own planet, Jupiter appears to have internal sources of heat. Some tens or hundreds of miles below the clouds it may be warm enough for water to exist in the liquid form; and given sufficient time, water plus hydrocarbons plus energy may add up to life.

A generation ago, practically all scientists would have dismissed the idea of life on Jupiter as absurd, owing to the "poisonous" nature of its atmosphere. But we now realize that this is a very naïve and self-centered viewpoint; modern biological theories suggest that the ancient earth, some four billion years ago, had a gaseous envelope much like Jupiter's — and it was in such an atmosphere that primitive life first arose. Only at a much later stage, when plants evolved, were the hydrogen-bearing compounds methane and ammonia replaced by oxygen; not until then could animal life appear. It is *oxygen* which is a deadly poison to primitive organisms — and, indeed, even to man if the pressure is increased too greatly.

Jupiter may, therefore, be in the very early stages of biological evolution; this could explain some of the extraordinary colors in its atmosphere, typical of many complex organic compounds. Whether this is fantasy or an exciting reality, we should know very soon, thanks to astronomical instruments in earth orbit. But it will be a very long time before we can actually send instruments — or men — down into that witch's brew of an atmosphere, stirred by storms and fertilized by bolts of lightning more powerful than any on earth.

The three other giants — Saturn, Uranus and Neptune — appear to be quite similar to Jupiter, except that they are considerably smaller and much less active. (But even the smallest, Neptune, is three and a half times the diameter of earth!) Saturn, however, has one unique and famous characteristic, its beautiful system of rings, which form a kind of multiple halo in the planet's equatorial plane. Although they appear solid when viewed through a telescope, the rings are actually composed of myriads of small particles in independent circular orbits. They are probably dust and ice; it may not be too far from the truth to say that Saturn is surrounded by a perpetual hailstorm.

The four giant planets possess twenty-nine known moons between them, and there must be many others still undiscovered. Jupiter (twelve satellites) and Saturn (ten) are almost small solar systems in themselves; indeed, some of their moons are comparable in size to the planets Mars and Mercury. However, because of their great distances, they appear as little more than pinpoints even in our most powerful telescopes. We can only assume that they are frozen, lifeless hunks of rock and ice.

Even so, they may have many surprises — and at least one of these worlds already presents us with a tantalizing mystery. Iapetus, the ninth moon of Saturn, is about six times brighter on one side of its orbit than on the other. There must be some remarkable formation — almost a giant natural mirror — flashing sunlight back at us from certain angles of illumination.

Saturn's largest moon, Titan, is sufficiently massive to hold an atmosphere — a tenuous one — of methane gas. *Any* atmosphere is useful for braking purposes, and methane is a particularly valuable substance, since it is twenty-five percent hydrogen. One day, our nuclear rockets may refuel here; Titan may be the key to the outer planets and beyond.

Until well into the next century, however, most of our manned explorations will be directed to smaller, more earthlike planets closer to the sun. Mercury, Venus and Mars all lie within a few months' flight time, even by today's slow-moving spacecraft, and they are all solid bodies, upon which it will be possible to make landings. Whether it will be desirable to do so is quite another matter.

Mercury, so close to the sun that the radiation it receives is ten times more powerful than on earth, may closely resemble the moon. Too small to retain an atmosphere, it is so hot that metals could melt in its equatorial regions. During a century of observation, astronomers had convinced themselves that the little planet keeps the same face always turned toward the sun, as the moon does toward the earth. Thus a whole science-fiction mythology had arisen, describing a world one side of which was frozen in perpetual night, while the other burned forever beneath a pitiless sun.

To the great embarrassment of astronomers, radar observations showed in 1965 that this simple, Dantesque picture is just not true. Mercury does know day and night, like any well-appointed planet; it spins on its axis once every fifty-nine days. This does little to alleviate its climate, but the extremes of heat and cold encountered there do not present a serious problem to space technology. Moreover, by a careful choice of latitude and time, it would be possible to find landing areas on Mercury where — for periods of several weeks — earthlike temperatures occurred.

There can be no doubt that men will one day visit Mercury, if only to install or service automatic monitoring equipment. Whether permanent bases will ever be established there is a question that only the future can decide. The reasonable assumption is that such a blistering lump of rock will have few resources and fewer attractions, except to a handful of devoted astrogeologists.

But nature is infinitely varied, and there may be much more on Mercury than we can possibly imagine today. For one thing, it seems to have an inexplicably high density — as if, for some peculiar reason, it is a repository of heavy metals. Commenting on this, the British mathematician R. A. Lyttleton recently informed the Royal Astronomical Society that "Mercury may well be the strangest planet in the solar system." This is a startling assertion; perhaps we should not write off this little world too quickly as an unimportant lump of rock.

However, not even those hardy optimists, the science-fiction writers, really expect to find any form of life on Mercury. Although it is not too difficult to imagine organisms that could survive there (particularly underground, where the temperature would be moderate), if the planet never possessed oceans, biological evolution could not have started. Yet again, we are talking of life as we know it; we should prepare ourselves for the discovery that we know very little about it indeed.

Of all the planets, the one for which we had the highest hopes was Venus. Almost a twin of earth in size, with a gravity very slightly weaker so that it would produce a mild buoyancy, Venus also comes closer to the earth (26,000,000 miles) than any other body, except for the moon and a few stray planetoids — wandering mountains about the size of Central Park. Though it is considerably nearer to the sun than our world, its permanent covering of brilliant white clouds reflects ninety percent of the solar heat back into space. For this reason, it was hoped that its temperature would not be excessively high. Nowhere else in the solar system seemed a more likely abode of life; the dazzling cloud cover at once evoked images of oceans and rain-drenched forests, perhaps trampled by great beasts like those that our own planet knew in the days of the dinosaurs.

Alas for these dreams! One by one we have been forced to relinquish them. First the spectroscope revealed that the atmosphere contains huge quantities of the suffocating gas carbon dioxide — which precluded any form of animal life, but did not eliminate plants. Then earth-based measurements of the radiation from the planet indicated that it was far hotter than anyone had imagined; so hot, in fact, that water could exist there only in the form of steam.

For a few years ingenious theories "explained away" these observations, but the last faint hopes for a merely tropical Venus were shattered by the Mariner II flyby of 1962, and finally laid to rest by the Russian Venera probes of 1969. As these descended into the atmosphere, they registered not only very high pressures (comparable to those a thousand or more feet down in our oceans) but temperatures of 800°F. Conditions on the surface of Venus appear roughly similar to those inside a blast furnace.

That still leaves Mars, beloved by romantic writers for almost a hundred years, largely thanks to the report in 1877 of a network of fine lines (the so-called canals) covering the surface of the planet like the map of an international airline system. It is now known that the canals are illusory; although there are some rather diffuse linear features, Mars in close-up is no more artificial-looking than the moon. In fact, as the brilliantly successful Mariners 4, 6 and 7 showed, it very closely resembles the moon, being covered with craters of all sizes up to at least a hundred miles in diameter.

Yet there is one all-important difference between Mars and the moon — and that is the presence of an atmosphere. That Mars has an atmosphere had been known for almost two centuries, because its surface features are occasionally obscured by mists or clouds. However, it is not an atmosphere which we could possibly breathe, for it contains virtually no oxygen. Even if it were pure oxygen, it would still be useless to us — for it is much thinner than the air on the summit of Mount Everest.

Nevertheless, any atmosphere, as long as it is not an actively poisonous one, is probably better than none. The tenuous Martian atmosphere acts as an effective screen against the small meteorites that might otherwise pepper the planet's surface, and it moderates the temperature extremes. It may also help to circulate the small (but possibly vital) amount of water vapor that exists on Mars.

The public attitude toward Mars has oscillated between extremes of pessimism and optimism, as a kind of overreaction to each new advance in knowledge. When the Mariner TV cameras showed a bleak, cratered landscape, there was an immediate assumption that the planet had now been proved to be lifeless. This was absurd; much better pictures of earth, from considerably shorter ranges, have shown no signs of life here! This question may not be settled even when we have landed robot probes on Mars, as Dr. Stuhlinger's experience in the Antarctic demonstrates. Once again we are faced with the difficult problem of proving a negative. If there *is* life on Mars, the proof may be obtained within a decade, or even less; but if there is not — it may take a century of careful examination to demonstrate this beyond doubt. Though large and active life forms could hardly be overlooked, it seems likely that anything surviving in such a rugged environment would be small, inconspicuous, and restricted to a few favored locations.

It might also have to be mobile to avoid the long, ferocious winters; although on a summer afternoon the equatorial temperature may rise to the seventies, the average temperature is below freezing point — and the minimum may be *two* hundred degrees below zero Fahrenheit.

But Mars is a small world — half the diameter of earth — and its year is almost twice as long; moreover, it has no seas to act as barriers to migration. Thus any creature which could travel a modest five miles *per day* need never experience winter.

It is said that a famous astronomer once received this message from William Randolph Hearst: "Is there life on Mars? Please cable one thousand words." He got the reply "Nobody knows" — repeated five hundred times.

Now at last there is a chance of a more informative answer. And if it should be "No," that will only be temporary. In what is, from the astronomical viewpoint, a mere flicker of time, there will be life on Mars — and perhaps most of the other planets and their satellites. It will have come from mother earth.

7. *Interplanetary Man*

While man's first footprints are still fresh upon the moon, it may seem slightly premature to discuss the exploration — and colonization — of the planets. Nevertheless, at the present rate of technological progress, such developments would appear to be inevitable during the next century. They may, indeed, be its prime concern — as the exploration of *this* earth has been a principal activity of the centuries since Columbus.

The solar system appears to have been admirably designed for the purposes of transportation — which may or may not impress the exponents of the "as God intended us to" school of thought. In the first place, although the planets move at considerable speeds, they all travel in the same direction — counterclockwise around the sun, for an observer at earth's North Pole. If any planet traveled in a retrograde orbit (i.e., clockwise) the problem of reaching it would be enormously increased.

To give a somewhat oversimplified example: earth and Mars move along their orbits at average speeds of 66,000 and 54,000 miles an hour. In changing from one planet to the other, therefore, it is necessary to make an adjustment of 12,000 miles an hour, which is a fairly modest figure even in terms of present-day rocketry.

Had they been traveling in *opposite* directions, however, the relative velocity to be overcome would have been the huge figure of 120,000 miles an hour, or ten times as much. And since the energy of a moving object depends on the square of its speed, this means that the difficulty of the transfer would be increased by a hundred — making Mars almost as inaccessible as the nearer stars.

A second simplifying factor is the relative flatness of the solar system; the planetary orbits all lie in very nearly the same plane, like the tracks of a circular race course. Only Mercury and Pluto — the two eccentrics at the opposite ends of the system — depart from this rule; their orbits are tilted at seven and seventeen degrees respectively to that of the earth, and even these inclinations are not likely to cause major problems.

A third reason for the — relative! — ease of interplanetary flight is much less obvious, and is not shown on any chart of the solar system. The planets are all held in their orbits by the gravitational grip of the sun, but that grip weakens very rapidly with distance. Moving outwards from the sun is like climbing up a steep hill; at first the slope is almost vertical, but presently it becomes less precipitous. And at last, it flattens out into a level plateau.

All the planets, even innermost Mercury, are far, far out on the fringes of the sun's gravitational field; by the time we get to earth, that field has been reduced to a mere ten-thousandth of its initial value. It therefore requires very little energy — or fuel — to travel from one planetary orbit to another; moving around the solar system is rather like gliding over a smooth but very slightly tilted sheet of ice. Once you set off in the right direction at the correct initial speed, the journey itself is effortless. Most of the work has to be done at the beginning of the trip, in escaping from the gravity of the earth.

It is easy to imagine a planetary system where this was not the case, and enormous amounts of energy would be needed to get from one orbit to another. We might thus be in the frustrating position of being able to reach the moon without difficulty — while the planets were virtually unattainable. So we should be thankful for the solar system we have. . . .

Though energy — which is simply the ability to do a measured amount of work — costs money, it is one of the cheapest commodities available. It can be purchased in many ways; in the modern world, the commonest and most convenient sources are the gasoline pump and the electric power outlet. When we look at various space missions, and see what they would cost purely on the basis of energy, the results are somewhat astonishing.

To lift a man all the way from the earth to the moon would require only ten dollars' worth of electrical energy. The cost of escaping from the moon itself is a ridiculous fifty cents; from the planets Mercury, Venus, Mars and the satellites of *all* the planets, only a few dollars. Naturally, it is more expensive to leave the giants; the worst case is that of Jupiter, whose titanic gravitational field requires three hundred dollars of energy for neutralization.

The fact that it cost not ten dollars each but nearly ten billion each to set the first men on the moon is some measure of our present ignorance and inefficiency. And here is another statistic: eight hundred pounds of kerosene and liquid oxygen, value thirty-five dollars, liberate enough energy to take a man from the earth to the moon. Yet the Saturn V burns a thousand *tons* per passenger.

In section 2, we indicated some of the methods by which today's absurd economics may be improved at least tenfold, and ultimately by even larger factors. At the same time, it is necessary to remind ourselves yet again that the future always contains unexpected and unpredictable developments, which invariably narrow the gap between the impossible and the easy.

All our ideas about travel beyond the earth are based upon rocket technology, and indeed we have no other means of propulsion in the vacuum of space. (Even the various nuclear and electric thrusters still depend upon the same principle of reaction or recoil.) However, it may well be that the rocket's history will parallel that of the balloon — which lifted mankind into a new element, but was eventually superseded.

A Saturn liftoff is the most magnificent spectacle yet contrived by man;

yet there must be better ways of achieving the same result, more compatible with nervous old ladies visiting their grandchildren on the moon, as well as with the peace and quiet of the countryside. The history of technology teaches us that the right tool always arrives at the right moment; witness how the transistor was ready when the Space Age dawned.

It may be a pure coincidence, but it seems slightly uncanny that success in the long search for gravitational waves was announced only a month before the first landing on the moon — by Dr. Weber of the University of Maryland. The old dream of controlling gravity may be a complete delusion — or it may foreshadow a basic industry of the twenty-first century. Anyone who is unwilling to admit this possibility should remember that when Hertz demonstrated the existence of electromagnetic waves in 1886, he saw no practical use for them. Now there is no man alive who is not affected by the radio networks that enmesh the globe — and are already stretching out to the planets. Perhaps the cycle is beginning again, and may lead to feats of space engineering as incredible to us as television would have been to the Victorians.

Even if such scientific breakthroughs do not happen (which is perhaps the most unlikely of all possibilities!) the entire solar system will become accessible to us during the coming century. Voyage times first measured in years will shrink to months, then to weeks, as nuclear propulsion systems become more efficient. On this planet, the evolution from the sailing ship to the intercontinental jet took three thousand years; the parallel evolution in space may take only a hundred.

And given the solar system — what will we do with it? At the very least, we will establish scientific bases on the more stable and hospitable bodies — such as Mars, Mercury, and the major satellites. Of these, the most important may be Titan, the giant moon of Saturn, which is almost as large as Mercury. However, it may not be unique as a source of methane (and hence hydrogen) for our nuclear rockets. Two of Jupiter's satellites, Ganymede and Callisto, are of about the same size as Titan, and may also serve as bases for the exploration of the outer planets.

Exploring the twenty-nine known (and probably many undiscovered) moons of the four great worlds Jupiter, Saturn, Uranus and Neptune is a task which may take generations; today, it is quite impossible to predict how rewarding, or otherwise, it will be. But exploring the giant planets themselves will be immeasurably more difficult; indeed, it may never be attempted by men, only by instruments. Even if the so-called "gas giants" possess solid surfaces, they may be so far down in their turbulent atmospheres that the pressures may be more crushing than in our deepest oceans. A spaceship capable of landing on (in?) Jupiter would have to be far stronger than any bathyscaphe designed to sound the Pacific trenches, and would have to be powered by energy sources completely beyond any technology we know today. And as an additional minor drawback, the intrepid crew would have to operate in a gravitational field that gave them two-and-a-half times their normal weight.

It would be rash to state that such feats will never be attempted, but more within the range of foreseeable engineering would be the exploration, by what have been called "buoyant stations" (i.e., balloons) of the atmospheres of the giant planets. Combination airship-spaceships, drifting or actively cruising at altitudes where the pressure was not impossibly high, may one day serve us as mobile manned bases on these strange worlds. From such floating platforms, we may send instruments, and beams of radiation, into the inaccessible depths thousands of miles below.

A lighter-than-"air" vehicle for the Jovian atmosphere poses some fascinating technical problems. Jupiter's atmosphere is mostly hydrogen — and as this is the lightest of all gases, what can we possibly put in our balloons so that they will float in such a medium? The answer can only be hydrogen itself — but *heated.* The recent surprising revival of hot-air ballooning as a sport may have important repercussions, half a century from now. It is quite conceivable that some youngster, soaring in today's skies, may be learning skills he will apply in wildly different surroundings, where the horizon is three times further away than on earth. . . .

We may also explore Venus by balloon long before we land on its dully glowing surface — probably first starting at the poles. Venus, unlike the earth, has its spin-axis almost at right angles to its orbit; having no axial tilt, it consequently has no seasons. The polar regions of the planet must therefore be, permanently, hundreds of degrees colder than the lower latitudes. If there are mountains here, water might exist on them — and some wildly optimistic scientists have even suggested the possibility of ice, which could reopen the whole question of Hesperian life.

However, it seems far more probable that any life on Venus (as on the moon) will be imported from earth. The astronomer-biologist Carl Sagan has suggested "seeding" the cloud layers with specially developed microorganisms, which could float in the turbulent upper atmosphere and feed on the immense quantities of carbon dioxide present. While doing this, they would release oxygen, and as their numbers multiplied there would be an ever-accelerating transformation of the atmosphere. In a relatively short time (perhaps centuries, perhaps even decades) the process which took geological ages on earth would have progressed far enough to make Venus hospitable to men.

Similar suggestions have been made regarding Mars, which may also possess vast quantities of oxygen locked up in its surface rocks. The material of the reddish Martian "deserts" may be iron oxide, and bacteria exist which can release its oxygen and leave the iron. This, in fact, is precisely how some of the great terrestrial iron deposits were formed. The bacteria involved have already had an immense economic impact on our own planet; one day we may set them to work again on other worlds.

These suggestions are by no means the most ambitious that have been put forward by reputable scientists. As early as 1948, the famous California Institute of Technology astronomer Fritz Zwicky — who was then heavily involved in the establishment of the American rocket industry — startled his colleagues by proposing the *reconstruction of the solar system.* He predicted that the time would come when man's increasing powers over his environment would allow him, literally, to move worlds, and thus to place the planets in orbits that would give them more favorable climates.

Even more mind-boggling ideas have been discussed by the British mathematician Freeman Dyson. He has argued that no intelligent species would tolerate indefinitely the fact that more than 99.9999999% of its sun's radiation is lost in space, and only a few billionths are intercepted by the planets. As the power requirements of a civilization increased, it might be necessary to dismantle a few planets and shield the sun completely with energy-absorbing screens.

This raises another question, first discussed by Tsiolkovsky during the 1920's in his pioneering studies of astronautics. *Do we really need planets?* When we have solved the purely technical problems of sustaining life in space, why should we continue to tolerate the perpetual drag of gravity, which cripples and sometimes kills us, and perhaps shortens our lives? The blissful weightlessness of free fall has now been experienced by a handful of astronauts — but has also been glimpsed by millions of skin-divers. It has its inconveniences, but these are easy to overcome. We evolved in the weightless environment of the sea; when we get into space, we may discover that this is where we really belong. . . .

Forty years ago, in an astonishing and recently reprinted book *The World, the Flesh and the Devil,* the British physicist J. D. Bernal suggested that eventually most of mankind would live in spherical space cities a few miles in diameter, moving in independent orbits around the sun. If this is correct, it is possible that far more people will live *off* the earth than the approximately one hundred billion who have lived on it since the dawn of time.

We must face the possibility — even the probability — that all our history on this world is but prelude to a far more complex future on an infinitely wider stage. On and among the planets we may see the founding of new societies, new cultures — perhaps even new species, as our descendants naturally or artificially adapt to alien environments.

There are some who may recoil in horror from these vertiginous glimpses of a cosmic future, or may feel that even to discuss them diverts attention from the immediate and desperate problems of our present age. That danger indeed exists; here — and only here — the tired old charge of "escapism" may sometimes be made against those whose eyes are fixed upon the stars.

Yet to let our minds wander from time to time in the centuries still far ahead may serve a useful purpose. It can help to put our present troubles in their true perspective — and it can remind us all what we stand to lose, if we do not solve them.

8. *The Shores of Infinity*

THE story of man has been one of expanding horizons; even today, there are a few Stone Age cultures where the limits of the unknown are only a dozen miles away.

First by land, thanks to conquerors and travelers like Alexander the Great and Marco Polo, the world widened. Yet not until the development of the sailing ship, and the arts of navigation, did Columbus, Magellan and their successors reveal the true lineaments of the globe. The fantastic cosmographies of the Dark Ages, with their mythical realms and monsters, lost their power over the minds of men. For the last four hundred years, every educated person has lived in essentially the world our parents knew.

But not the world *we* know. In one brief decade, our imaginations have had to encompass the moon. We can pinpoint the very moment when it ceased to be a heavenly body and became a place; it was November 23, 1966, when Lunar Orbiter 2 radioed back to earth the historic photograph of the Copernicus crater — and for the first time we looked not upon, but *across,* the landscape of another world.

In a few more years, we will experience the same revelation with Mars, Mercury, Venus. . . . Before the end of this century, the entire solar system will have become the background of our lives, as robot probes penetrate to its farthest reaches, and men prepare to follow them. Human thought will undergo another change of scale, comparable to that which occurred during the first Age of Discovery, half a millennium ago.

But for all the obvious parallels, there will be a profound difference between these two eras. Columbus and his fellow navigators were filling in the details of a world which, though large, they knew to be finite. The men of the twenty-first century will always be aware that, far beyond their widest voyaging, the stars and galaxies are scattered across a volume of space which is unimaginably large, and may indeed be infinite.

There is a familiar household object which gives a good idea of the scale of the solar system, and of the gulfs beyond. Take an ordinary twelve-inch LP record, and imagine that the hole in the center is the earth's orbit — 180,000,000 miles wide. Then Pluto lies on the rim, and all the outer planets are scattered across the disc; Saturn is at the edge of the label.

As yet, we have just started to explore the hole in the middle, which contains the sun, Mercury and Venus, and have sent a few probes a tenth of an inch outwards toward Mars. The remainder is still unknown, though already within our reach if we are prepared to spend years in silent coasting through space.

On this scale, the very nearest of the stars — and therefore the closest possible system of other planets — is three-quarters of a mile away. Our sun is a barely visible speck of dust, one of a hundred thousand million which on the average are about a mile apart. The whole collection of suns (the Galaxy) forms a disc about five times the diameter of earth. Even with this drastic reduction, our model has got a little out of hand. . . .

It is not surprising, therefore, that we do not know if there are any other solar systems besides our own. Across gulfs which shrink the sun itself to a feeble star, even a giant planet like Jupiter is utterly invisible. Nevertheless, there is indirect evidence that a few of the neighbor stars do have dark, planet-sized companions, and today most astronomers think it likely that the majority of suns possess solar systems. From this, it is an easy step to assume that planets bearing life — and intelligence — are commonplace throughout the universe.

This is probable; the alternative — that this tiny earth is the *only* inhabited planet in a cosmos of at least 10,000,000,000,000,000,000,000 suns — seems the wildest fantasy one can imagine. But at this stage of our knowledge there is no proof whatsoever that any life, or any intelligence, exists beyond the solar system. Some scientists, rashly ignoring the lessons of the past, have denied that such proof will ever be obtained; they feel that the distances involved present an insuperable barrier to knowledge. And they ridicule the idea that living creatures — men or other beings, should they exist — can ever span the abyss between the stars.

The development of space flight itself shows how dangerous it is to make such negative predictions. Moreover, in this case they can be shown to be false, at least to a high degree of probability. There may indeed be absolute bars to interstellar *travel* — but if so, they must be based on factors not yet known. What is already certain is that there are no serious difficulties in establishing interstellar *communication* — always assuming, of course, that there is someone to talk to at the other end. . . .

The quite amazing development of radio astronomy during the last thirty years has given us all the technology required. The gigantic antennas designed to catch faint, natural signals from the depths of space, and the sensitive receivers to amplify them, are the very tools needed for this task. In barely more than half a century since Marconi flashed his first signals across the Atlantic (a feat then widely declared to be impossible!) we have developed equipment with which we can talk to the stars.

No one has yet tried to do this, though there have been limited attempts to listen for intelligent messages from space. If these are ever discovered, it is likely to be as a by-product of some radio-astronomical investigation, because no one is going to tie up millions of dollars' worth of equipment on a random search that may last for decades — or centuries — with no certainty that it will ever meet with success. The detection of the extraordinary pulsars occurred in just such a fashion; at first, their trains of accurately timed radio pulses appeared to fit exactly the specifications of intelligent signals. Now it seems that they can be explained as a natural, though very surprising, phenomenon; the sharply defined pulses may be produced by spinning "neutron stars" — dense bodies only a few miles in diameter, yet weighing as much as the sun.

To be quite certain that we are receiving intelligent signals, we must prove that they carry some kind of message, even if it is not one that we can interpret. So far, there is no indication that the pulsars — or any other sources of radiation in space — are doing this.

That discovery might come at any moment, and would have a shattering impact upon human philosophy, religion and perhaps even politics. The mere knowledge that another intelligence existed somewhere else in the universe would affect our thoughts and our behavior in a myriad subtle ways. At the very least, it might cause our quarrelsome species to close its ranks.

Beyond this, the possibilities are so numerous that speculation is limited only by the laws of logic. Just a few of the questions that might arise are: does the message contain any useful information? (It would be possible to transmit a cosmic encyclopaedia, and an advanced, benevolent civilization might do so.) Should we attempt to answer? (It might not be worth it; if the message came from the Andromeda galaxy, it would have been on its way since the first ape men experimented with clubs. On the other hand, if it came from one of the nearer stars, our reply could be received in a decade or so.) And perhaps most interesting of all: is the source approaching? ("We shall be landing on the White House Lawn/Red Square in thirty minutes . . .")

The theme of cosmic confrontation is, of course, the classic stand-by of the science fiction writers, but it is now time that we took it seriously. (In the past it has sometimes been taken too seriously, as in 1938 when Orson Welles's *War of the Worlds* radio play spread panic across the eastern U.S. seaboard.) Some writers have expressed perhaps premature thanks for the existence of "God's quarantine regulations," which may allow us to communicate across the interstellar distances, but which will prohibit any form of physical contact.

It is true that the distances involved in flight to the stars are about a million times greater than those which must be crossed to reach the planets; but this does not imply a proportionate increase in difficulty. It is a surprising fact that a Saturn V could send a payload of many tons clear out of the solar system — to reach, for example, Sirius. But it would take a few hundred thousand years to get there.

However, this is at the very beginning — the log-canoe stage — of our space technology, which will advance beyond recognition in the centuries to come. There is no theoretical reason why speeds which are a substantial fraction of the velocity of light may not be ultimately achieved, at least by robot probes on one-way missions. This would make the closer stars perhaps twenty years away, and it would be surprising if, during the next few centuries, we do not attempt to send successors of today's Mariners and Voyagers to the nearest stellar systems.

But what of *manned* flight? It has been said that interstellar travel is not an engineering problem, but a biological one. Certainly, biological techniques — known or foreseeable — give a number of possible solutions.

One is the multi-generation starship, a mobile, self-contained worldlet which might cruise for centuries, until the descendants of the original voyagers made planetfall in a new solar system. Another way of arriving at the same result would involve suspended animation (hibernation); if this proves to be impossible with human beings, then we might send frozen ova, which were automatically fertilized a couple of decades before the end of the voyage. The children thus born could be reared by robot nurses, and in due course introduced to their distant heritage of human knowledge and history.

If this particular idea sounds nicely calculated to bring a gleam of maniacal delight to the eyes of Dr. Strangelove — can anyone doubt that we would attempt it, if the sun was about to explode and we had the time and the technology thus to make our race immortal? The galaxy is full of detonating stars — how many doomed races may already have tried such desperate experiments? As already remarked, our island universe, the Milky Way system, contains a hundred thousand million stars. Equally significant is the fact that it has existed for at least five thousand million years — that is, more than a million times the duration of human history. Over such expanses of space and time, anything that is technically possible has probably been achieved not once, but over and over again. There may be innumerable cosmic arks, making their lumbering way between the stars. . . .

It is possible that even interstellar travel may not be particularly time consuming — at least, from the viewpoint of the travelers themselves. As is widely known, though imperfectly understood, Einstein's theory of relativity predicts that as velocities approach that of light (186,000 miles a second, or 670,000,000 miles an hour) time itself appears to slow down.

This prediction has been experimentally verified with high-speed atomic particles, and has fascinating consequences for space travelers. It means (in theory at least) that journeys of any distance are possible during the span of a human life, and indeed in as short a period of time as may be desired, *if* there is enough power to give the necessary speed, and the crew can withstand the acceleration involved. To give a concrete example (taken from Shklovskii and Sagan's book *Intelligent Life in the Universe*) one can imagine a spaceship setting out for the Andromeda galaxy at a constant acceleration of one gravity, so that the crew would experience their normal weight. If this acceleration could be maintained, the crew would be twenty-eight years older when the ship arrived at Andromeda.

Now this is very surprising — because light takes about two million years to make the same journey, and nothing can exceed the speed of light! Yet no contradiction is involved. From the point of view of outside observers, the spaceship never quite attains the speed of light, and so the voyage lasts a full two million years. But to the travelers — and all their clocks, since everything in the ship is equally affected — only twenty-eight years would have elapsed. They would have no way of telling that anything peculiar had happened to them, unless they turned around and went straight home. Then, fifty-six years older, they would land on an earth where four million years had passed. . . .

If anyone asks: "How long did the voyage *really* take?" the answer is that both figures are correct — it all depends on the observer. One day, this "time-dilatation" phenomenon may cause little more surprise than the fact that it can be noon in New York, when it is 5 P.M. in London.

A trip to Andromeda is a rather extreme case, but journeys to the neighboring stars, out to a few dozen light years away, do not appear quite so far-fetched; for example, the bright star Vega (distance twenty-six light years) could be reached in about six years of ship time and the twelve-year-older crew could return home fifty-two years after they had left. This would give them at least a sporting chance of seeing their children again. . . .

Unfortunately, the levels of power and energy needed for these interesting feats are of such a magnitude that we do not know, even in theory, how they may be attained. The ultimate form of nuclear propulsion, involving the *total* conversion of mass into thrust with a hundred percent efficiency (which is certainly impossible) would be utterly inadequate, and power outputs comparable to those of the sun would be needed for the more ambitious relativistic missions.

Because of this, some scientists have argued that velocities approaching that of light can never be attained, no matter how far present or foreseeable methods of propulsion are improved. This conclusion is probably correct — but it is about as useful and relevant as a demonstration that no wooden, coal-burning airplane can ever make a supersonic crossing of the Atlantic.

It is absurd to imagine that the energy sources, and the methods of employing them, which we can envisage here at the very dawn of the Neotechnic Age are those that will still be employed a thousand years from now. Even at the beginning of *this* century, who would have dreamed of the two hundred million horsepower that lift the Saturn V? It is doubtful, in fact, if so much power was then available to the entire human race.

To state that the energy of a whole sun would be needed to drive a star ship does not prove the impossibility of such a vehicle. It merely means that we may have to wait a few centuries before we can build one — which, perhaps, is desirable for a number of excellent reasons.

Nor should we be discouraged by the undoubted fact that nuclear energy — the most concentrated source of power now known — is wholly inadequate for high-speed star faring. Recent astronomical discoveries strongly suggest that nuclear power is pretty feeble stuff, compared to some of the forces being let loose in the cosmos. When we observe "quasars" or radio galaxies liberating energies equivalent to the simultaneous explosion of a billion stars, we may well wonder if we are glimpsing some new order of creation. Fifty years ago, studies of the sun gave us our first hint of the powers locked in the hydrogen atom, which we have now released — though not yet tamed — here on earth. One day we may likewise learn the secrets of the quasars; and if we survive that knowledge, we will be on our way to the stars.

Whether we shall be setting forth into a universe which is still unbearably empty, or one which is already full of life, is a riddle which the coming centuries will unfold. Those who described the first landing on the moon as man's greatest adventure are right; but how great that adventure will really be we may not know for a thousand years.

It is not merely an adventure of the body, but of the mind and spirit, and no one can say where it will end. We may discover that our place in the universe is humble indeed; we should not shrink from the knowledge, if it turns out that we are far nearer the apes than the angels.

Even if this is true, a future of infinite promise lies ahead. We may yet have a splendid and inspiring role to play, on a stage wider and more marvelous than ever dreamed of by any poet or dramatist of the past. For it may be that the old astrologers had the truth exactly reversed, when they believed that the stars controlled the destinies of men.

The time may come when men control the destinies of stars.

L'envoi

FIVE hundred million years ago, the moon summoned life out of its first home, the sea, and led it onto the empty land. For as it drew the tides across the barren continents of primeval earth, their daily rhythm exposed to sun and air the creatures of the shallows. Most perished — but some adapted to the new and hostile environment. The conquest of the land had begun.

We shall never know when this happened, on the shores of what vanished sea. There were no eyes or cameras present to record so obscure, so inconspicuous an event. Now, the moon calls again — and this time life responds with a roar that shakes earth and sky.

When a Saturn V soars spaceward on nearly four thousand tons of thrust, it signifies more than a triumph of technology. It opens the next chapter of evolution.

No wonder that the drama of a launch engages our emotions so deeply. The rising rocket appeals to instincts older than reason; the gulf it bridges is not only that between world and world — but the deeper chasm between heart and brain.

Acknowledgments

As in the case of an Apollo mission, this book is the product of many hands and many minds. Just as the Apollo 11 astronauts had to have — and eagerly acknowledged — the assistance of hundreds of people at Mission Control and elsewhere, indeed all over the world, *First On the Moon* necessarily had its own ground support. Some of the writing is in the third person, partly because this was the easiest way to explain the immediate historical background which was relevant to the flight of Apollo 11, and partly because it was clearly impossible for Mr. Armstrong, Colonel Collins and Colonel Aldrin to be everywhere at once. Yet in a very real sense this *is* the Apollo 11 astronauts' personal book; it represents their own view of the grandest voyage yet in history. It is also a "source book." The half million words (at least) of interviews and recorded conversations from which this narrative is drawn reflect the generous cooperation of many astronauts, and of many others in the NASA community.

Most of the interviewing was done by Dora Jane Hamblin and Elizabeth Dunn of the staff of *Life* magazine, but other acknowledgments are in order: to Harold Keen of San Diego, California; to David Chandler of New Orleans; to Tommy West of Houston, reporting from recovery carrier *Hornet*; to Ernest Shirley of Sydney, Australia; and to Joan Downs of *Life* magazine and Enid Colfer Farmer, who along with Elizabeth Dunn recorded the inward tensions of the astronauts' families as they experienced the flight of Apollo 11. We owe special thanks to Mrs. Armstrong, Mrs. Collins and Mrs. Aldrin for putting up with us, and to James Schefter of Nassau Bay, Texas, who spent so very many hours playing back the tapes on which were recorded conversations between Columbia, Eagle and Mission Control. Some tapes were played as many as twenty times to identify which astronaut had said what; and because of Mr. Schefter's diligence, we believe that the ground-to-air talk which appears on the preceding pages is here published correctly for the first time. We are grateful to Carmen Casanova, Edward B. Clarke Jr., Joanne

Cowen, Laura Garcia, William P. Harley, Ethel McNicol and Catherine Radich, who typed the bulk of the manuscript in off-duty hours; to Susan Hickson, our secretary and assistant editor, who organized our files and meticulously stitched hundreds of corrections onto the final copy, and to Michelle Cliff, who tracked down scores of instant answers to knotty questions. A particular kind of gratitude is reserved for our "backup pilot," Colonel Thomas P. Stafford, who read the manuscript for technical errors and made valuable editorial suggestions.

The Notes which follow are fewer in number than might be considered customary for a book of this length. The reason is that attribution is usually self-evident in the text; where this is the case, no note seemed necessary. Responsibility for any error of historical fact or allusion is shared by the undersigned.

Gene Farmer
Dora Jane Hamblin

Notes

Chapter 1

1. Barbara Tuchman, *The Guns of August* (New York, Macmillan, 1962), p. 26.
2. Richard E. Byrd, *Alone* (New York, Putnam, 1938), p. 4.
3. Hugh Sidey, *John F. Kennedy, President* (New York, Atheneum, 1963), p. 122.
4. The Rt. Hon. George Brown, M.P., in conversation with Gene Farmer, October, 1957.
5. President John F. Kennedy as quoted by Dr. Jerome B. Wiesner, "Go for Apollo," *Newsweek* (July 7, 1969), p. 42.
6. Chester Wilmot, *The Struggle for Europe* (London, The Reprint Society, 1954), p. 243.

Chapter 2

1. Christopher Morley, *Human Being* (New York, Doubleday, 1932), Chapter 11.
2. Jules Verne, *From the Earth to the Moon,* translated by Edward Roth (New York, Dover, 1960), pp. 192–193, 197.
3. Bernard Weinraub, "Tourists Crowd Cocoa Beach as Apollo Countdown Begins," *New York Times,* July 11, 1969.
4. The minimum of two thousand feet was dictated by the fact that the spacecraft had to achieve that altitude before the ground radar locked on. Subsequently Apollo 12 was launched (November 14, 1969) with considerably less ceiling, and under thunderstorm conditions. This may not have been a wise decision.
5. *Investigation into the Apollo 204 Accident,* first session; Volume I, p. 67, testimony of Dr. Maxime Faget.

Chapter 3

1. Dora Jane Hamblin, "Men and a Stupendously Powerful Machine," *Life* magazine (December 20, 1968), pp. 16–22.
2. According to Boeing, the 747 assembly hangar encloses about 160,000,000 cubic feet, as compared to the Vehicle Assembly Building's 129,482,000.
3. She received a letter of acceptance during the flight of Apollo 11.

Chapter 4

1. Theodore H. White, *Fire in the Ashes: Europe in Mid-Century* (New York, Sloane, 1953), p. 60.
2. Comparisons supplied by Alexander R. Ogston, senior technical aviation advisor, Esso International Inc., New York.
3. Throughout the spring and early summer of 1969 American space experts felt reasonably sure that the Soviets had developed a more powerful booster than the Saturn V. In the autumn American intelligence sources picked up evidence that the Soviets' "big boy," probably developing ten million pounds of thrust, had exploded on its launch pad. The booster had probably been developed to build a space station in orbit; this could have been done with Soviet boosters already available, but in the opinion of Sir Bernard Lovell, F.R.S., "the number of launches necessary . . . would be economically prohibitive." Letter to Gene Farmer dated December 8, 1969.
4. When hurricane Camille devastated the Mississippi gulf coast on August 17, 1969, NASA made the Mississippi Test Facility, which was fifteen miles inland and suffered only minor damage, available as a refugee center, providing medical supplies and land clearing equipment and serving as the only communications for the entire area.
5. Sir Bernard Lovell, F.R.S., in a letter to Gene Farmer dated September 18, 1969.
6. David Sheridan, "How an Idea No One Wanted Grew Up to be the LEM," *Life* magazine (March 14, 1969), p. 20.
7. Christopher C. Kraft Jr., assistant director for flight operations, in a memorandum dated April 4, 1966 to the assistant director for flight crew operations.
8. Mr. Kraft speaks here of two successful dockings. Although the Neil Armstrong–David Scott flight, Gemini 8, was the first American "hard dock," two Gemini flights, flown by Frank Borman and James Lovell in one spacecraft, and by Walter Schirra and Thomas Stafford in the other, had maneuvered to within one foot of each other in December 1965.

Chapter 5

1. The tape was made by Michael Kapp, general manager, A. and R., Capitol Records.
2. Dora Jane Hamblin, "Spacecraft Anonymous," *Life* magazine (October 18, 1968), p. 115.
3. Hamblin, "Spacecraft Anonymous," *Life* magazine (October 18, 1968), p. 113.
4. Roy Neal in an interview with Ron De Paolo of the *Life* magazine Los Angeles Bureau, July 1, 1969.
5. Neil Armstrong was in fact one of the first two civilians to become an astronaut. The other was Elliot M. See Jr., killed in a plane crash, without ever having flown in space.

Chapter 6

1. Arthur M. Schlesinger Jr., *A Thousand Days: John F. Kennedy in the White House* (Boston, Houghton Mifflin, 1965), p. 419.
2. Michael Collins in conversation with Dora Jane Hamblin, March 1969.
3. Jules Verne, *All Around the Moon*, translated by Edward Roth (New York, Dover, 1960), pp. 251, 289.
4. John Moreau, "The Space Chef Gives a Custom Service," *New York Post*, July 16, 1969.
5. In Italian press coverage of Apollo 11, Collins was frequently identified as "the Roman." When the crew of Apollo 11 made its world tour in the autumn of 1969, Collins was summoned to his birthplace at 16 Via Tevere. In the presence of the mayor of Rome and other dignitaries, he helped unveil a three-foot marble plaque: "In this house was born 31 October 1930 Michael Collins, intrepid astronaut of the mission Apollo 11, first man on the moon [sic]."
6. He may have been right. According to the Baker Planetarium in Houston, there was a two-hour period during the night of July 16 when, in the flat Nassau Bay area, it was possible for Apollo 11 to be visible to the naked eye.
7. In *The French* (New York, Putnam, 1969) Sanche de Gramont states (p. 109) that Collins was "raised in France." This bit of mythology presumably was born in Chambley.

Chapter 7

1. Dr. Thomas Paine, in a conversation with Gene Farmer, Washington, D.C., July 9, 1969.
2. The press, foreign and domestic, did have a problem, since from the earliest days of Mercury the astronauts had been under contract to *Life* magazine for their personal stories — and the stories of their immediate families.
3. *Montclair Times*, Centennial Edition, June 13, 1968.
4. There is a story in the Air Force that Buzz got his nickname by "buzzing" the Air Force Academy at Colorado Springs, but this is definitely incorrect.
5. *Montclair Times*, Special Buzz Aldrin Day Supplement, September 11, 1969.
6. Frank Borman, who was a year ahead of Buzz Aldrin at West Point, was Coach "Red" Blaik's football manager in 1949, and Blaik later drew a subtle distinction between the two men: "Aldrin always evidenced his desire to excel. Borman excelled without showing his hand." Colonel Earl H. Blaik in a letter to Gene Farmer dated April 14, 1969.
7. On the eve of the flight of Apollo 11 MIT was unwilling to discuss the circumstances surrounding Buzz Aldrin's thesis. Professor Walter Wrigley, chairman of the committee which passed on Aldrin's thesis, told *Time* correspondent Barry Hillenbrand, "All theses are, of course, in new and unexplored areas and therefore by nature controversial."
8. The four who died before the flight of Apollo 11 were Air Force Captain Theodore C. Freeman; Air Force Major Charles A. Bassett II; Navy

Lieutenant Commander Roger B. Chaffee; and Marine Corps Major C. C. Williams.
9. Handwritten text made available by Colonel Edwin E. Aldrin Sr., of Brielle, New Jersey.
10. Bruce Catton, *A Stillness at Appomattox* (New York, Doubleday, 1953), p. 379.

Chapter 8

1. The debate on this point was still going on even after the success of Apollo 11, and probably will continue to go on for years.
2. By 1969 a number of scientists had begun to question the "dead" theory. Apollo 11 brought back no evidence of lunar organisms, but more lunar landings would be required before the case could be considered closed.
3. Ralph B. Baldwin, *The Face of the Moon* (University of Chicago Press, 1949). Dr. H. Percy Wilkins, F.R.A.S., and Patrick Moore, F.R.A.S., *The Moon* (New York, Macmillan, 1956). *Encyclopaedia Britannica*, Vol. 15, p. 813. "Astronomical Notes from Berkeley," *Sky and Telescope*, Vol. 22 (October 1961), pp. 184–191. Gene Marine, "Politics Haunts Astronomers at Berkeley," *Bulletin of Atomic Scientists*, Vol. 17, No. 8 (October 1961), pp. 345–346. Also, Roland Flamini, who covered for *Time* magazine the 1967 meeting of the International Astronomical Union congress in Prague.
4. Virgil I. Grissom, *We Seven* (New York, Simon and Schuster, 1962), p. 149.
5. The Soviet Union did not request rock samples, presumably because this request would have set a precedent for a reciprocal request from the United States. British geologists were allotted the greatest share — 2.3 percent of the total earth weight of 25 kilograms.
6. Roderick Peattie, *The Inverted Mountains* (New York, Vanguard, 1948), pp. 57–60, 120–121. John McGregor, *Southwestern Archaeology* (Urbana, University of Illinois Press, 1965), pp. 295–302, 381–382, 419–421, 475. Joseph Miller, *Arizona* (New York, Hastings House, 1966), p. 286. National Park Service–Southwestern Monuments Association, publication number 6-66-20M, 9th edition. U.S. Department of the Interior, 1968, 306–123/119. National Park Service–Southwestern Monuments Association, publication number 5-68-20M, 9th edition. When the lava cooled, the Indians came back and for more than a century this area was a mixture of Indian culture — Pueblos from the northeast, Hohokam irrigation farmers from central Arizona, Mogollon groups from the south, Cohoninos from the west. A severe drought which began in 1215 drove them all out for good by 1225. When the Spanish came through the region between 1583 and 1605 they encountered small bands of Indians in the mountains near Flagstaff. These probably were hunting and gathering parties of either Havasupais from the Grand Canyon or Yavapais from the Verde Valley. No Indians were reported between the Hopi villages and the San Francisco mountains. The Navajo, who are seen in the national monument area today, did not move into the region until about 1870.
7. Armstrong and Aldrin were due to go to Flagstaff that day in May 1969;

at the last minute they had to cancel out, and the resident astrogeologists went through a scheduled exercise without them. Just before noon, hunched over a plotting board at Astrogeology's own "mission control" in downtown Flagstaff, Arnold Brokaw saw a closed-circuit television picture go dead. After a moment a hand-lettered sign appeared on the screen. Someone had drawn an impressive bolt of lightning and the single word: BLAM! A violent storm had sent the geologists scurrying to the shelter of their vans. A voice crackled over the radio from the field: "It's just as well the astronauts didn't come!" Dora Jane Hamblin, reporting from Flagstaff, 1969.

8. Dr. Shoemaker, chairman of the California Institute of Technology's Division of Geological Sciences, thought the scientific thrust of Apollo inadequate and after the flight of Apollo 11 indicated his intention to leave the program after the third moon landing in 1970. UPI dispatch, October 9, 1969.

Chapter 9

1. Michel Van Langren (1600–1675) was actually the Spanish court astronomer. It is unclear whether he named the Sea of Fertility, since his map work was swallowed up by the Italian Riccioli when Riccioli published *Almagestum Novum* in 1651.
2. *New York Times,* May 26, 1969.
3. Originally the LM was spelled as well as pronounced "LEM." The pronunciation of the abbreviation "LM" remains the same.
4. Commander (now Captain) Conrad received a good deal of critical mail for his choice of "Intrepid." But the source of the name was as much American as British. At least four American ships of war have been called Intrepid, but Pete Conrad had no warlike intentions when he went to the moon as commander of Apollo 12.
5. In fact, Jules Verne's spaceship was named the Columbiad.
6. By this time there was wide awareness that President Nixon's prelaunch dinner with the Apollo 11 astronauts was almost certain to be called off. The reporters who covered the July 5 press conference contributed their share of "needling" in a needlessly embarrassing situation by spending nine dollars to buy three hundred surgical masks, which they put on as the astronauts entered the room.
7. Both the descent and ascent stages of the LM flown by McDivitt and Schweickart on Apollo 9 remained in earth orbit. On Apollo 10 the descent stage was fired into solar orbit and the ascent stage into such a low lunar orbit that it probably crashed on the moon within a few hours. At the time Apollo 12 was launched nobody could be sure what had happened to the ascent stage of the Apollo 11 LM; the crew of Apollo 12 merely hoped not to run into it while cruising around the moon.
8. *Encyclopedia of Space* (New York, McGraw-Hill, 1968), p. 635.
9. Neil Armstrong thus joined the Weber Booster Club, organized by Weber Aircraft in 1957. The club is made up of airmen who have been "boosted out of an intolerable situation by a Weber escape system." As of 1969 the club numbered more than 450 members. After his ejection from the LLRV

in May 1968, Armstrong was presented with the membership plaque on which was mounted the "ditch" handle from the ejection seat, a membership card and a key chain medallion with the club's insignia, an armchair floating beneath a billowing parachute. NASA test pilot Joseph Algranti also gained membership after his successful ejection from the LLTV in December 1968.

Chapter 10

1. Quoted from Mircea Eliade, *Myths, Dreams and Mysteries* (New York, Harper and Row, 1961), pp. 103–106. Text of paper made available by the Reverend Dean Woodruff.
2. Duke actually came from Lancaster, South Carolina, and attended the United States Naval Academy. But he did do graduate work at MIT.
3. The discrepancy between the figure recalled by Mr. Ryan and the figure quoted by General Lindbergh is probably accounted for by the cost of the Wright-Whirlwind engine which — according to General Lindbergh — Franklin Mahoney agreed to supply at cost. See *The Spirit of St. Louis* (New York, Scribner, 1953), pp. 69–81. Adding everything up, the flight of *The Spirit of St. Louis* probably cost around fifteen thousand dollars.
4. At a state dinner for the Apollo 11 astronauts, in Los Angeles following the mission, President Nixon awarded a mass citation to the 400,000 members of the Apollo team who had labored on the ground — contractors, assemblers, industry, labor, etc. Steve Bales, the twenty-six-year-old guidance officer who correctly analyzed the alarms of Eagle's onboard computer and made the final 'go' decisions, was selected to receive the award on behalf of all the others.
5. *New York Times* Sunday Magazine, November 9, 1969.

Chapter 11

1. Mission Control in Houston recorded Neil Armstrong as saying: "That's one small step for man" — without the article "a." Tape recorders are fallible. When Charles ("Pete") Conrad, the flight commander of Apollo 12, stepped onto the moon on November 19, 1969, he paraphrased the quotation: "Whoopee, man, that may have been a small step for Neil, but that's a long one for me."
2. Harriet and Donn Eisele were divorced in July 1969. Colonel Eisele has since remarried.
3. *Houston Post*, August 9, 1969.

Chapter 12

1. After two weeks of failure, and well after Eagle had left the moon, a laser fired from the University of California's Lick Observatory hit the target and measured the distance to be 226,970.9 miles with a time uncertainty which could mean a margin of error of three hundred feet.
2. Pictures taken of Aldrin on the moon, by Armstrong, showed the knees of his spacesuit so dirty that many wondered if he had fallen. Aldrin did

not recall slipping to his knees on the lunar surface. He said, "I think some of it came from kicking up the dirt."

3. An amendment proposed by Representative Richard L. Roudebush in November 1969 and attached to a NASA appropriations bill called for the planting of the American flag and no other on each subsequent lunar landing mission. The amendment was passed in both the House and the Senate.
4. President Nixon was criticized for "taking up the astronauts' time" by making the telephone call to the moon. In the opinion of the editors of this book, the criticism was misplaced. The Apollo 11 astronauts had plenty of time to do their jobs. The idea of placing the telephone call was the President's — and, in our view, a good one. (G.F., D.J.H.)
5. Vesicular is an adjective which describes a surface as having vesicles, i.e., small cavities. Vesicular cavities are usually found in basaltic lava, where they are produced by the expansion of vapor in a molten mass.
6. The "cold" effect was remedied for the flight of Apollo 12. Hammocks were put in the lunar module Intrepid — along with some blankets. Pete Conrad and Al Bean slept better than Armstrong and Aldrin did. They were also able to extract their lunar core samples without the difficulty experienced by Eagle's crew.
7. The best estimate of Britain's Sir Bernard Lovell, F.R.S., is that Luna 15 crashed into the moon's Sea of Crises, about five hundred miles from where Armstrong and Aldrin landed in Eagle. The purpose of Luna 15 — i.e., whether or not a "soft" or "hard" landing was intended, and whether or not there was a failure in telemetry, remains obscure. So does the purpose of three simultaneous Soviet space flights in October 1969 — Soyuz 6, 7 and 8. In Sir Bernard's opinion, the Soviet computing systems are still probably inadequate to deal with unforeseen discrepancies in flight patterns: "We know that they are considerably behind the West, particularly in flexibility, and the importance of this has been adequately demonstrated in the recent Apollo flights." Letter to Gene Farmer dated December 8, 1969.

CHAPTER 13

1. The rendezvous radar is put in SLEW to turn the antenna mechanically or make it swing back and forth.
2. More correctly called the King of the Belgians.
3. The game was rained out and postponed, and the President did not see it.
4. Dora Jane Hamblin, "The Lunar Laboratory," *Life* magazine (July 4, 1969) pp. 54–58.
5. There was reason for hurry in the matter of getting the rocks back to the lunar receiving laboratory; if in fact they did possess special properties the scientists did not want to give the properties a chance to erode in a terrestrial environment.
6. Identification of the exact landing site did not become official until the early morning of July 29, after the onboard film had been received and processed in the lunar receiving laboratory. Eagle landed in the Sea of Tranquility at zero degrees, 41 minutes and 15 seconds north latitude,

and 23 degrees, 25 minutes and 45 seconds east longitude. In the flight plan the projected landing site had been zero degrees, 42 minutes and 50 seconds north latitude, and 23 degrees, 42 minutes and 28 seconds east longitude.

7. Asked for an explanation of the weird noises, the astronauts looked at each other and grinned. Armstrong said: "We are guilty again. We sent the whistles . . . [*laughter*] . . . and bells. We had our little tape recorder which we used to record our comments during the flight in addition to playing music in the lonely hours, and we thought that we'd just share some of that with the people in the control center."

Chapter 14

1. Robert Ardrey, *The Territorial Imperative* (New York, Atheneum, 1966), pp. 300–301, 340–341.
2. Admiral McCain's father, Rear Admiral John S. McCain, was commander of a land-based air South Pacific force during the Second World War. Admiral John Sidney McCain Jr.'s son, John S. McCain III, Lieutenant Commander USN, was believed in 1969 to be a prisoner of war in North Vietnam.
3. Walter Lord, *A Night to Remember* (New York, Holt, 1955), pp. 136–147.
4. Enid Colfer Farmer, representing *Life* magazine during Apollo 11.
5. The first three groups of astronauts, called within the organization "the original seven," "the nine" and "the fourteen," are of course the men the public became most aware of because they did most of the space flying. Of "the seven," only Gus Grissom had been killed at the time of Apollo 11. Of "the nine," Ed White and Elliot See were dead. Of "the fourteen," four were lost: Roger Chaffee, Ted Freeman, Charlie Bassett and C. C. Williams. Major Givens had become an astronaut in 1966. Two of the eight widows had remarried at the time this book went to press. Faith Freeman is now Mrs. Victor Ettredge of Nassau Bay, Texas; her husband, formerly with the Air Force, is an engineer for NASA. Martha Chaffee is now Mrs. William Canfield of Houston, Texas; her husband is a real estate investor.
6. Armstrong had Wapakoneta, Ohio, and Aldrin Montclair, New Jersey; Collins had his "day" in New Orleans, Louisiana, the home state of the congressman who had appointed him to West Point.
7. On January 29, 1970, Colonel Borman announced that he would leave both NASA and the Air Force in midyear to join the wealthy Texan H. Ross Perot in starting a new foundation to promote public debate on national affairs.